ADVANCES IN

CHROMATOGRAPHY

Volume 24

ADVANCES IN CHROMATOGRAPHY

Volume 24

Edited by

J. CALVIN GIDDINGS
EXECUTIVE EDITOR

DEPARTMENT OF CHEMISTRY
UNIVERSITY OF UTAH
SALT LAKE CITY, UTAH

ELI GRUSHKA
GAS CHROMATOGRAPHY AND LIQUID CHROMATOGRAPHY

DEPARTMENT OF INORGANIC AND ANALYTICAL CHEMISTRY
THE HEBREW UNIVERSITY OF JERUSALEM
JERUSALEM, ISRAEL

JACK CAZES
MACROMOLECULAR AND INDUSTRIAL CHROMATOGRAPHY

FAIRFIELD, CONNECTICUT

PHYLLIS R. BROWN
BIOCHEMICAL CHROMATOGRAPHY

DEPARTMENT OF CHEMISTRY
UNIVERSITY OF RHODE ISLAND
KINGSTON, RHODE ISLAND

MARCEL DEKKER, Inc., New York and Basel

Library of Congress Cataloging in Publication Data
Main entry under title:

Advances in chromatography. v. 1-
1965-
New York, M. Dekker
v. illus. 24 cm.
Editors: v.1- J. C. Giddings and R. A. Keller.
1. Chromatographic analysis—Addresses, essays, lectures.
I. Giddings, John Calvin, [date] ed. II. Keller, Roy A., [date]
ed.
QD271.A23 544. 65-27435
ISBN 0-8247-7049-8

Marcel Dekker, Inc.
270 Madison Avenue, New York, New York 10016

Current printing (last digit):
10 9 8 7 6 5 4 3 2 1

Printed in the United States of America

Contributors to Volume 24

Monica L. Bell Chemistry Department, University of Mississippi, University, Mississippi

Julie G. Bower Department of Chemistry, University of Houston, Houston, Texas

Keith D. Bower Department of Chemistry, University of Houston, Houston, Texas

Henri Colin Laboratoire de Chimie Analytique Physique, Ecole Polytechnique, Palaiseau, France

Stanley N. Deming Department of Chemistry, University of Houston, Houston, Texas

John N. Driscoll HNU Systems, Inc., Newton Highlands, Massachusetts

Keith R. Eberhardt Statistical Engineering Division, Center for Applied Mathematics, National Bureau of Standards, Washington, D.C.

Georges A. Guiochon Laboratoire de Chimie Analytique Physique, Ecole Polytechnique, Palaiseau, France

Haleem J. Issaq Program Resources, Inc., NCI-Frederick Cancer Research Facility, Frederick, Maryland

Yoichiro Ito Laboratory of Technical Development, National Heart, Lung, and Blood Institute, Bethesda, Maryland

Karen Kafadar* Statistical Engineering Division, Center for Applied Mathematics, National Bureau of Standards, Washington, D.C.

George Karaiskakis Physical Chemistry Laboratory, University of Patras, Patras, Greece

Nicholas A. Katsanos Physical Chemistry Laboratory, University of Patras, Patras, Greece

Ante M. Krstulović[†] Laboratoire de Chimie Analytique Physique, Ecole Polytechnique, Palaiseau, France

Ira S. Krull The Barnett Institute of Chemical Analysis, Northeastern University, Boston, Massachusetts

Ping J. Lin[§] Chemistry Department, University of Mississippi, University, Mississippi

Jon F. Parcher Chemistry Department, University of Mississippi, University, Mississippi

Michael E. Swartz The Barnett Institute of Chemical Analysis and Department of Chemistry, Northeastern University, Boston, Massachusetts

---

Current affiliations:
*Quality Department, Hewlett-Packard Company, Stanford Park Division, Palo Alto, California
[†]L.E.R.S.-Synthelabo, Paris, France
[§]Department of Chemical Engineering, Feng Chia University, Taichung, Taiwan

Contents

Contents of Other Volumes

Volume 1

Volume 2

Volume 3

Volume 4

Volume 5

Volume 6

Volume 7

Volume 8

Volume 9

Volume 10

Volume 11

Volume 12

Volume 13

Volume 14

Volume 15

Volume 16

Volume 17

Volume 18

Volume 19

Volume 20

Volume 21

Volume 22

Volume 23

ADVANCES IN

CHROMATOGRAPHY

Volume 24

1

Some Basic Statistical Methods for Chromatographic Data

Karen Kafadar* and Keith R. Eberhardt / Statistical Engineering Division, National Bureau of Standards, Washington, D.C.

---

*Current affiliation: Hewlett-Packard Company, Stanford Park Division, Palo Alto, California

I. INTRODUCTION

A. The Need for Statistics in Measurement Processes

The application of statistical methods has been increasing over the past decades, particularly in the fields of engineering and physical and chemical sciences. Many statistical concepts were formulated originally for biology or agricultural experiments, but their importance has reached all branches of science, since measurement methods for determining physical, biological, or behavioral relationships always involve some variability. Statistics is the science of collecting, analyzing, and interpreting numerical data. An important goal of statistics is to assess the magnitude of fluctuations in the data from various sources, thereby gaining understanding of a measurement process.

At first, scientists or engineers may be reluctant to admit that their measurement processes contain errors. An inducement to repeat the measurements often is sufficient to demonstrate that even the finest measuring equipment does not operate error free. If the measurement process consists of several steps, using different measuring equipment (balances, chromatographs, etc.), it is important to assess the magnitude of the variation in the measurements from *each* source, so that the final variability in the responses can be examined in light of what is known about the variability of each step.

Frequently, measurements may be costly or time-consuming, so the amount of available data will be small relative to the population of measurements one would like to describe. Thus, statistical methods are essential when generalizing from a small sample to a larger population. Statistical inference, properly applied, aids in this generalization. Furthermore, a well-planned, statistically designed experiment will maximize the amount of information per measurement, thereby reducing the cost of unnecessary work.

B. Applications to Measurement Data

When a sample is analyzed to determine the concentration of a particular analyte, the main target is generally the concentration of the analyte in a larger batch of material from which the sample was obtained, rather than merely the sample itself. To ensure that the analytical result from the sample corresponds closely with the target quantity, one must examine the variation introduced in the measurement by the sampling process, by the analytical procedure, and by the chemist who performs the analysis. An additional complication in this problem is that the response actually measured is not concentration but, rather, a related quantity like peak area or peak height. A proper statistical model for the peak responses must be formulated to ensure that a meaningful uncertainty is associated with the reported concentration.

From the results of such an experiment, one can expect to gain information about uncertainties in the measurements which may be useful for further investigations. Future experiments may be designed more efficiently if the uncertainties from various sources can be identified. For example, to determine the average concentration of an analyte in a lot of 1000 vials, a number of sample vials may be measured. Each measurement result is affected by analytical error as well as vial-to-vial variability. If previous experiments have established the magnitude of the analytical error, then the greater effort in the chemical analyses may be devoted to the measurement or detection of inhomogeneity among the sample vials. Such information will be important to users of any part of the lot of vials. Statistical sampling methods can be used to ensure that such information can be attributed correctly to the entire lot.

C. Additional Considerations

Frequently, a great benefit in a statistical approach is the recognition of additional factors affecting the response. This forces a more complete evaluation of the aims and problems of the investigation and thereby leads to a more clearly defined and practical experiment than might otherwise have been performed. For example, in running a series of sample solutions on the liquid chromatograph, it may become obvious that the response of the external standard to which the sample responses are being compared is drifting during the day, but that the drift is less noticeable after a certain number of operating hours. One may then either adjust the resulting concentrations for this drift, or specify a design for the procedure and order of the measurements to balance out the effects of the drift, or, often best, both. External environmental conditions such as temperature, pressure, and humidity may likewise affect the response to a degree that is substantial beyond the inherent analytical error. Again, an experiment may be designed to balance out or compensate for the effects of these factors on the response. An identification of the sources of variability is essential in any chemical investigation.

It is not unlikely that one or more outlying measurements may appear far from the bulk of the data. Sometimes these outlying values represent natural random variation in the response, or they may be the results of errors due to a variety of reasons, such as a misrecorded value, a door slam, or an incorrect procedure. The proper use of statistical methods in such cases is essential for the correct interpretation of the data. Inappropriate treatment of outliers could result in misleading conclusions. As with any procedure based on mathematical foundations, the validity of the underlying assumptions must be checked, and appropriate steps must be taken in cases of doubt.

D. Topics Covered in This Article

This article is intended to illustrate some useful statistical concepts for chemists who perform chromatographic experiments. To this end, while not a complete introduction to this field, Sec. II reviews basic concepts, linear additive models, the estimation of the mean and variance for the simplest linear additive model, and estimation in higher-order models involving several factors. Those who are familiar with these notions may wish to scan Sec. II for notational conventions used in this article, perhaps devoting a little more attention to Sec. II.D. In Sec. III, the concepts described in Sec. II are discussed specifically with reference to data obtained via chromatography. Several other topics are introduced by way of example: treatment of outliers, response drift in external standards, the decomposition and assessment of various sources of error, and the use of information from multiple internal standards and multiple analyte peaks. These examples are based on actual experiments conducted at the National Bureau of Standards, but the concepts may be applied usefully to measurement processes of various kinds.

II. BASIC STATISTICAL CONCEPTS

A. Probability Models

Measurements of any sort, such as peak areas, are recorded as numbers. The application of statistics to data requires an assumed model which governs these data. Although such models usually are based on theoretical assumptions to derive meaningful results, this section will avoid complicated theoretical discussions and merely indicate the important statistical ideas that are applicable by the chemist.

To apply statistical methods, one must relate data, which are variable (subject to numerous sources of variation), to some quantity of interest, which is fixed but unknown (such as the long-run mean of the measurement process), by means of a model. In this context, a *probability model* is a mathematical description of the relative frequency of all possible outcomes which could arise from a measurement process. For example, one might specify, by means of a mathematical formula, the probability that a measurement will be recorded as 4.2 cm (e.g., between 4.15 and 4.25 cm). Such mathematical formulas are called *probability density functions* and may be tailored to meet the specific circumstances. The Gaussian ("normal") density function (Fig. 1) is the most familiar probability model, but there are many others as well.

B. Models for Measurement Processes

If more than one measurement is obtained on a given process, part of the model may include a term for the dependence among measurements. A practical definition of *independence* says that independence holds when, given the sample and the measurement technique, the probability density function of any future measurements is unaffected by the outcome of the past measurements. For example, if knowledge that the first

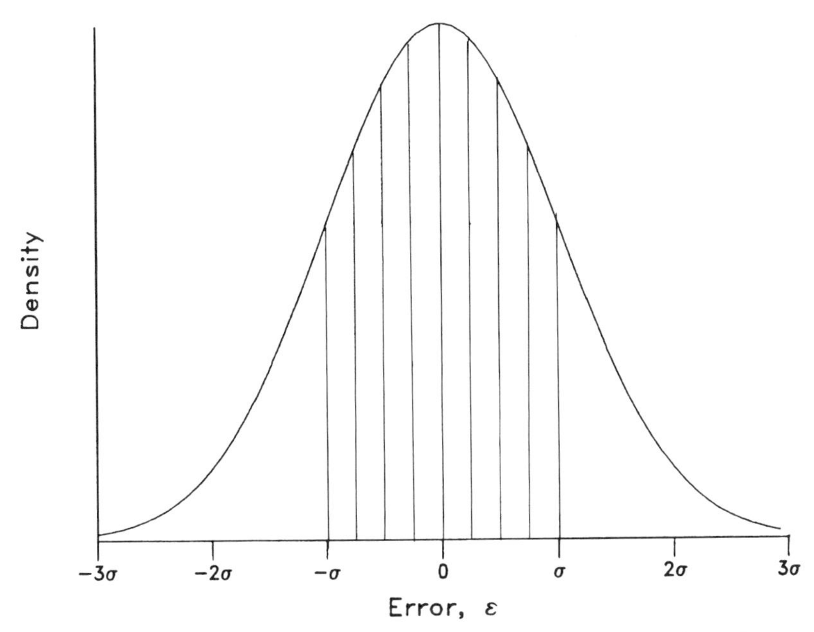

Figure 1 A graph of the Gaussian (normal) density function, $\phi(x) = \exp(-x^2/2\sigma^2)/\sigma\sqrt{2\pi}$, illustrating the determination of the probability that a random variable so distributed falls within 1 standard deviation (σ) of the mean [$= \int_{-\sigma}^{\sigma} \phi(x)\, dx = 0.6824$].

measurement on a well-defined sample was 4.2 cm helps not at all in predicting the value of the next measurement, then this is said to be an example of an independent measurement process.

Independence simplifies our models. If n measurements are independent, then the specification of only the univariate model for one of them is sufficient for the evaluation of the probability of all outcomes of the ensemble of all n measurements in the process. Thus, the starting place for the simplest models is a description of the sequence of independent measurements, all having the same probability density function.

Consider the following model for a measurement process. Denote by x_i the outcome of the ith measurement in the process. Suppose that the assumption of independence holds and that each measurement can be modeled as the sum of two terms: a fixed but unknown parameter or constant (usually the mean), and some random depature from this mean that may be different with each measurement. Thus,

$$x_i = \mu + \varepsilon_i, \quad i = 1, 2, \ldots, n$$

The average value μ is often unknown and is a fixed quantity to be estimated in the presence of the random "error" ε_i, whose magnitude is of interest. Since ε_i is a random quantity, it is described by means of a probability model of all possible outcomes, as described above. For example, the error may follow the Gaussian probability density function shown in Fig. 1: On the average, the departure from μ is zero, much of the time ε_i takes values between -1σ and $+1\sigma$ (i.e., with probability 0.682), and almost always ε_i takes values between -3σ and $+3\sigma$ (i.e., with probability 0.997).

Several modifications to this setup are possible. The error may not be zero on the average (e.g., if the measurements are recorded consistently higher than the assumed or estimated average). The errors may be dependent, requiring a more complex model, or have an asymmetric distribution, or the error may be more or less precise, corresponding to a change in the scale parameter (here called σ). Even more serious may be misspecification of the model for the error itself: instead of being Gaussian, the density may have heavier tails, such as that shown in Fig. 2.

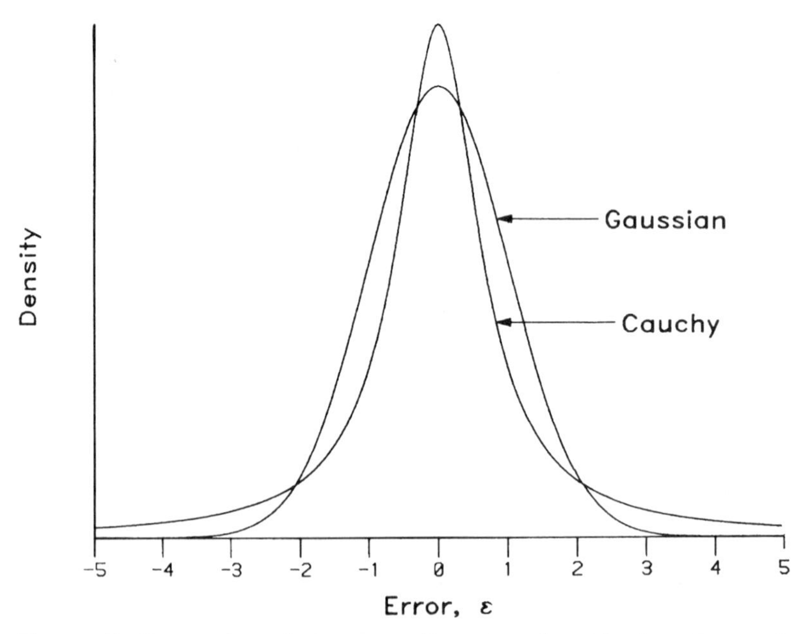

Figure 2 Comparison between a Gaussian density function and a density function having longer tails than the Gaussian (the Cauchy density function is illustrated, scaled so that the area between -0.67 and 0.67 is 0.5 for both distributions).

Even if the density function is not Gaussian, *location* and *scale parameters* analogous to μ and σ may be defined. The next section will address the estimation of these parameters.

C. Estimators for Location and Scale in Simple Models

For the simplest linear model of a measurement process, each measurement may be represented as the sum of two terms:

$$x_i = \mu + \varepsilon_i, \quad i = 1, 2, \ldots, n \tag{1}$$

where μ is parameter of constant value, usually the unknown mean, and $\varepsilon_1, \varepsilon_2, \ldots, \varepsilon_n$ are independent random errors, whose probability density function has mean 0 and variance σ^2. Estimates of the parameters μ, the average of the measurements, and σ, the standard deviation of the errors, are of interest. Another useful quantity, for nonnegative measurements, is the relative standard deviation (coefficient of variation) σ/μ. For example, $\sigma = 4$ may seem large if $\mu = 5$ (80% relative standard deviation), but moderately small if the average value is 100 (4% relative standard deviation). In the following subsections, three methods of estimating location and scale will be defined and given special notation. In later sections, generic estimates of μ and σ will be denoted by $\hat{\mu}$ and $\hat{\sigma}$.

1. Sample Mean and Sample Variance

Given a random sample $x_1, x_2, \ldots, x_n$ according to the model in Eq. (1), a simple estimator of the population mean is the *sample mean*

$$\bar{x} = \frac{\Sigma x_i}{n}$$

and an estimate of σ^2 is the *sample variance* s^2:

$$s^2 = \frac{\Sigma(x_i - \bar{x})^2}{n - 1}$$

When the data are Gaussian, it may be shown that $\bar{x}$ and s^2 are "optimal" estimators for μ and σ^2, in that, on the average, they estimate μ and σ^2 accurately and with high precision. When the data are *not* Gaussian, however, $\bar{x}$ and s^2 may be highly inefficient estimators of μ and σ^2. This is particularly true for heavy-tailed distributions, which are prone to produce outlying observations, far from the center of the data. In such cases, estimation methods which concentrate on the central part of the data are preferable. These methods generate more efficient estimators as well as valid confidence intervals. Two such procedures are given in the following subsections.

2. Trimmed Mean and Winsorized Variance

When the underlying probability density function of the data has somewhat longer tails than the Gaussian, the ordinary sample mean may not be as precise an estimate of the population mean as the trimmed mean. The *g-trimmed mean* is computed by first sorting the data from smallest to largest, trimming (i.e., deleting) the g smallest and g largest values and averaging the remaining values. Formally, the data are sorted (i.e., renumbered) in ascending order as $x_{(1)}$, $x_{(2)}$, . . . , $x_{(n)}$; then the g-trimmed mean is defined as

$$\bar{x}_{Tg} = (x_{(g+1)} + x_{(g+2)} + \cdots + x_{(n-g)})/(n - 2g)$$

Typically, g is chosen so that g/n is between 0.10 and 0.20 (i.e., 10 to 20% of the data at each end are ignored).

Using $\bar{x}_{Tg}$ to estimate μ requires an associated estimate of the variance σ^2, given by the *g-Winsorized variance* s_{Tg}^2.

$$s_{Tg}^2 = [(g + 1)(x_{(g+1)} - \bar{x}_{Tg})^2 + (x_{(g+2)} - \bar{x}_{Tg})^2 + \cdots + (x_{(n-g-1)} - \bar{x}_{Tg})^2 + (g + 1)(x_{(n-g)} - \bar{x}_{Tg})^2]/(n - 2g - 1)$$

The estimated standard deviation associated with the g-trimmed mean is given by the square root of this quantity, s_{Tg}.

An example may help to illustrate the use of these estimators. Suppose that a sample solution, together with its container, is weighed on a balance six times and that the recorded weights are 30.2, 30.8, 30.3, 30.6, 30.4, and 39.5 g. Then

$$\bar{x} = 32.0 \text{ g and } s = 3.7 \text{ g}$$

For a g-trimmed mean using g = 1,

$$x_{(1)}, x_{(2)}, \ldots, x_{(6)} = 30.2, 30.3, 30.4, 30.6, 30.8, 39.5$$

$$\bar{x}_{T1} = (30.3 + 30.4 + 30.6 + 30.8)/4 = 30.5 \text{ g}$$

$$s_{T1} = \{[2(0.2)^2 + (0.1)^2 + (0.1)^2 + 2(0.3)^2]/3\}^{1/2} = 0.3 \text{ g}$$

In the original data, the last observation should have been 30.5 but was recorded incorrectly as 39.5. Having been informed of the error, one might have calculated the corrected sample mean and standard deviation:

corrected $\bar{x}$ = 30.5 g
corrected s = 0.2 g

Often, contamination or non-Gaussianity is far less obvious than in the above case, but the example serves to illustrate the importance of using estimators that can provide reliable estimates of μ and σ in the presence of one or a few outliers.

3. Median and Median Absolute Deviation

For more insurance against the possibility of heavy-tailed distributions, the *sample median* and an associated scale estimate are recommended as estimators of μ and σ. These are defined as

$$\text{Median } x_M = \text{middle observation in sorted data (n odd)}$$

$$\text{Median } x_M = \text{average of two middle observations (n even)}$$

and

$$s_M = 1.5\text{MAD} = 1.5\text{med}|x_i - x_M|$$

The multiplier 1.5 for the MAD (*median absolute deviation* from the median) is recommended because the MAD typically underestimates the standard deviation. In Gaussian populations, the average value of the MAD is about two-thirds of σ, and the factor 1.5 inflates the MAD about the right amount for a wide variety of populations.

With the data of Sec. II.C.2, these estimates are very simple to calculate by hand for small samples:

$$x_M = (x_{(3)} + x_{(4)})/2 = 30.5 \text{ g}$$

$$s_M = 1.5\text{med}\{0.3, 0.2, 0.1, 0.3, 9.0\} = 1.5(0.25) = 0.4 \text{ g}$$

Like the g-trimmed mean and g-Winsorized variance, these estimators are robust, in that they are insensitive to a small fraction of possibly outlying data.

4. Confidence Limits for the Mean

Given a particular underlying distribution for the measurements, every estimator has its own probability distribution. For any estimator of the target parameter, the variability of the estimator about its target value is important. Ideally, the estimator comes very close to the true value of the parameter in a high percentage of the data sets on which it is used. The mathematical formulation of this problem is embedded in the following question: What is the probability that an estimator, say $\hat{\mu}$, of a parameter, lies within c units of the true value of the parameter, say, μ? Mathematically,

$$\text{Prob}\{|\hat{\mu} - \mu| \leqslant c\} = ? \tag{2}$$

Notice that this question is different from asking about probabilities of events concerning individual observations x_i, but the probability density function of $\hat{\mu}$ is related to the assumed underlying model for the data.

Generally, the statistician chooses a value for the probability in Eq. (2), say, 0.95, and then determines the value of c so that the statement holds. Since $\{|\hat{\mu} - \mu| \leqslant c\}$ is equivalent to $\{\hat{\mu} - c \leqslant \mu \leqslant \hat{\mu} + c\}$, Eq. (2) is the probability that the *confidence interval*, $\hat{\mu} \pm c$, contains the parameter μ. The practical interpretation for the confidence interval is the following: If this procedure for calculating $\hat{\mu}$ is carried out on many sets of n observations, then one can expect that 95% of all the resulting confidence intervals will cover the true mean (5% of the intervals so generated will miss μ entirely).

Values of c for various values of the probability (2) can be determined for the three estimators defined in the previous three sections. The confidence limits on the mean μ are then given by $\hat{\mu} \pm c$. Denoting the probability in Eq. (2) by $1 - \alpha$ (by convention), c is given by the following formulas:

1. Using the sample mean, the allowance c in the confidence limits for μ is given by

   $$c = t_{(n-1,\ \alpha/2)} s/\sqrt{n},$$

   where $t_{(n-1,\alpha/2)}$ is the abscissa on the Student's t density function on n-1 degrees of freedom such that the area under the right tail is $\alpha/2$. The values from Student's t are used because the probability in Eq. (2) arises from the area under a density function of this form, when $\bar{x}$ is calculated in this way from n independent Gaussian observations. When the distribution is heavy tailed, the confidence interval usually will be conservative, sometimes excessively so. Values of Student's t are easily available for many tail areas and degrees of freedom, depending upon the values of n from the data and α chosen from Eq. (2) (e.g., Ref. 1).
2. Using the g-trimmed mean, the allowance c for the $100(1 - \alpha)$% confidence limits on μ is given by

   $$c = t_{(f,\alpha/2)} s_{Tg}/\sqrt{(n - 2g)}, \quad f = (n - 2g - 1)$$

   where $t_{(f,\alpha/2)}$ again refers to Student's t density function, here with f degrees of freedom (see Refs. 1 and 2).
3. Using the median, 95% confidence limits on μ are given by $(x_M \pm 2.4 s_M/\sqrt{n})$. Limits for other confidence levels are discussed in Ref. 3.

D. Estimators for Higher-Order Models

When the data can be grouped according to one or more criteria, the simple linear model (1) may be inadequate for expressing the variation in each observation. It is then necessary to consider higher-order models that involve parameters associated with the additional classifications. The total variation in the data may then be decomposed according to the various classifications, and the important factors will be distinguished by large contributions to the total variation in the data. This type of decomposition of the total variation into its various components is called an *analysis of variance*.

An example will illustrate how a set of observations may be decomposed according to its sources of variation. Suppose that four measurements are taken on each of six vials of a solution. If there is no difference among the vials, then Eq. (3) may describe the response adequately:

$$x_i = \mu + \varepsilon_i, \quad i = 1, 2, \cdots, 24 \tag{3}$$

However, if a difference among the vials is suspected for any reason (e.g., filled on different days, or filled from an inhomogeneous batch of material), then "vial" may a *factor* having six *levels*, and the model should include terms (usually called *effects*) to distinguish the six vials from one another:

$$x_{ij} = \mu + \alpha_i + \varepsilon_{ij} \tag{4}$$

where i = 1, 2, . . ., 6 vials and j = 1,2,3,4 measurements per vial. The effect α_i represents the difference between the concentration of the ith vial and the mean concentration of the lot.

Now suppose that the chemist took two measurements in the morning and two in the afternoon on each of the six vials. Then the model should include a factor to distinguish between the two time periods:

$$x_{ijk} = \mu + \alpha_i + \beta_j + \varepsilon_{ijk} \tag{5}$$

where i = 1,2, . . ., 6 vials, j = 1,2 periods, and k = 1,2 measurements. The effect β_j represents the departure from μ due to the jth time period (morning or afternoon), and ε_{ijk} again assumes the role of the "remainder" that is unexplained by the other two factors.

The measurement errors in each of the above models are assumed to be random, independent variables and therefore to have a probability density function. If the six sample vials are drawn randomly from a lot of vials, the vial effects ($\alpha_1, \alpha_2, \ldots, \alpha_6$) also may be considered random variables from the population of all vials, which is itself modeled by a probability density function. The probability density functions of the α's and ε's are assumed to be centered at zero, so that μ represents the overall average concentration for all vials. The effects due to the time periods (β_1, β_2) could be treated as fixed, unknown constants. The magnitude of the difference $\beta_1 - \beta_2$ represents the differential bias between periods.

The goals of a general analysis of variance are (1) to assess the degree to which the effects of the factors represented by the α's and β's may exceed or be comparable to the variance of the error terms, ε's and (2) to provide appropriate estimates of variability for the estimated effects. This section presents two methods for estimating the parameters in the single-factor model (4) and for assessing the contribution to the overall variance of the observations, relative to that from the "unexplained" error terms.

1. Analysis of Variance: One Factor

The general form of model (4), an example of a *linear additive model*, is

$$x_{ij} = \mu + \alpha_i + \varepsilon_{ij} \tag{6}$$

where i = 1, 2, . . ., p groups and j = 1, 2, . . ., n_i measurements per group. The $N = \Sigma n_i$ measurements may be displayed in a table:

Group 1	Group 2	. . .	Group p
x_{11}	x_{21}	. . .	x_{p1}
x_{12}	x_{22}	. . .	x_{p2}
⋮	⋮	. . .	⋮
x_{1n_1}	x_{2n_2}	. . .	x_{pn_p}

A graphical display of the data is obtained by plotting the values x_{ij} versus the group number i, as in Fig. 3. This reveals any unusual features in the data very quickly.

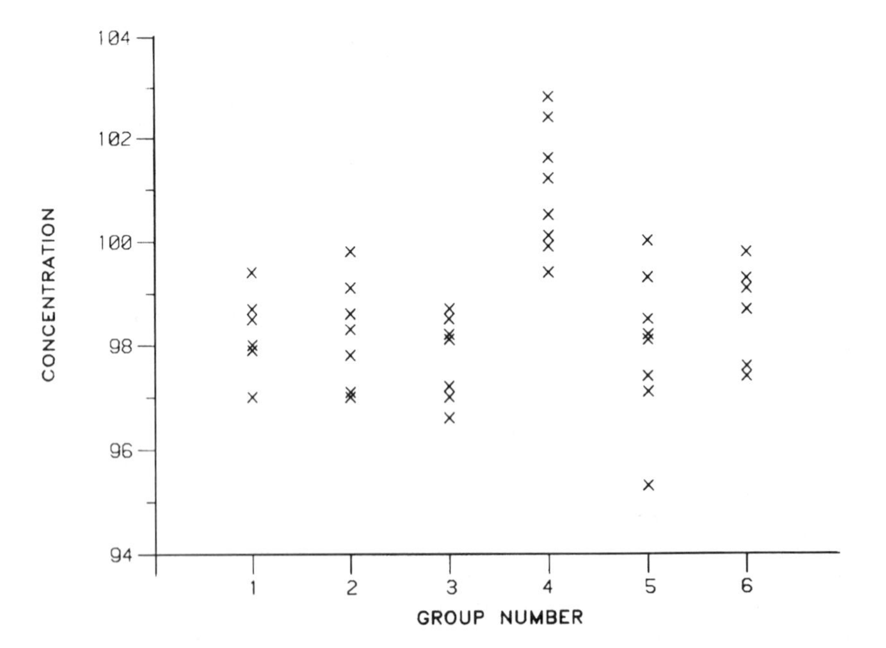

Figure 3 Illustrative graphical display of data from one-way analysis with six groups, in which the data from group 4 are consistently high.

The overall mean μ may be estimated by the grand average of all measurements:

$$\hat{\mu} = \bar{x}_{\cdot\cdot} = \Sigma\Sigma x_{ij}/N \tag{7}$$

and the effect due to the ith group, α_i, may be estimated by

$$\hat{\alpha}_i = \bar{x}_{i\cdot} - \bar{x}_{\cdot\cdot} = \sum_j x_{ij}/n_i - \bar{x}_{\cdot\cdot} \tag{8}$$

The dot subscript notation indicates that the data values have been averaged over the corresponding index.

The analysis of variance procedure begins by considering the total variation in all $N = \Sigma n_i$ measurements about the grand mean. The overall standard deviation, corresponding to the definition given in Sec. II.C, would be calculated as

$$s = \left(\frac{\Sigma\Sigma(x_{ij} - \bar{x}_{\cdot\cdot})^2}{N - 1} \right)^{1/2}$$

A well-known identity re-expresses the numerator sum of squares as

$$\Sigma\Sigma(x_{ij} - \bar{x}_{\cdot\cdot})^2 = \Sigma\Sigma(x_{ij} - \bar{x}_{i\cdot})^2 + \Sigma n_i(\bar{x}_{i\cdot} - \bar{x}_{\cdot\cdot})^2 \tag{9}$$

The first sum on the right-hand side of Eq. (9) sums the squared deviations of the measurements about their *group means*; i.e., it gives a measure of how the observations *within* the groups vary among themselves. This term is called the *within-group sum of squares*. Analogously, the second sum on the right-hand side sums the squared deviations of each *group mean* about the *overall mean*, so it gives a measure of how different the groups are from one another. This term is called the *between-group sum of squares*. Since these two sums of squares involve different combinations of the data, they are made commensurate by dividing each by a term called the *degrees of freedom* (d.f.). The result is called a *mean square*. Table 1 summarizes the calculations in this one-way analysis of variance and introduces some notation for the important quantities derived.

Table 1 One-Way Analysis of Variance

Variation	Sums of squares	d.f.	Mean squares
Between groups	$SS_B = \Sigma n_i(\bar{x}_{i\cdot} - \bar{x}_{\cdot\cdot})^2$	p-1	$MS_B = SS_B/(p-1)$
Within groups	$SS_W = \Sigma\Sigma(x_{ij} - \bar{x}_{i\cdot})^2$	N-p	$MS_W = SS_W/(N-p)$
Total	$SS_T = \Sigma\Sigma(x_{ij} - x_{\cdot\cdot})^2$	N-1	

The differences among the *means* $\bar{x}_{1\cdot}, \bar{x}_{2\cdot}, \ldots, \bar{x}_{p\cdot}$, as reflected in MS_B, are assessed in relation to the variation among the data within groups. If the means are not very different (MS_B small), relative to the differences within the groups themselves (MS_W large), then the ratio MS_B/MS_W would be small. This condition leads to the conclusion that the α's in Eq. (6) are all identically zero. But if the observations are very tightly clustered around their own group means (MS_W small), and the means are quite different (MS_B large), then this ratio is large and suggests that the groups themselves arise from populations having different mean values, represented by nonzero α's. In addition, if the data are Gaussian, the distribution of the value of this ratio MS_B/MS_W may be judged, because this ratio itself is distributed, when all α's are zero, according to a tabulated function called the F distribution. Such tables, together with explanations on their usage, are readily accessible (e.g., see Ref. 1). Fortunately, this distribution is also a good approximation under much more general conditions, when all α's are zero, which require only randomization of the contributions to error.

2. Median Polish

As shown in Sec. II.C for estimating a single location parameter, the median provides a robust estimate of the mean of the measurements in the presence of outlier-prone data. Likewise, for the one-factor model in Eq. (6), an alternative analysis to the analysis of variance presented above is a degenerate form of the median polish algorithm [4], wherein sample means are replaced by sample medians. Corresponding to these estimates, median polish leads to

$$\hat{\mu} = \operatorname*{med}_{ij}\{x_{ij}\} = M$$

$$\hat{\alpha}_i = M_i - M, \quad M_i = \operatorname*{med}_{j}\{x_{ij}\}$$

Thus, the quantity μ is estimated by the grand median of all N observations rather than by the grand mean of all observations as in Eq. (7). Similarly, α_i is estimated as the difference between the median of the ith group and the grand median.

No exact statistical theory analogous to that for the analysis of variance table exists for median polish. However, the between and within components of variation can be compared informally as follows. These components may be measured by defining quantities analogous to the terms in the analysis of variance table:

$$MM_B = \bar{n}(1.5\operatorname*{med}_{i}|M_i - M|)^2, \quad \bar{n} = \Sigma n_i/p$$

$$MM_W = \operatorname*{med}_{i}\{u_i^2\}, \quad u_i = 1.5\operatorname*{med}_{j}|x_{ij} - M_i|$$

Their ratio, MM_B/MM_W, may be thought of as a "pseudo-F" statistic, whose magnitude may be regarded as analogous to the ratio MS_B/MS_W

from the analysis of variance. The approximation involved in referring such "pseudo-F" values to a conventional F table has not been studied carefully, but such reference seems likely to provide useful insight (see Sec. III.C).

Median polish is useful as a comparison with the conclusions derived from the classical analysis of variance. If the estimates of the parameters μ and α_i, or the ratios of the components of variation, are very different, the data should be examined to determine the cause of the differences (e.g., outliers, or asymmetric or non-Gaussian distribution of the data). Examples in Sec. III will illustrate the practice of comparing the results from two analyses.

The median polish technique can be extended to apply to models like that in Eq. (5) having two factors, and, further, to models having three or four factors. An example of a median polish with two factors is given in Sec. III.D.

III. APPLICATIONS AND EXTENSIONS

The aforementioned techniques are illustrated most clearly on actual data. No statistical procedure can be applied blindly without taking into account all aspects of the problem. While some applications are straightforward, the following examples rely on a combination of simple and more advanced techniques. The first example deals with the problem when some high-performance liquid chromatography (HPLC) data involve a time component in the external standard that affects the estimation of the main parameter of interest. The second example illustrates the use of robust estimators in the face of potential outliers, the recognition of which often requires a high degree of familiarity with the data. The third example is an extension of the first, as it includes other factors to be considered when determining total variability. Higher-order models are discussed in the fourth example involving data on polychlorinated biphenyl (PCB) concentration.

A. HPLC Analysis of Halocarbons

A Standard Reference Material at the National Bureau of Standards, known as SRM 1639, was certified for amounts of seven halocarbons in a methanol solution. The certification of these highly volatile compounds required three steps. First, the analytes and methanol were weighed on a balance and then carefully combined. This step provided gravimetrically determined concentrations of each analyte. Then the mixture was put into ampuls in such a way so as to minimize evaporation of any component. An analysis of a set of sample vials via HPLC was required to assure homogeneity among the vials at this second step. Finally, gas chromatography was used as a second, independent method of determining the concentration of each analyte.

The HPLC phase of the procedure necessitated the chemical analysis of many sample vials, so the chemist wanted to monitor the HPLC apparatus to ensure that any significant drift could be taken into acount. Thus, an HPLC analysis on an external standard was run several times throughout the day, and peak areas for each of seven different analytes were recorded.

A statistical analysis based on the logarithms of the peak areas conforms to model assumptions about the distribution of the error terms more closely than an analysis based on the untransformed data [5]. In an HPLC analysis using external standards, the concentration of an analyte is determined from the equation

$$C(a) = [P(a)/P(s)]C(s) \qquad (10)$$

where C and P denote concentration and peak area, respectively, and a and s denote analyte and standard, respectively. Taking logarithms of both sides of Eq. (10) results in additivity:

$$\log[C(a)] = \log[P(a)] - \log[P(s)] + \log[C(s)] \qquad (11)$$

Typically, log[C(s)] is known, but log[P(a)] and log [P(s)] are measured during the chemical analysis and are thus subject to experimental error. Histograms of logarithms of peak areas reveal more nearly symmetric distributions than those of the raw peak areas. Both additivity of the model and symmetry of the distributions argue for a statistical analysis based on logarithms. (For many chemists, $\log_{10}$ is more familiar, but $\log_e$ is perfectly acceptable.)

To detect instrumental drift, it is appropriate to plot, for each analyte, x_i = logarithm of the peak area of analyte for run i versus t_i = time of ith run (in minutes since 11:00 A.M.). Seven such plots are possible, corresponding respectively to the analytes $CHCl_3$, $CHBrCl_2$, $CHBr_2Cl$, $CHBr_3$, C_2HCl_3, CCl_4, and C_2Cl_4. Naturally, the logarithms of the peak areas in this standard will have different levels and degrees of variability for the different analytes. To compare all seven plots on one graph, Fig. 4 plots normalized values $(x_i - M)/s_M$ versus time t_i, where M is the median over time of the data on a particular analyte, and s_M is the scale estimate 1.5MAD, described in Sec. II.C.

Having normalized the logarithms of the peak areas for their different locations and spreads, Fig. 4 reveals that the seven analytes are all roughly linearly increasing with time with a more or less common rate of drift. However, the data at time 11:50 A.M. are suspiciously low. The chemist noted in his lab book that the slope sensitivity (SS) of the peak area integrator was different at this time: SS = 25, whereas SS = 10 for all other runs. While this is a minor change, it does give reason for excluding these data. Notice that inclusion of the low values from 11:50 A.M. would result in somewhat higher estimates of the drift.

A very simple technique for estimating the drift for each analyte involves a calculation of a slope based on two median points in the outer

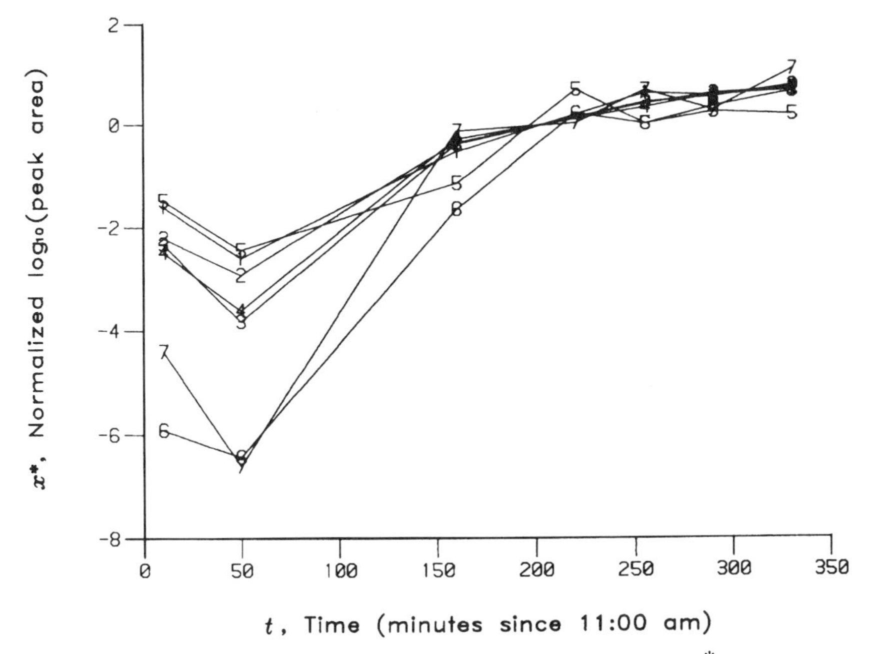

Figure 4 Data from HPLC homogeneity study: plot of x_i^* versus the ith time t_i, where $x_i^* = (x_i - M)/s_M$ for $x_i = \log_{10}$(peak area at time i of given analyte), $M = \text{med}\{x_i\}$, and $s_M = 1.5\text{MAD}\{x_i\}$. Seven analytes are represented: 1 = $CHCl_3$, 2 = $CHBrCl_2$, 3 = $CHBr_2Cl$, 4 = $CHBr_3$, 5 = C_2HCl_3, 6 = CCl_4, and 7 = C_2Cl_4.

thirds of the data. For example, the six normalized values corresponding to the first analyte are sorted:

Time t_i:	10	160	220	255	290	330
Medians (outer thirds):	85				310	
Data x_i:	-1.624	-0.572	—	0.603	0.572	0.692
Medians (outer thirds):	-1.098				0.632	

Slope = (differences in x's)/(difference in t's) = 1.730/225 = 0.00769

Notice that this estimate of slope is especially convenient in this problem because it does not depend on the missing x data corresponding to $t_i = 220$. The above calculations show that the normalized logarithm of the HPLC response (logarithm of the peak area) appears to drift at a rate of 0.00769 unit/min for the analyte $CHCl_3$. Proceeding in this way for all seven analytes, we have seven estimates of the normalized drift parameter. Relevant calculations are given in Table 2. The first five analytes exhibit roughly the same amount of drift, while the last two drift somewhat more rapidly. The median estimate of the normalized drift is 0.00921 unit/min. Formally, the model states that

Table 2 Calculations for Drift in HPLC Standard

Minutes since 11:00 A.M.	Peak area for analyte						
	$CHCl_3$	$CHBrCl_2$	$CHBr_2Cl$	$CHBr_3$	C_2HCl_3	CCl_4	C_2Cl_4
A. Raw data							
10	38567	14661	15411	12358	36627	5917	22588
50	36938	14053	14671	11788	35699	5815	22180
160	40372	16332	16411	13540	36934	6822	23396
220	—	—	—	—	38827	7967	23429
255	42492	17125	16877	13915	38138	7220	23558
290	42435	17343	16947	14070	38398	7307	23484
330	42656	17455	17077	14166	38353	7384	23645
M = median (log peak area)	4.617	4.223	4.221	4.138	4.581	3.859	4.370
B. $\log_{10}$ (peak area) - median (over time) = y_i - M							
10	-0.0307	-0.0572	-0.0334	-0.0456	-0.0176	-0.0864	-0.0159
50	-0.0494	-0.0756	-0.0547	-0.0661	-0.0287	-0.0940	-0.0238
160	-0.0108	-0.0103	-0.0061	-0.0059	-0.0139	-0.0246	-0.0006

220	—	—	—	—	0.0078	0.0028	0
255	0.0114	0.0103	0.0061	0.0059	0	0	0.0024
290	0.0108	0.0158	0.0079	0.0107	0.0030	0.0052	0.0010
330	0.0131	0.0186	0.0112	0.0137	0.0024	0.0097	0.0040
s_M = 1.5 MAD	0.0189	0.0258	0.0143	0.0183	0.0117	0.0146	0.0036
C. Normalized log (peak area) $y_i^* = (y_i - M)/s_M$							
10	-1.624	-2.218	-2.334	-2.488	-1.505	-5.907	-4.438
50	-2.615	-2.931	-3.828	-3.607	-2.459	-6.423	-6.664
160	-0.572	-0.400	-0.425	-0.324	-1.194	-1.686	-0.171
220	—	—	—	—	0.667	0.193	0
255	0.603	0.400	0.425	0.324	0	0	0.667
290	0.572	0.612	0.551	0.586	0.253	0.356	0.285
330	0.692	0.721	0.783	0.747	0.209	0.667	1.114
Normalized drift estimate (excludes 11:50 A.M. data)							
	0.00769	0.00878	0.00910	0.00921	0.00703	0.01914	0.01335

$$x_i^* = (x_i - M)/s_M = A + Bt_i + (\text{error})$$

and that B has been estimated as 0.00921. Hence the drift for each analyte in its unnormalized scale is Bs_M units/min. (A standard error for this slope estimate may be obtained by a procedure discussed in Chap. 8 of Ref. 6.)

This drift estimate now becomes important when standards can be run only infrequently throughout the day. Knowing the rate of drift permits an adjustment to the logarithm of the standard peak area to correspond to the time at which samples are run. For example, suppose a reference standard is run at 3:00 P.M., and three sample vials are run at 3:45 P.M., 4:30 P.M., and 5:15 P.M.. Apart from a constant factor corresponding to the logarithm of the concentration of standard, Eq. (11) leads to

log(concentration of analyte in sample vial) =
log(peak area from sample) - adjusted log(peak area from standard) + constant

i.e., the logarithm of the peak area for the standard has been adjsuted to match the value it might have been had it been run at exactly the same time as the sample. Since 45 min elapsed between the measurement on the standard and that of the first vial, the logarithm of the peak area from the standard has been adjusted by adding 0.00921 unit/min $\times$ 0.01890 (= s_M) = 0.00783 unit, so that both standard and sample are based on approximately the same time. Likewise, the logarithm of the peak area from the standard for the second and third sample vials have been adjusted by 0.00921 $\times$ 0.01890 $\times$ 90 = 0.01567 unit and 0.00921 $\times$ 0.01890 $\times$ 135 = 0.02350 unit, respectively.

This example illustrates the following:

Simple plotting is important.
Proper normalization of variables facilitates comparison.
Simple statistics such as the median and the MAD are useful in the normalization.
Logarithms of peak area ratios should be adjusted to match the times at which samples are run. This important adjustment requres a sufficient number of runs on the standards to estimate the drift rate accurately.

The next two sections will use these same HPLC data to illustrate additional statistical techniques. Because the factor corresponding to the logarithm of the concentration in the standard will be constant for all data, only the logarithm of the peak area ratio need be considered. It will be assumed that the values have been properly time adjusted, as outlined above.

B. Outliers

This second example will be simplified to illustrate the computation of simple location and scale estimators described in Sec. II.C, and the differences among these estimators in the presence of outliers. Ten sample vials of SRM 1639 were chemically analyzed for C_2HCl_3 via HPLC (Sec. III.A) as part of a study to verify the homogeneity of the material. Using the notation in Sec. II.C, let

$$\begin{aligned} x_i &= \log(\text{concentration of } C_2HCl_3 \text{ in ith vial}) \\ &= \log(\text{peak area ratio}) + \log(\text{concentration of standard}) \\ &\quad + \log(\text{error term}) \\ &= \mu + K + \varepsilon_i \end{aligned} \tag{12}$$

The logarithm of the concentration of the standard, K, is known and is the same for all 10 measurements, so μ is the main target for estimation. [Equivalently, μ = log(average concentration in samples/concentration in standard).] The 10 values are shown in Table 3. One of the values, 0.04227, is somewhat smaller than the others. The calculations for the three methods of estimation given in Sec. II.C are shown below the data in Table 3. The trimmed mean and the median are somewhat larger compared to the sample mean, and their associated scale estimates are smaller than the sample standard deviation. These differences are caused by essentially the one low value. This illustrates an important use of robust estimators: comparisons of the results may highlight potential anomalies in the data. It is important to follow up on any suspicious data by inquiring into the chemical measurements. In this example, one chemical explanation may be related to the time at which this observation was run.

At this point, a word of caution should be raised concerning a potentially illegitimate use of robust estimators. Strictly speaking, it is not "fair" to compute several estimates and to choose the "best" answer. Using one estimator, a 95% confidence interval will fail to cover the true value of the parameter 5% of the time. However, if three such intervals are computed and only the shortest is reported, then the probability that the reported interval fails to cover the true value is larger than 5%. How much larger? A simple approximation based on the so-called Bonferroni inequality may be derived by the following line of reasoning.

The probability that the shortest of three confidence intervals fails to cover the true value is less than or equal to the probability of the union of three events: interval 1 fails <u>or</u> interval 2 fails <u>or</u> interval 3 fails. This in turn is less than or equal to the sum of the individual probabilities for each interval. Thus the probability that the shortest of three 95% confidence intervals will fail to cover the true parameter value can be no more than $3 \times 5\% = 15\%$.

If the data of Table 3 were to be summarized by reporting the shortest of the confidence intervals in panel D, the confidence level

Table 3 Three Estimation Methods for Calculating C_2HCl_3 Concentration in MeOH (SRM 1639)

A. The data [10 values of log(peak area ratio)]

0.05357, 0.05536, 0.05508, 0.05048, 0.05431, 0.05291, 0.04227, 0.04614, 0.05142, 0.04965

B. Sorted data

0.04227, 0.04614, 0.04965, 0.05048, 0.05142, 0.05291, 0.05357, 0.05431, 0.05508, 0.05536

C. "Plot" of data

```
                X
                X                 .04• = 0.040 - 0.044
                X                 .04+ = 0.045 - 0.049
        X       X      X          .05• = 0.050 - 0.054
   X    X       X      X          .05+ = 0.055 - 0.059
 -------------------------
  .04•  .04+   .05•   .05+
```

D. Three estimators for μ and σ

1. Sample mean and sample standard deviation:
   $\bar{x} = 0.05112$, $s = 0.00420$ (d.f. = 9)
   95% confidence interval for average log(peak area ratio)
   $= 0.05112 \pm 2.262(0.00420/\sqrt{10}) = 0.05112 \pm 0.00300 = (0.04812, 0.05412)$
2. 10% trimmed mean (g = 1) and Winsorized standard deviation:
   $\bar{x}_{T1} = 0.05170$, $s_{T1} = 0.00382$ (d.f. = 7)
   95% confidence interval for average log(peak area ratio)
   $= 0.05170 \pm 2.365(0.00382/\sqrt{8}) = 0.05170 \pm 0.00319 = (0.04850, 0.05489)$
3. Median and MAD:
   Median = 0.05216; 1.5 MAD = 0.00350
   95% confidence interval for average log(peak area ratio)
   $= 0.05216 \pm 2.4(0.00350/\sqrt{10}) = 0.05216 \pm 0.00266 = (0.04950, 0.05482)$

could be reported honestly (and conservatively) as 85%, but not 95%. In order to ensure retaining at least 95% confidence when computing three confidence intervals for the same data, each confidence interval should be computed at the $100(1 - \alpha/3)\% = 98.3\%$ level.

Outliers are not unusual; unfortunately, reporting them is far less common. Sometimes an investigator is all too quick to disregard a slightly discrepant observation as an outlier. In a scientific approach to measurement, it must be realized that "realistic performance parameters require the acceptance of all data that cannot be rejected for cause" [7]. Alternative estimators are important tools for identifying outliers and searching for their causes.

C. Assessing Various Sources of Error

As described in Sec. III.A, part of the certification process for SRM 1639 required the assessment of the homogeneity among the ampuls. Although the mixture was put into ampuls in a carefully monitored environment to prevent evaporation of any component, perfect control was impossible. After this procedure, it was necessary to ascertain the vial-to-vial homogeneity via the HPLC process.

Three sets of 10 sample vials each were chemically analyzed. The three sets related to the beginning, middle, and end of the ampul sequence. (Data in Sec. III.B were the 10 measurements in the second set of vials.) Using the model (6), we have

$$x_{ij} = \mu + \alpha_i + \varepsilon_{ij}$$

$$j = 1, 2, \ldots, n_i = \text{number in ith set and } i = 1,2,3 \text{ sets},$$

where μ is the mean log [ratio of concentrations C(s)/C(a)], α_i is the departure from μ due to the ith set and ε_{ij} is the random error term. The number of vials in each set, n_i, was targeted at 10, but only 9 vials were analyzed in the first set. The final numbers in each group are $n_1 = 9$, $n_2 = 10$, and $n_3 = 11$. The data are given in Table 4 and are plotted in Fig. 5. (For convenience, the data have been listed in ascending order in each set.)

Curiously, every group exhibits at least one unusually low value. These measurements were on samples that were run on one particular day, and unusual conditions may have affected the equipment during this time. With or without these values, however, Fig. 5 shows that the logarithms of the peak area ratios tend to exhibit about the same variability. In Table 4, the total sum of squares is decomposed into the between and within sums of squares (3.1×10^{-5} and 3.9×10^{-4}, respectively), yielding a small, nonsignificant ratio of mean squares $1.08 = 1.5 \times 10^{-5}/1.4 \times 10^{-5}$ when compared to the F distribution. Notice that the analogous mean-squared quantities from the median polish are comparable in magnitude (0.73×10^{-5} and 0.99×10^{-5}, respectively); their ratio, analogous to the F statistic, is 0.74. Based on these data, there is no evidence of inhomogeneity among the sample vials put into ampuls at the beginning, middle, or end of the ampul-filling process.

In estimating the relative precision for these data, it would be inappropriate to estimate the relative standard deviation in the usual way as $\hat{\sigma}/\hat{\mu}$ (Sec. II.C), because these data have been transformed via logarithms to make them more nearly symmetric. However, standard propagation of error formulas [8] show that 2.3 times the (raw) standard deviation of $\log_{10}$-transformed data actually estimates the *relative*

Table 4 Logarithms of peak area ratios for Three Sets of Ampuls in SRM 1639 Homogeneity Study (HPLC of C_2HCl_3 in MeOH)

A. Data [log(peak area ratios)]

Set 1	Set 2	Set 3
0.04314	0.04227	0.04425
0.04594	0.04614	0.04855
0.04718	0.04965	0.05128
0.05081	0.05048	0.05292
0.05086	0.05142	0.05293
0.05149	0.05291	0.05308
0.05256	0.05357	0.05397
0.05296	0.05431	0.05426
0.05474	0.05508	0.05486
	0.05536	0.05493
		0.05588
Mean		
0.04996	0.005112	0.05245
Standard deviation		
0.00376	0.00420	0.00338
Median		
0.05086	0.05216	0.05308
1.5 MAD		
0.00376	0.00350	0.00267

Total sum of squares = 4.16×10^{-4}
Grand mean = 0.05126
Grand standard deviation = 0.00379
Grand median = 0.05273
Grand 1.5MAD = 0.00295

B. Analysis of variance

Source	Sum of squares	d.f.	Mean squares	F ratio
Between sets	1.15×10^{-5}	2	1.54×10^{-5}	1.08
Within sets	3.86×10^{-4}	27	1.43×10^{-5}	
Total	4.16×10^{-4}	29		

C. Analysis by medians

Source	d.f.	"Mean squares"	"Pseudo-F"
Between sets	2	7.31×10^{-6}	0.74
Within sets	27	9.92×10^{-6}	
Total	29		

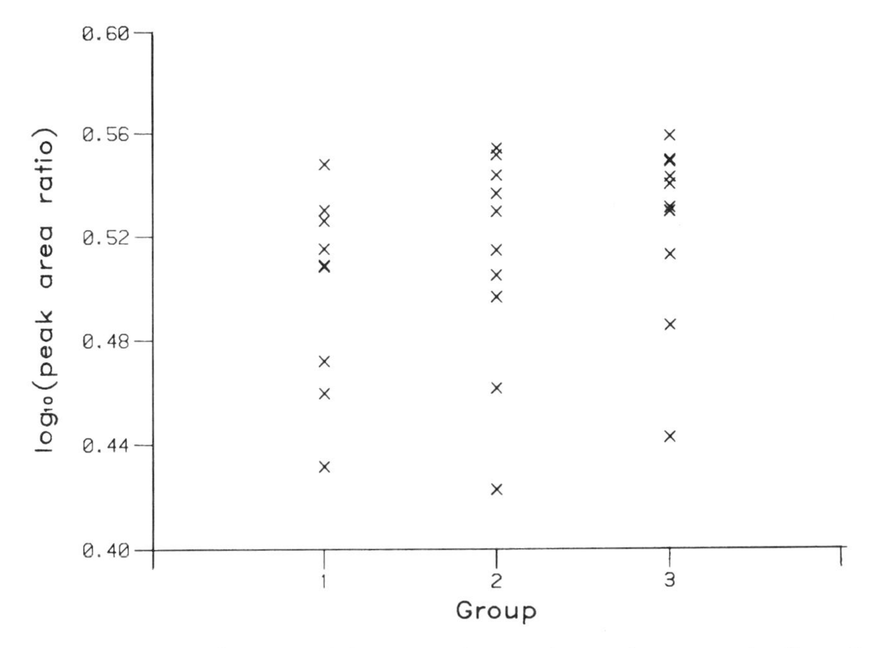

Figure 5 Data from HPLC homogeneity study on the concentration of C_2HCl_3 in the Standard Reference Material for halocarbons in methanol. Three sets of ampuls were analyzed via HPLC to yield peak area ratios. The logarithms of these ratios are plotted against the group number, where 1 = "early," 2 = "middle," 3 = "late" in the ampul-filling process.

standard deviation of the untransformed data. Thus, if σ^* and μ^* refer to the standard deviation and mean of the untransformed data, and σ is the standard deviation of the $\log_{10}$-transformed data, then

$$2.3\sigma \simeq \sigma^*/\mu^*$$

(The factor 2.3 = log 10 is the conversion factor from natural to common logarithms. This factor is unnecessary if the data are transformed via natural logarithms.)

For this study, the estimated standard deviation in $\log_{10}$ units is 0.0052, so the estimated *relative* standard deviation, in the *original* units, is 2.3 X 0.0052 = 0.012, or 1.2%. This is the apparent analytical error in the HPLC procedure for measuring the concentration ratio, (concentration of analyte in sample)/(concentration of analyte in standard). Assuming that the concentration in the standard is known with negligible uncertainty, 1.2% is also the relative standard deviation of the procedure for measuring concentration in the sample.

D. Summarizing Results from Multiple Internal Standards and Analyte Peaks

A grave danger in statistical techniques is their misapplication, either by using inappropriate procedures or by neglecting fundamental assumptions which accompany them. Sometimes different results from different estimation methods can signal the presence of model violations. The discrepancy may not be obvious, and blind faith in the assumptions can mislead the experimenter in both the estimates of model parameters and their uncertainties. In other circumstances, such as in Sec. III.C, despite the appearance of suspected outliers, both the analysis by means and the analysis by medians lead to the same conclusion.

Another serious misuse of statistical procedures concerns nonindependence. As stated in Sec. II.B, it is reasonable to assume that separate, distinct measurements have independent errors associated with them. So the response of a sample from HPLC or gas-chromatographic equipment may be treated as independent from the response obtained on a different sample run on another occasion. However, for an analyte such as a PCB mixture, the response consists of not one, but several peaks, all referring to the same "analyte" under investigation, so the concentration can be determined in several ways, corresponding to the different peak areas. But these calculated concentrations are *not* independent, so it is not appropriate to calculate a standard error of $\hat{\mu}$ (e.g., $\bar{x}$ or x_M or any other estimator) by naive application of the formulas given in Sec. II.C. In fact, such a calculated standard error often grossly underestimates the true uncertainty of $\hat{\mu}$, because the data are highly correlated. It is important to recognize that all these values really refer to only one run for one sample, which is insufficient for estimating a population mean at all. If several independent samples are run and the information for each sample is summarized by one value, then the variation among these independent summary values may be used to calculate a realistic uncertainty.

The example which illustrates this principle arose from some gas-chromatographic measurements on NBS Standard Reference Material 1581 for the concentration of polychlorinated biphenyls (PCBs) in motor oil. The material was prepared in a large batch and then carefully blended and dispensed in 5-ml ampuls. Six sample vials were selected at random for chemical analysis using three independently prepared calibration solutions. Each measurement of the analyte gave 10 distinct peaks. In addition, each sample was "spiked" with three internal standards (IS1, IS2, IS3), which were in fact three PCB isomers that were not present in the original mixture. Since analyte concentration can be calculated from any combination of one internal standard and one analyte peak, the analyte concentration can be calculated in 30 different ways for each run. But the 30 values are not independent.

However, the values from each replication using the different calibration solutions can be treated independently from each other. Thus a technique for summarizing the 30 dependent values would yield a summary value which would be independent of the summary values based on other samples.

A statistical analysis then proceeds as follows:

1. For each calibration run, the 30 dependent values (10 analyte peaks X 3 internal standards) are summarized to yield an overall summary value, plus some information about the relative values of the effects due to the choice of analyte peaks and the internal standards.
2. The independent summary values corresponding to the different calibration runs are analyzed according to the appropriate statistical model.

The PCB data illustrating this procedure are described in detail in Ref. 5. First, three calibration solutions were prepared, denoted A, B, and C. Each calibration solution was used twice for calculating concentrations in two samples (the logarithms of which were averaged) on three separate occasions. Each occasion resulted in 30 values. Let Z_{mjln} be the random variable which denotes the logarithm of the concentration based on the mth calibration solution (m = A, B, C) the jth replication or occasion (j = 1, 2, 3), the lth analyte peak, and the nth internal standard (n = 1, 2, 3). The data are plotted in Fig. 6.

Step 1 summarizes the 30 values $\{Z_{mjln}\}$ over l = 1, 2, . . . , 10 and n = 1, 2, 3, by modeling each observation according to the following statistical model:

$$Z_{mjln} = \mu^{(mj)} + \alpha_l^{(mj)} + \beta_n^{(mj)} + \varepsilon_{ln}^{(mj)} \tag{13}$$

where μ represents the overall average value, α_l represents the departure from μ due to calculating the concentration using the lth analyte peak, β_n represents the departure from μ using the nth internal standard, and ε_{ln} represents the unexplained residual from this model. The superscripts are reminders that this additive model holds for each replication (j) of each calibration solution (m), so there are really nine such models. Note that the errors in model (13) are not independent.

As shown in Sec. II.D.1, the parameters in model (13) may be estimated using sample means, or, as shown in Sec. II.D.2, using the more robust median polish. Since chemical reasoning suggests that the concentrations based on some analyte peaks may be less reliable than those based on other peaks, the parameters in model (13) were estimated using median polish. For two-way tables, this procedure is explained fully in Ref. 4 and is applied specifically to these data in Ref. 5.

Median polish is carried out for all nine sets (j = 1, 2, 3 replications on each of m = A, B, C calibration runs). The result is nine

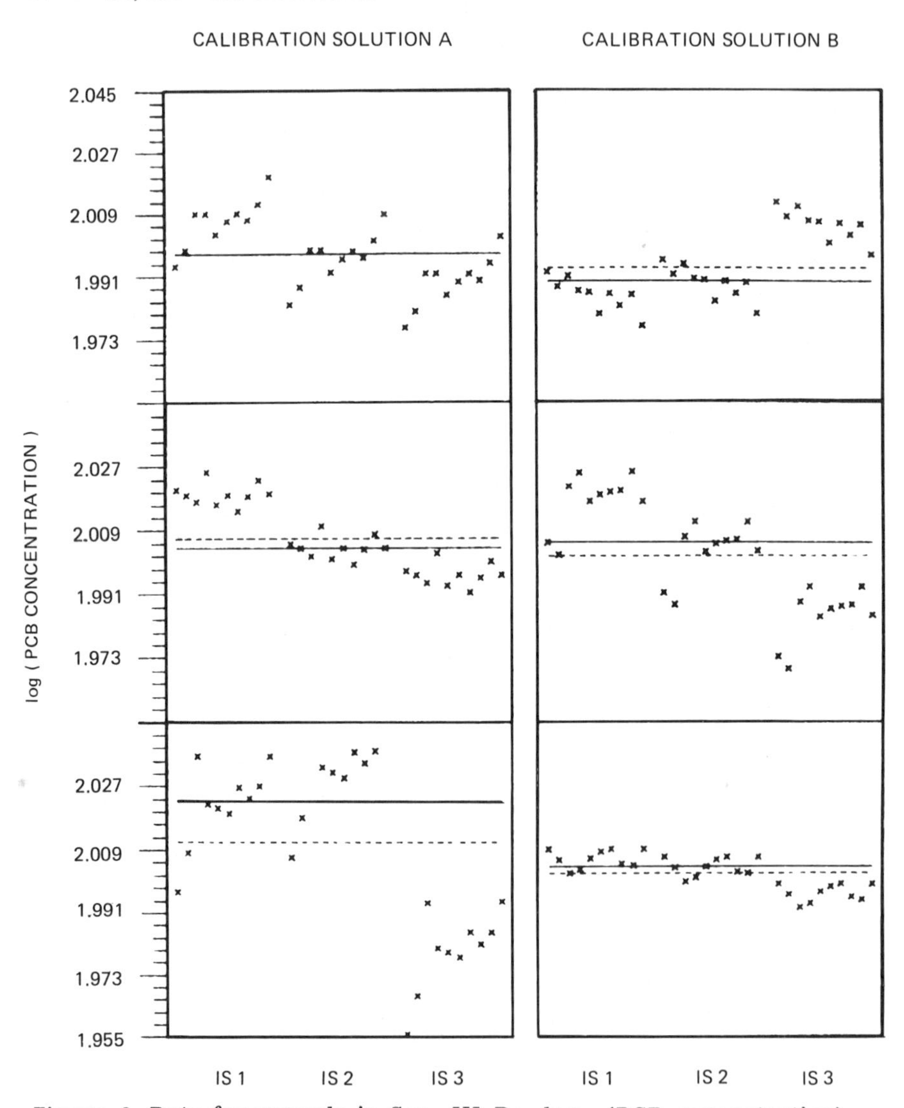

Figure 6 Data for example in Sec. III.D: $\log_{10}$(PCB concentration) using 10 analyte peaks and 3 internal standards, relative to one of three calibration solutions (left = A, middle = B, right = C). Three replications are shown in each set. The dashed line is the mean of 30 values and the solid line is the estimated typical value from two-way median polish.

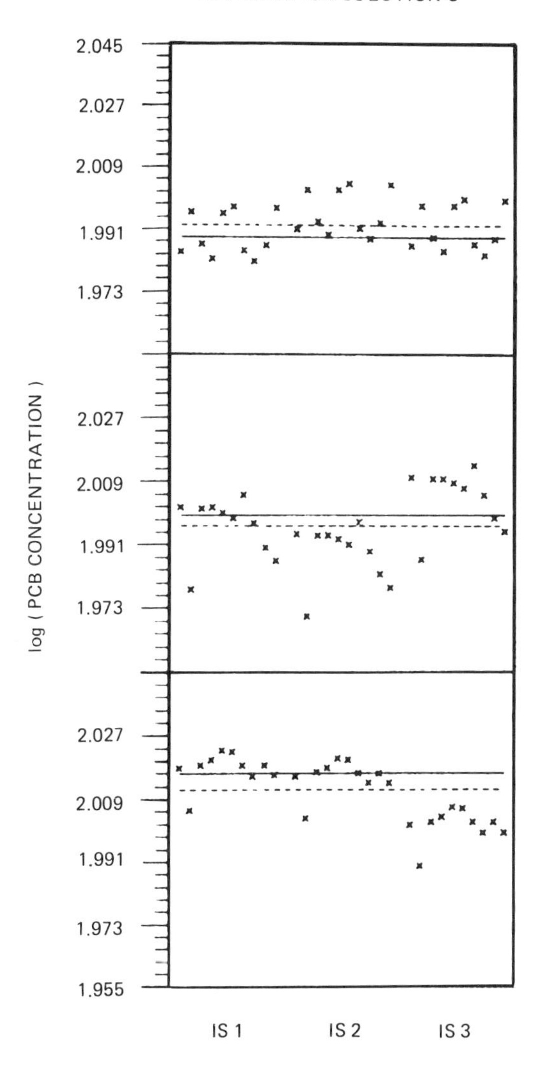

Figure 6 (continued)

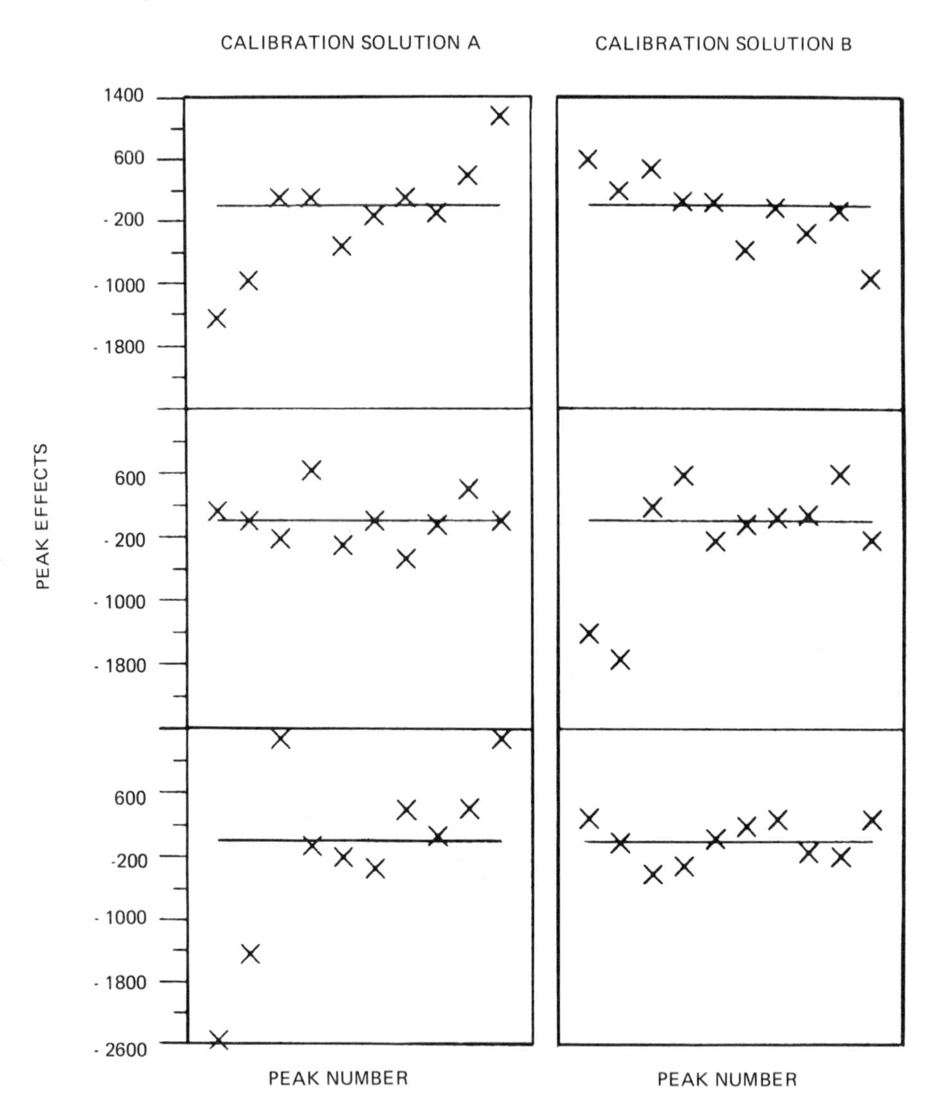

Figure 7 Peak effects (multiplied by 1000) from an additive two-way model of the data in Fig. 6. Plotted in order of peak number, there are three replications, based on three calibration solutions, or nine sets of effects.

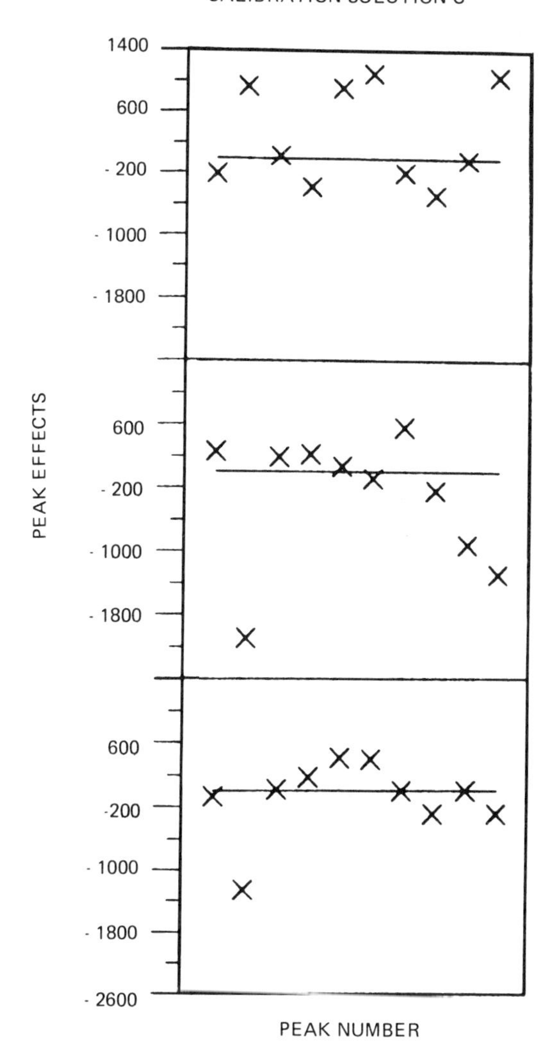

Figure 7 (continued)

overall summary values $\hat{\mu}^{(mj)}$, nine sets of $\{\hat{\alpha}_l^{(mj)}\}$, and nine sets of $\{\hat{\beta}_n^{(mj)}\}$. At the end of each analysis of the model (13), the analyte peak effects $\{\hat{\alpha}_l^{(mj)}\}$, $l = 1, 2, \ldots, 10$, are plotted as a function of peak number (Fig. 7). Notice that some peaks have unusually low peak effects (e.g., the second peak in calibration solution C). Notice also that there are no statistical tests performed on this model, since the data are not independent; the main interest is the summary value and some notion of the analyte peak effects. The summary value based on the median polish is shown in Fig. 6 as a solid line; that based on the analysis of variance (grand mean) is shown as a dashed line. The difference between them is obvious when one group of values carries the sample mean off with it.

Now the nine independent summary values from step 1 can be represented by the following statistical model:

$$\mu^{(mj)} = \eta + \xi_m + \delta_{mj} \tag{14}$$

where η = is the overall common value of the logarithm of the concentration, ξ_m is the effect due to the mth calibration solution, and δ_{mj} is the remaining unexplained error term. The δ_{mj} are independent, since each of the nine occasions involved a separate run of the gas-chromatographic equipment. Now model (14) looks just like the model in Eq. (6), and the corresponding analysis of variance table is given in Table 5.

The importance of this example is the recognition of the difference between dependent data, for which common statistical estimators such as s^2 are overoptimistic, and independent data, for which such procedures are appropriate. For readers having further interest in these data, Ref. 5 goes into greater depth and illustrates other statistical principles as well.

IV. SUMMARY

This article has reviewed some basic underlying notions of statistics that are applicable particularly to measurements taken via gas and liquid chromatography. Such data require the formulation of models, which may be revised as further data and insight become available. "We must be prepared to use many models, and find their use helpful for many specific purposes, when we already know that they are wrong—and in what ways" [9]. From such models, we can separate the important factors from the inconsequential ones, and examine the remainder for any additional information and model revision if necessary.

Various methods of estimation were presented for simple models, and standard errors for these estimates provide indications of their uncertainties. An important step, often overlooked in a statistical

Table 5 One-Way Analysis of Variance for Nine "Typical Values" of log (Concentration of PCB) in SRM 1581

		Calibration Solution		
Replication		A	B	C
1		1.99723	1.98942	1.98833
2		2.00399	2.00542	1.99936
3		2.02271	2.00415	2.01660
Source	d.f.	Sums of squares	Mean squares	F-ratio
Between calibrations	2	0.0001151	0.0000576	0.378
Within calibrations	6	0.0009127	0.0001521	
Total	8	0.0010278		

95% Student's t confidence interval for grand mean (8 d.f.): (1.99431, 2.01174) (log scale) or (98.7, 102.7) in the original (untransformed) units (μg/g)

analysis of data, is the simple plot. A graphical analysis provides useful insights concerning locations, spreads, and relationships among the data values. "There is no excuse for failing to plot and look" [4].

Finally, applications to real problems illustrate the usefulness of these techniques. They highlight the importance of a careful statistical analysis to obtain valid conclusions.

ACKNOWLEDGMENTS

We would like to thank Dr. S. N. Chesler and R. G. Christensen of the Center for Analytical Chemistry (NBS) for supplying the data in Sec. III and Dr. H. H. Ku, Professor J. W. Tukey, Professor I. Olkin, Dr. W. Carmichael, M. G. Natrella, and R. G. Christensen for their comments on this article.

REFERENCES

1. M. G. Natrella, *Experimental Statistics*, NBS Special Publication Handbook 91, U.S. Government Printing Office, Washington, D.C., 1962.

2. J. W. Tukey, and D. H. McLaughlin, Sankhya *25A*, 331–352 (1963).
3. P. Horn, J. Am. Stat. Assoc. *78*, 930–936 (1983).
4. J. W. Tukey, *Exploratory Data Analysis*, Addison-Wesley, Reading, Mass., 1977.
5. K. Kafadar and K. R. Eberhardt, Nat. Bur. Stand. J. Res. *88*, 37–46 (1983).
6. F. Mosteller and J. W. Tukey, *Data Analysis and Regression: A Second Course in Statistics*, Addison-Wesley, Reading, Mass., 1977.
7. P. E. Pontius, NBS Technical Note #288, U.S. Government Printing Office, Washington, D.C., 1966.
8. H. H. Ku, Nat. Bur. Stand. J. Res. *70C*, 263–273 (1966); also reprinted in NBS Special Publication 300, Vol. 1, U.S. Government Printing Office, Washington, D.C., 1969, pp. 331–341.
9. J. W. Tukey, Notes to Mathematics 596, by J. R. Thompson and D. Brillinger, Princeton University, Princeton, N.J., 1963.

2

Multifactor Optimization of HPLC Conditions

Stanley N. Deming, Julie G. Bower, and Keith D. Bower / University of Houston, Houston, Texas

I. INTRODUCTION

Although research is still being directed toward further improvements in liquid-chromatographic stationary phases, most practicing analysts are able to adjust only the composition of the eluent to improve their separations. Thus much current research is appropriately directed toward a fundamental understanding of how changes in eluent composition affect the chromatographic separation process.

The great versatility of high-performance liquid chromatography (HPLC) is, in fact, largely attributable to the rich variety of factors that can be adjusted in the mobile phase. Solvent composition, polarity, dielectric constant, hydrogen bonding capability, concentration of surface-active molecules, pH, and ionic strength are only a few of the many mobile phase factors (variables) that can affect sample retention times.

The intents of this article are to show why multifactor studies are required for a broad understanding of real chromatographic systems and to discuss several approaches that have been used for the systematic multifactor optimization of chromatographic methods.

II. SINGLE-FACTOR EFFECTS

When investigating multifactor systems, many researchers use a primitive approach in which each factor is investigated individually, one factor at a time. The philosophy underlying this approach is based on the classical belief that if a researcher wants to find out how one factor affects a system, then the research should hold all other factors constant and vary the one factor of current interest [1].

As an example of this approach, suppose the investigator is interested in finding out how the factor pH affects the retention time of a hydrocinnamic acid ($C_6H_5CH_2CH_2COOH$) sample in a reversed-phase "ion-pair" liquid-chromatographic system [2]. The investigator might prepare four eluents, each containing 780 ml of distilled water, 10 ml of 1 M acetic acid to be used as a buffer, 1.5 mmol of octylamine hydrochloride (the so-called "ion-pairing agent" or "ion-interaction reagent," IIR), a few drops of $HClO_4$ or NaOH to adjust the pH, and sufficient methanol to bring the total volume to 1.00 liter. The pH of the four eluents might be adjusted to 3.6, 4.4, 5.2, and 6.0. The investigator could then use each of these eluents with a reversed-phase column to measure the retention time of a hydrocinnamic acid sample. The results might be those shown in Fig. 1. If the investigator now asks the simple question, What is the effect of pH on retention time?, then, based on the results shown in Fig. 1, the answer would probably be, Not much.

As another example of this single-factor approach to investigating chromatographic systems, the investigator might ask, What is the effect of ion-interaction reagent concentration on the retention time? Again, the investigator might prepare a series of four eluents, but this time the investigator would hold the pH constant (at 3.6, say) and vary the concentration of octylamine hydrochloride IIR: 0.0, 1.5, 3.0, and 5.0 mM. The results might be those shown in Fig. 2. What is the effect of ion-interaction reagent on retention time? Again, the answer would probably be, Not much.

III. MULTIFACTOR EFFECTS

So far, there is nothing wrong with the "good science" that has been carried out in these two sets of experiments. Difficulties could, how-

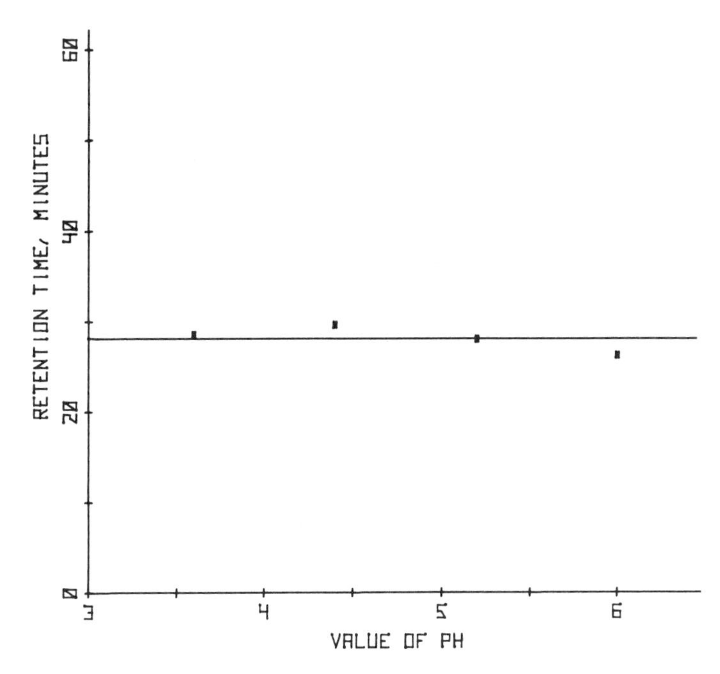

Figure 1 Single-factor effect of pH on the retention time of hydrocinnamic acid. The concentration of IIR is 1.5 mM. (Data from Ref. 2.)

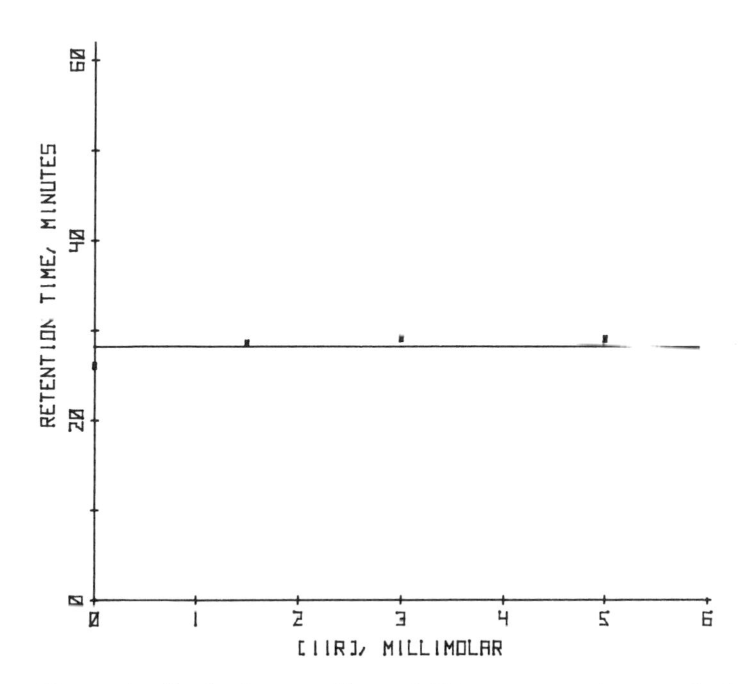

Figure 2 Single-factor effect of IIR concentration on the retention time of hydrocinnamic acid; pH = 3.6. (Data from Ref. 2.)

ever, be encountered if the investigator believed that the effect of pH was always "not much," independent of the concentration of IIR, or that the effect of the concentration of IIR was always "not much," independent of the pH. What many researchers fail to realize is that primitive questions such as, What is the effect of pH on retention time? or What is the effect of ion-interaction reagent concentration on retention time? yield answers that are conditional: *The answer depends upon the values of the other factors involved!*

Figure 3 is intended to clarify this concept. The two sets of connected dots show where in "factor space" the investigator has carried out the two sets of experiments shown in Figs. 1 and 2. We have seen in Fig. 1 what the effect of pH is when the concentration of IIR is 1.5 mM. We have seen in Fig. 2 what the effect of the concentration of IIR is when the pH is 3.6. Figure 3 shows clearly that we *do not* yet know, for example, what the effect of pH might be when the concentration of IIR is 5.0 mM, or what the effect of the concentration of IIR might be when the pH is 6.0.

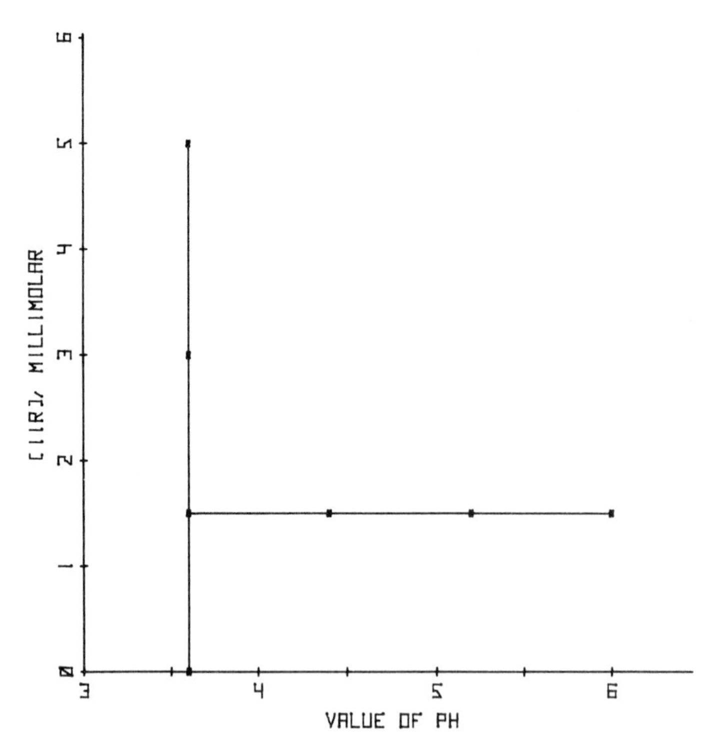

Figure 3 Combination of pH and IIR concentration for experiments shown in Figs. 1 and 2.

If the effect of one factor depends upon the value of another factor, the factors are said to *interact* mathematically. In most chemical systems, factor interaction is the rule rather than the exception [3]. Thus it is important to plan experiments in a way that will reveal these interactions if they exist. To investigate factor interactions, the experimental design must investigate more than "straight-line paths" such as the two shown in Fig. 3. It is beyond the scope of this paper to discuss the fundamentals of experimental design; such material may be found in other sources [4]. Experimental designs that are useful for discovering factor effects and their interactions are the "factorial designs" [5], including "central composite" [6], "Box–Behnken" [7], and "simplex mixture" [8] designs. An example of a "two-factor four-level" full factorial design is shown in Fig. 4. This design might be used to carry out nine more chromatographic experiments to supplement the seven experiments shown in Fig. 3.

The usefulness of factorial-type designs is shown in Fig. 5, in which retention time is plotted as a function of pH *for different values of* IIR concentration, and in Fig. 6, in which retention time is plotted as a function of IIR concentration *for different values of* pH. (Figures

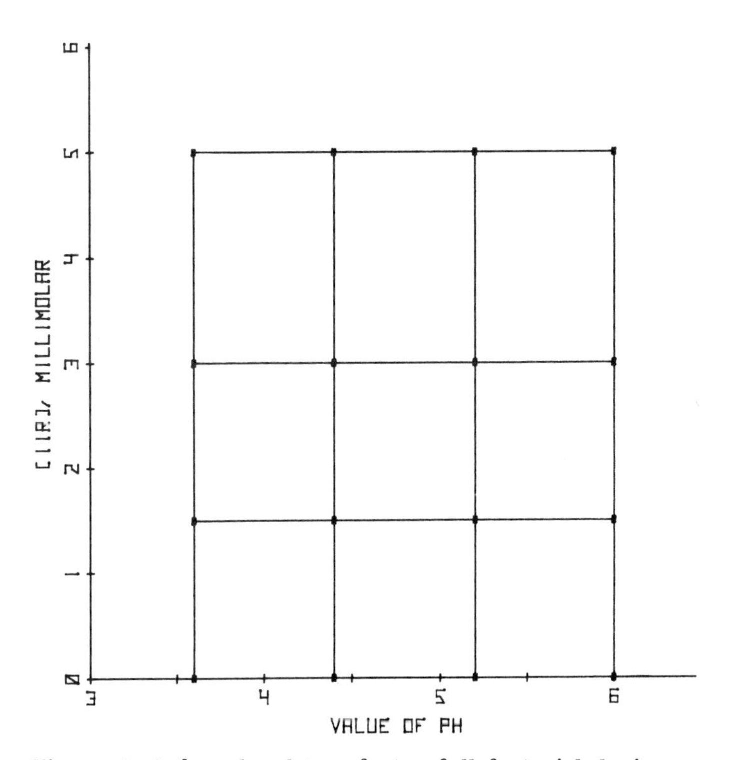

Figure 4 A four-level two-factor full factorial design.

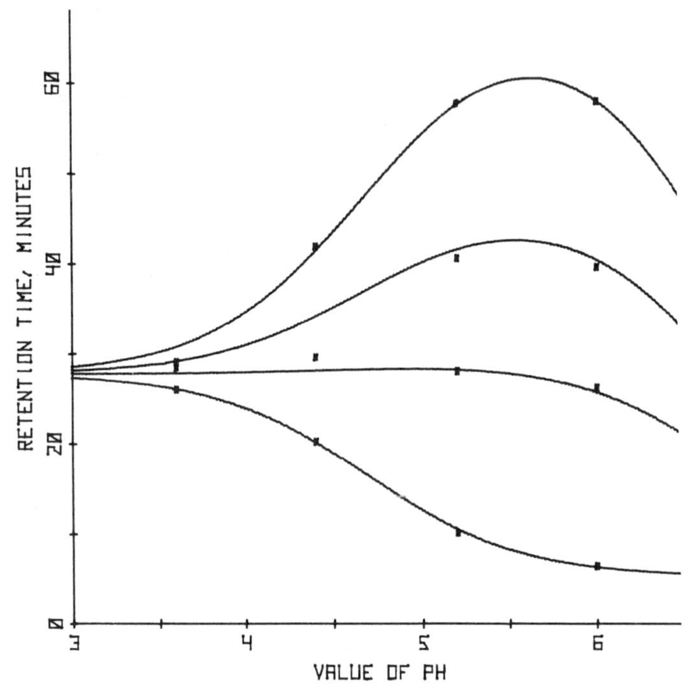

Figure 5 Effect of pH on retention time of hydrocinnamic acid at different values of IIR concentration. Lower curve to upper curve: 0.0, 1.5, 3.0, and 5.0 mM IIR. (Adapted from Ref. 2.)

5 and 6 contain the same information viewed from two different orthogonal directions.) It is seen that the effect of pH does depend upon the concentration of IIR and that the effect of the concentration of IIR does depend upon the pH [2].

IV. MODELS

Information gained from multifactor experimental designs such as the one shown in Fig. 4 is often very useful for obtaining a much better mechanistic understanding of the separation process. Although the illustrations shown here represent an extreme case, they do nevertheless make a point; if the only information an investigator has is that in Figs. 1 and 2, the investigator might be tempted to come to the general conclusions that pH does not have an effect on hydrocinnamic acid's retention time (wrong) and that the concentration of IIR does not have an effect on hydrocinnamic acid's retention time (also wrong).

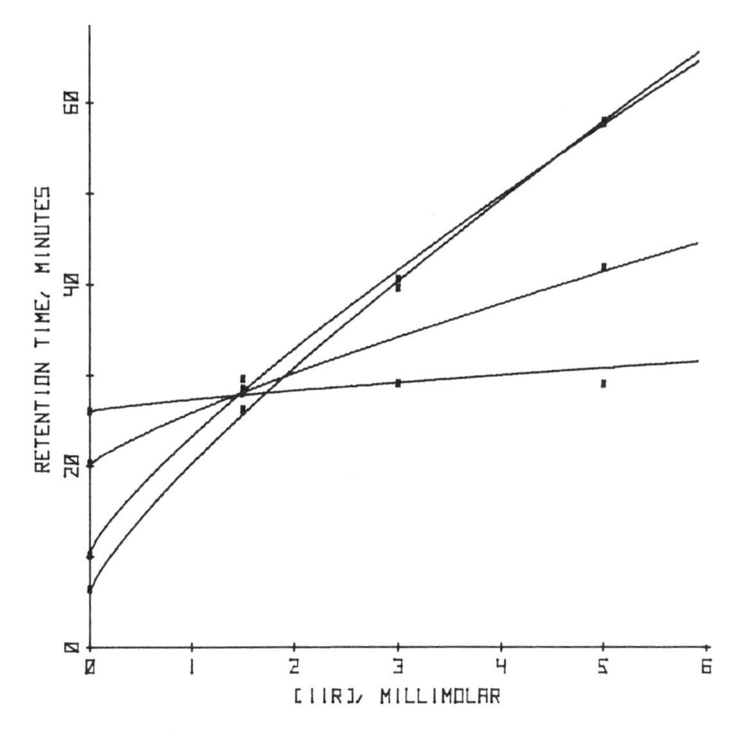

Figure 6 Effect of IIR concentration on the retention time of hydrocinnamic acid at different values of pH. At zero IIR concentration (upper curve to lower curve); pH = 3.6, 4.4, 5.2, and 6.0. (Data from Ref. 2.)

It is clear from the multifactor results shown in Figs. 5 and 6 that both pH and IIR concentration have important effects on the retention time of hydrocinnamic acid. Thus any model that attempts to explain the chromatographic behavior of hydrocinnamic acid in a reversed-phase "ion-pair" system must include the effects of both pH and IIR concentration in a way that includes the mathematical interaction between them.

A semiempirical model that explains the results shown in Figs. 5 and 6 is

$$t_R = f_{HA}t_{HA} + f_A t_A + f_A f_{HS} b[\text{IIR}]^{1/n} \tag{1}$$

where t_R is the observed retention time of hydrocinnamic acid; t_{HA} is the retention time of hydrocinnamic acid in its protonated conjugate acid form (i.e., at low pH in the absence of any IIR); t_A is the retention time of hydrocinnamic acid in its unprotonated conjugate base form (i.e., at high pH in the absence of any IIR); f_{HA} is the fraction of hydrocinnamic acid existing in the conjugate acid form, $f_{HA} = [H^+]/([H^+] + K_{HA})$, where K_{HA} is the acid dissociation constant of hydrocinnamic acid; f_A is the fraction of hydrocinnamic acid existing in the conjugate base form, $f_A = K_{HA}/([H^+] + K_{HA})$; f_{HS} is the fraction of octylamine existing in the protonated conjugate acid form, $f_{HS} = [H^+]/([H^+] + K_{HS})$, where K_{HS} is the acid dissociation constant of octylammonium ion; and b and n are empirical constants arising from the use of a Freundlich adsorption isotherm. The third term of the model, $f_A f_{HS} b[IIR]^{1/n}$, contains the mathematical interaction between pH and IIR concentration. Further details of this model and its mechanistic basis may be found in the literature [2]. This model was fitted to the experimental data points shown in Figs. 5 and 6; the lines in Figs. 1, 2, 5, and 6 were drawn from the fitted model.

An alternative view of the information in Figs. 5 and 6 is the response surface shown in Fig. 7. The interaction between pH and IIR concentration is clearly evident. Figure 8 shows the behavior of eight other sample compounds: trans-cinnamic, phenylacetic, trans-p-coumaric, trans-ferulic, trans-caffeic, and vanillic acids; phenylethylamine (a weak base); and phenylalanine (a zwitterionic compound). The following discussion assumes that these eight compounds plus hydrocinnamic acid are the components of a nine-component sample.

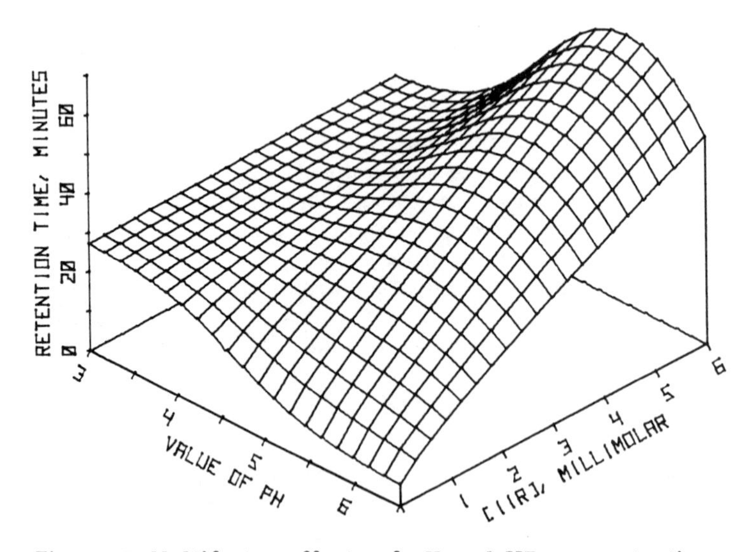

Figure 7 Multifactor effects of pH and IIR concentration on the retention time of hydrocinnamic acid. (Adapted from Ref. 2.)

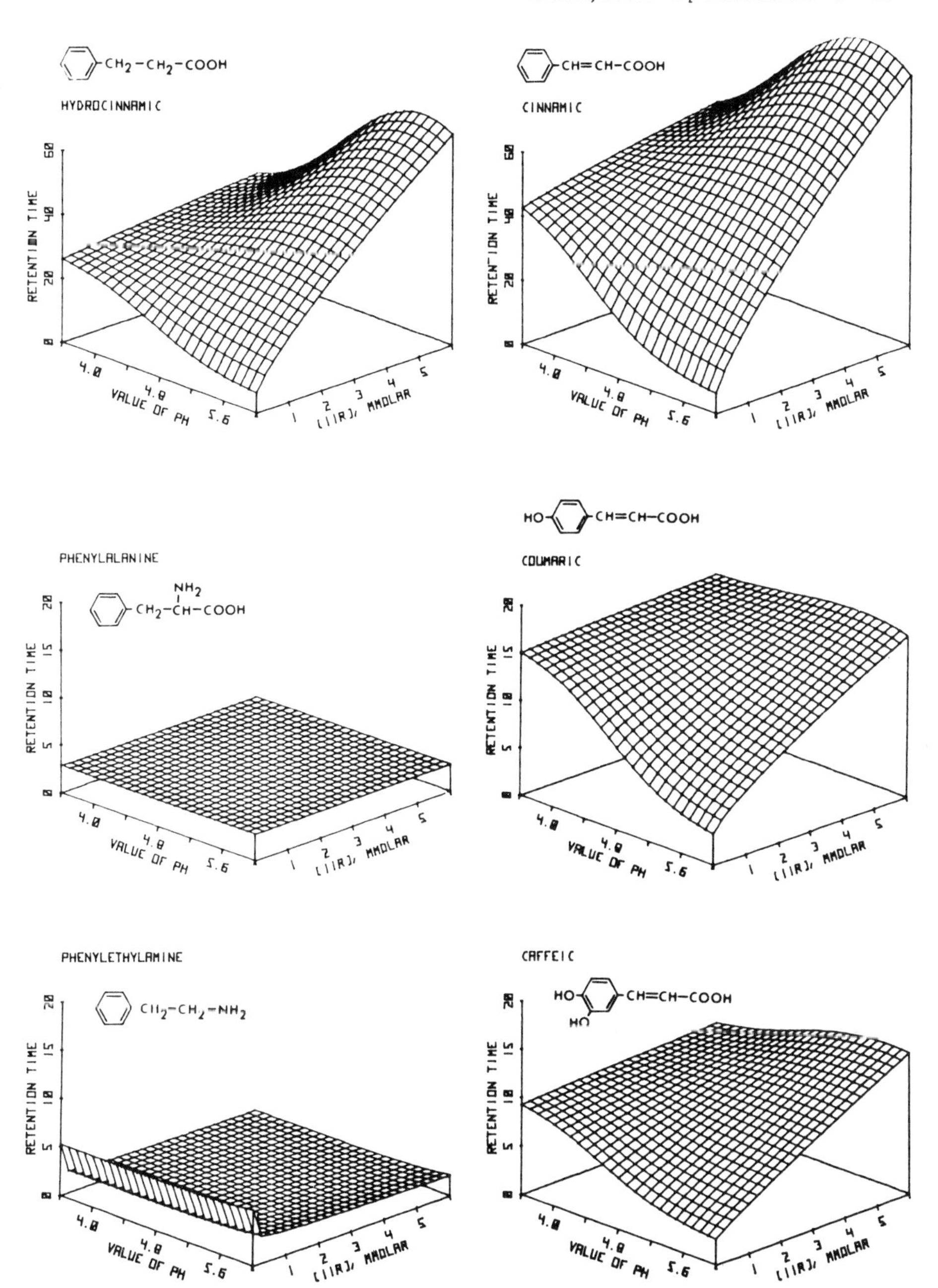

Figure 8 Combined effects of pH and IIR concentration on the reversed-phase liquid-chromatographic behavior of nine sample components. (From Ref. 2.)

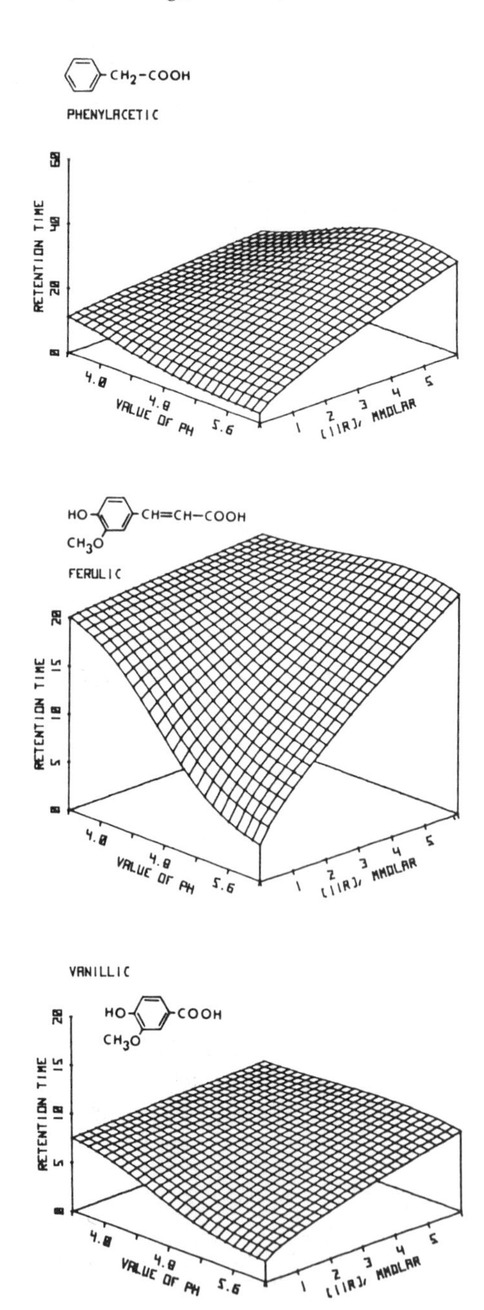

Figure 8 (continued)

V. MULTIFACTOR OPTIMIZATION

A goal of liquid-chromatographic methods development is to obtain adequate separation of all components of interest in a reasonable analysis time. This goad is achieved (or approached as closely as possible) by adjusting accessible chromatographic factors (pH, IIR concentration, etc.) to give the desired response.

Optimization of chromatographic systems is unusually difficult because of the ever-present possibility of "elution order reversal." This is illustrated in Fig. 9, in which the retention time response surfaces for trans-ferulic acid and phenylacetic acid have been superimposed. The intersection of the two surfaces occurs at combinations of pH and IIR concentration that give identical retention times for these two components; i.e., under these conditions of pH and IIR concentration, trans-ferulic and phenylacetic acids will not be separated. For those combinations of pH and IIR concentration that occur to the "left" of this intersection, phenylacetic acid will elute first followed by trans-ferulic acid. For those combinations of pH and IIR concentration that occur to the "right" of this intersection, the elution order will be reversed: trans-ferulic acid would elute first, followed by phenylacetic acid. Thus there are two broad regions of experimental conditions that will give acceptable separation of the two compounds; there is a narrow region between them that will give unacceptable separation.

These ideas can be made quantitative by applying to the present multifactor case the "window diagram" technique pioneered by Laub and Purnell [9] for single-factor optimization. Because relative retention (alpha) is a truer measure of separation than is the difference in retention times, the two-dimensional "alpha diagram" shown in Fig. 10 can be produced by dividing the higher surface shown in Fig. 9 by the lower surface at all combinations of pH and IIR concentration. (Actually, the capacity factor surfaces must first be formed by subtracting the time equivalent of the void volume from each surface in Fig. 9 and dividing each of the resulting surfaces by the time equivalent of the void volume; the ratios of these capacity factor surfaces then give Fig. 10.) The two regions giving acceptable separation are evident in Fig. 10 as the higher parts of the surface; the unacceptable region occurs in the "valley" of Fig. 10.

Other measures of chromatographic performance can also be used for the vertical axis: separation, resolution, percent overlap, and so on.

If all other pairs of compounds in the nine-component chromatographic sample are treated in the same way, then a number of diagrams similar to Fig. 10 will be produced. (Each of these surfaces corresponds to a single line in Laub and Purnell's window diagrams.) If all of these alpha surfaces are superimposed, and if only those parts of the surfaces that are first encountered above the pH—IIR plane are plotted, the resulting surface shown in Fig. 11 is the two-dimen-

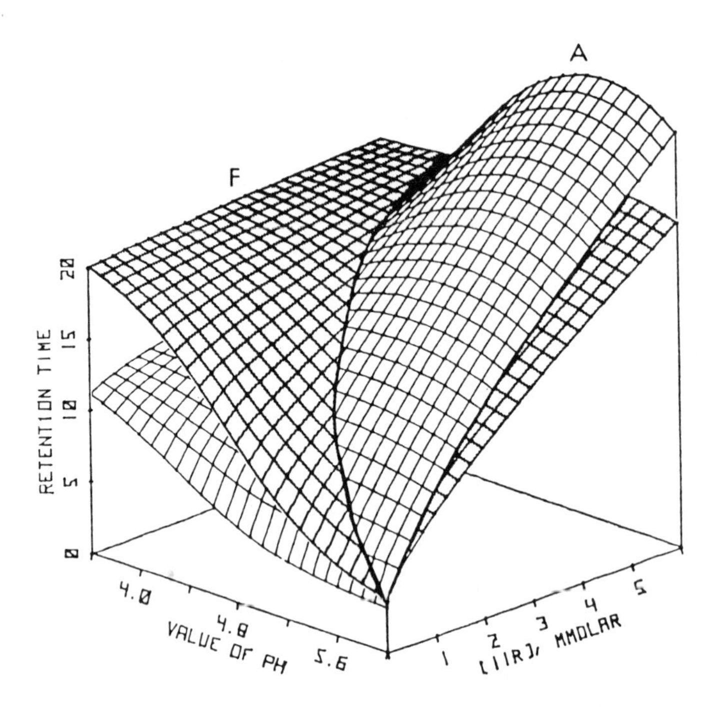

Figure 9 Predicted retention behavior of trans-ferulic acid (F) and phenylacetic acid (A) as functions of pH and IIR concentration. (From Ref. 20.)

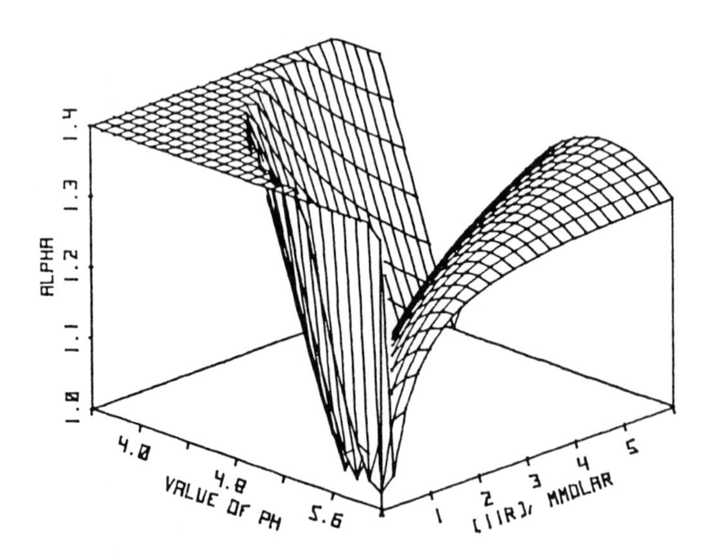

Figure 10 Predicted relative retention (alpha) values for trans-ferulic acid and phenylacetic acid as a function of pH and IIR concentration. Values of alpha greater than 1.4 are set equal to 1.4. (From Ref. 20.)

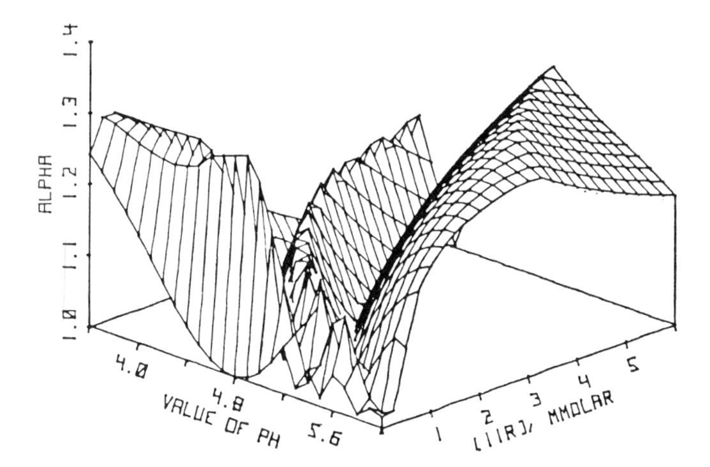

Figure 11 Two-dimensional window diagram for a nine-component mixture. (From Ref. 20.)

sional equivalent of the Laub and Purnell windows: a response surface that looks like a mountain range.

The interpretation of Fig. 11 is as follows: Any point on the surface represents the best separation of the worst-separated pair of compounds at that particular combination of pH and IIR concentrations. (That statement merits reading several times.) If the investigator is willing to define optimum chromatography as the best relative retention of the worst-separated pairs of compounds, then optimum chromatography occurs at the top of the tallest "mountain" in Fig. 11. If the investigator were to prepare an eluent with the corresponding pH and IIR concentration, then the best separation possible for this system should be obtained.

Just how closely the actual chromatographic performance matches that predicted by the window diagram techniques will depend both upon the accuracy of the models used in constructing the window diagrams and upon the precision and accuracy of the data that have been used to fit these models. Thus it is important that high-quality experimental data be obtained and that good chromatographic models be used.

Remember, however, that any optimum obtained in this way is still *conditional*! If a third factor were included, say, the percentage of methanol in the eluent, then the shape of Fig. 11 would, in general, change as this third factor is varied. The "optimum chromatography" might get better—or worse—as the percentage of methanol is changed. Three- and higher-factor systems are challenging in this regard and offer even more power for obtaining improved separations. Much research remains to be done to develop efficient experimental designs, models, and search algorithms for these three- and higher-factor systems.

VI. RUGGEDNESS

Figure 11 offers three regions that give approximately the same locally optimum results: the region near pH = 4.0, [IIR] = 0.05 mM (near the front left side); the region near pH = 4.4, [IIR] = 3.0 mM (along the ridge in the middle); and the region near pH = 5.8, [IIR] = 3.0 mM (near the right side). Although all three optima would yield approximately the same quality of separation, the region on the right is preferable because it is more "rugged."

Ruggedness is a measure of a system's lack of sensitivity to small changes in operating conditions. The region in the middle of Fig. 11 is very sensitive to small changes in pH and in the concentration of IIR. If a technician were to prepare an eluent that was in error by a small fraction of a pH unit, the resulting separation would probably be useless: At least two peaks would be nearly merged. The same situation exists for the region at the left of Fig. 11—small changes in pH or IIR concentration could cause major degradation of the quality of the separation. The region at the right of Fig. 11 is clearly more rugged and would be preferable for that reason. Figure 12 shows the chromatogram that was obtained in this region at pH = 5.8, [IIR] = 3.2 mM.

VII. OTHER APPROACHES TO MULTIFACTOR OPTIMIZATION

References 10–34 represent a selection of papers taken from the recent chromatographic literature that give slightly different approaches to multifactor optimization (e.g., the use of solvent strength and solvent selectivity [19]), discuss concepts involved in multifactor optimization (e.g., relationships between chromatographic response functions and performance characteristics [32]), or review different aspects of multifactor optimization [25,26]. The content of these articles is evident from their titles.

Of special interest is the "overlapping resolution map" (ORM) technique introduced by Glajch et al. [19]. In most applications using ORMs, four solvents have been used to optimize the separation of a set of similar compounds. The four solvents are typically methanol (MeOH), acetonitrile (ACN), tetrahydrofuran (THF), and water. The fourth sovlent, water, is used to adjust the solvent strength of each of the other solvents so the resulting mixtures (MeOH–water, ACN–water, and THF–water) are approximately isoeluotropic. Once these solvent mixtures have been defined, water is no longer a variable. Thus there are only three variables: MeOH–water, ACN–water, and THF–water.

However, in any experimental design involving mixtures [8], there is always the constraint that the sum of the mole fractions of all components must be unity. In the present case,

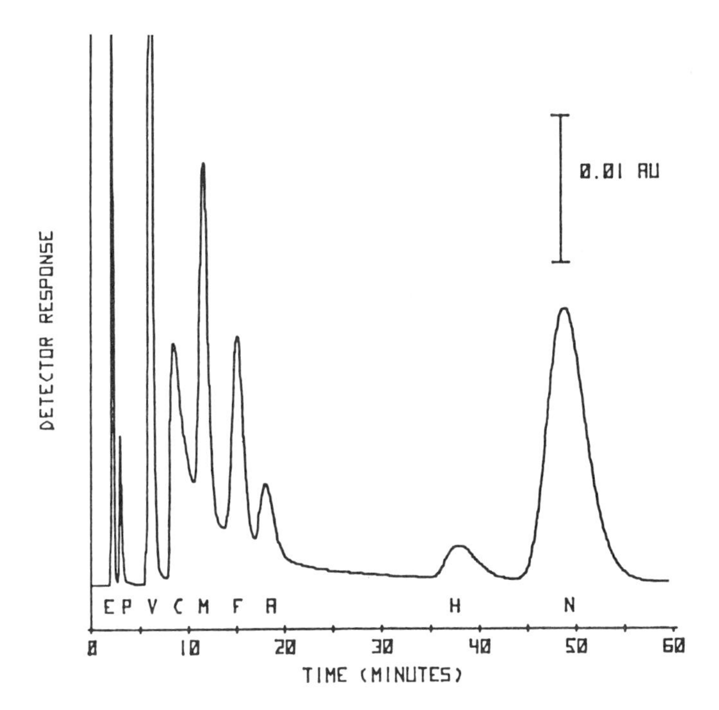

Figure 12 Chromatogram of a nine-component mixture at pH 5.8 and [IIR] = 3.2 mM (ultraviolet detector, 254 nm; E = phenylethylamine, P = phenylalanine, V = vanillic acid, C = trans-caffeic acid, M = trans-p-coumaric acid, F = trans-ferulic acid, A = phenylacetic acid, H = hydrocinnamic acid, N = trans-cinnamic acid). (From Ref. 20.)

$$X_{MeOH-water} + X_{ACN-water} + X_{THF-water} = 1 \qquad (2)$$

Thus there are only two *independent* variables (the third is fixed by the values of the first two). This situation is shown graphically in Fig. 13. Only those experiments that fall on the triangular oblique plane are permitted.

An alternative view of this situation is shown in Fig. 14. This view is generated by looking down on the MeOH−water/ACN−water plane from the THF−water axis. Although the mole fraction of THF−water is not shown, it may be obtained from a knowledge of $X_{MeOH-water}$ and $X_{ACN-water}$ by difference from Eq. (2). A two-factor three-level factorial design has been superimposed over the feasible region. Only six of the nine experiments (those lying within the feasible region and its boundaries) can actually be carried out. For purposes involving degrees of freedom in model fitting, one additional experiment

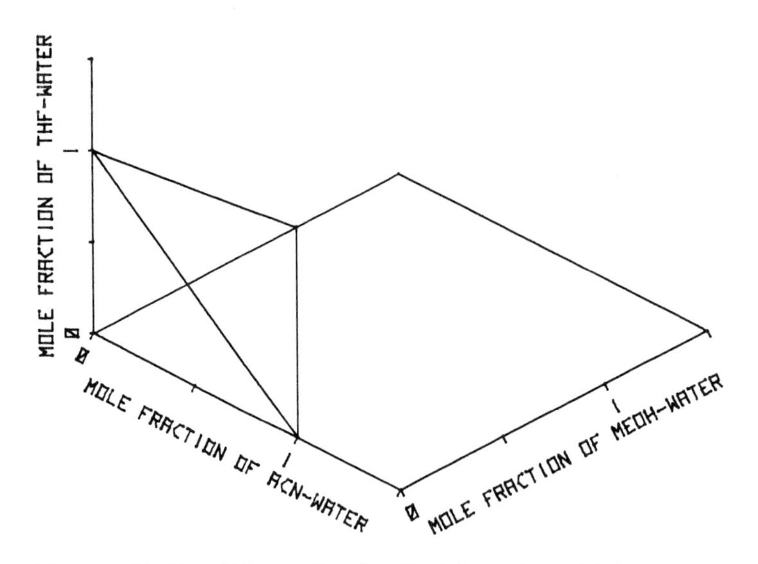

Figure 13 Feasible region for MeOH−water, ACN−water, and THF−water, given the constraint that the sum of the mole fractions must equal unity.

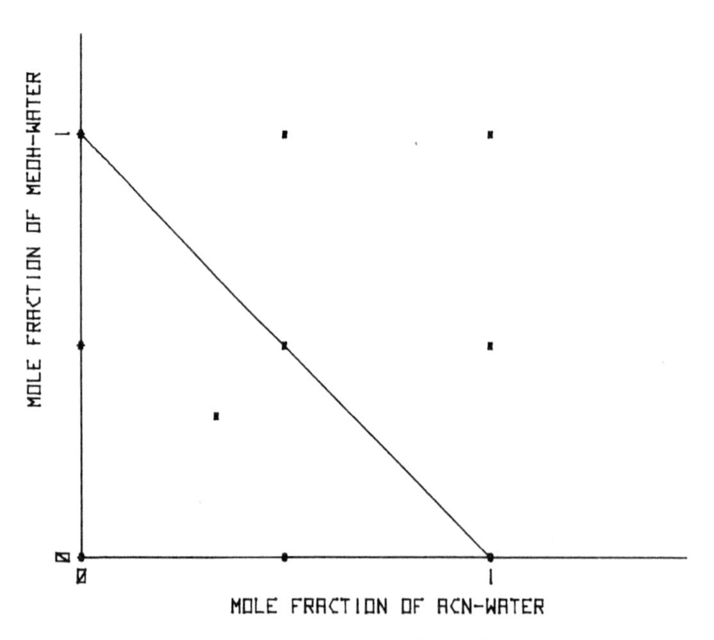

Figure 14 An alternative view of the feasible region for MeOH−water, ACN−water, and THF−water, given the constraint that the sum of the mole fractions must equal unity. A two-factor three-level full factorial experimental design is superimposed.

is carried out in the "middle" of the feasible region; this gives a total of seven experiments.

Figure 15 gives a physical chemistry ternary phase representation of the information in Fig. 14. The top of the triangle corresponds to MeOH—water only ($X_{MeOH-water} = 1$); the bottom left side corresponds to THF—water only; and the bottom right side corresponds to ACN—water only. Experiments are carried out at the indicated locations to determine the retention times of each component of interest. The results for each component are fitted to a full second-order polynomial model (including a mathematical interaction effect):

$$t_R = b_0 + b_1X_1 + b_2X_2 + b_{11}X_1^2 + b_{22}X_2^2 + b_{12}X_1X_2 \qquad (3)$$

where the b's are parameters of the model, and X_1 and X_2 represent the mole fractions of MeOH—water and ACN—water, respectively.

The resulting smooth parabolic response surfaces are treated in a way similar to that of the window diagrams discussed in the previous section. The resolution maps can be overlapped to find the set of chromatographic conditions that give the best overall resolution. The original paper contains additional details [19].

VIII. SEQUENTIAL SIMPLEX OPTIMIZATION

Sequential simplex optimization [1] has been used to optimize multifactor chromatographic systems (see, e.g., Ref. 16, 30, and 34). Although the sequential simplex technique is a powerful multifactor op-

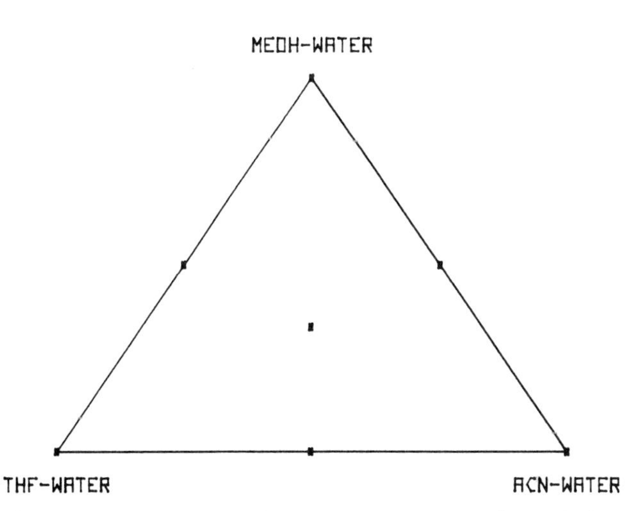

Figure 15 An alternative representation of the information in Fig. 14.

mization technique, its use in chromatographic systems is limited because of the existence of multiple optima caused by elution order reversal (see Fig. 11). If started from any region of factor space (e.g., any region of pH and IIR concentration), the simplex will converge to one of the several local optima. However, there is no guarantee that it will converge to the global optimum (the best of the local optima).

Thus in chromatographic systems where elution order reversal is possible, window diagram techniques should be used first. If the resulting chromatographic performance needs further improvement, a small simplex can then be used to "fine-tune" the system and obtain the maximum chromatographic performance possible from the system.

IX. FUTURE DIRECTIONS

The success of multifactor optimization techniques rests on the availability of good multifactor predictive models. Equation (1) seems to work well for many systems in which pH and IIR concentration are varied. Equation (3) seems to work well when relatively "soft" factors such as solvent mole fractions are varied.

In the future, a better multifactor mechanistic understanding of the chromatographic separation process will be required to take full advantage of the rich variety of factors that can be adjusted in the mobile phase.

REFERENCES

1. S. L. Morgan and S. N. Deming, Anal. Chem. *46*, 1170–1181 (1974).
2. R. C. Kong, B. Sachok, and S. N. Deming, J. Chromatogr. *199*, 307–316 (1980).
3. G. E. P. Box and K. B. Wilson, J. R. Stat. Soc. B *13*, 1–45 (1951).
4. G. E. P. Box, W. G. Hunter, and J. S. Hunter, *Statistics for Experimenters: An Introduction to Design, Data Analysis, and Model Building*, Wiley, New York, 1978.
5. W. G. Cochran and G. M. Cox, *Experimental Designs*, 2nd ed., Wiley, New York, 1957.
6. M. G. Natrella, *Experimental Statistics*, National Bureau of Standards Handbook 91, U.S. Government Printing Office, Washington, D.C., 1963.
7. G. E. P. Box and D. W. Behnken, Technometrics *2*, 455–475 (1960).
8. J. A. Cornell, *Experiments with Mixtures: Designs, Models, and the Analyis of Mixture Data*, Wiley, New York, 1981.

9. R. J. Laub and J. H. Purnell, J. Chromatogr. *112*, 71–79 (1975).
10. L. R. Snyder, J. Chromatogr. Sci. *15*, 441–449 (1977).
11. H. Engelhardt and H. Englass, J. Chromatogr. *158*, 249–259 (1978).
12. R. J. Laub and J. H. Purnell, J. Chromatogr. *161*, 49–57 (1978).
13. R. J. Laub and J. H. Purnell, J. Chromatogr. *161*, 59–68 (1978).
14. L. R. Snyder, J. Chromatogr. Sci. *16*, 223–234 (1978).
15. R. A. Hartwick, C. M. Grill, and P. R. Brown, Anal. Chem. *51*, 34–38 (1979).
16. M. W. Watson and P. W. Carr, Anal. Chem. *51*, 1835–1842 (1979).
17. J. -P. Thomas, A. Brun, and J. -P. Bounine, J. Chromatogr. *172*, 107–130 (1979).
18. J. R. Gant, J. W. Dolan, and L. R. Snyder, J. Chromator. *185*, 153–177 (1979).
19. J. L. Glajch, J. J. Kirkland, K. M. Squire, and J. M. Minor, J. Chromatogr. *199*, 57–79 (1980).
20. B. Sachok, R. C. Kong, and S. N. Deming, J. Chromatogr. *199*, 317–325 (1980).
21. H. Colin and G. Guiochon, J. Chromatogr. Sci. *18*, 54–63 (1980).
22. B. Sachok, J. J. Stranahan, and S. N. Deming, Anal. Chem. *53*, 70–74 (1981).
23. H. Poppe and E. H. Slaats, Chromatographia *14*, 89–94 (1981).
24. L. A. Jones, R. W. Beaver, and T. L. Sohmoeger, Anal. Chem. *54*, 182–186 (1982).
25. I. E. Frank and B. R. Kowalski, Anal. Chem. *54*, 223R–243R (1982).
26. R. E. Majors, H. G. Barth, and C. H. Lochmuller, Anal. Chem. *54*, 323R–363R (1982).
27. P. E. Antle, Chromatographia *15*, 277–281 (1982).
28. W. Wegscheider, E. P. Lankmayr, and K. W. Budna, Chromatographia *15*, 498–504 (1982).
29. P. J. Schoenmakers, A. C. J. H. Brouen, H. A. H. Billiet, and L. deGalen, Chromatographia *15*, 688–696 (1982).
30. J. C. Berridge, J. Chromatogr. *244*, 1–14 (1982).
31. J. L. Glajch and J. J. Kirkland, Anal. Chem. *55*, 319A–336A (1983).
32. W. Wegscheider, E. P. Lankmayr, and M. Otto, Anal. Chim. *150*, 87–103 (1983).
33. M. Otto and W. Wegscheider, J. Chromatogr. *258*, 11–22 (1983).
34. J. H. Nickel and S. N. Deming, Liquid Chromatogr. *1*, 414–417 (1983).

3

Statistical and Graphical Methods of Isocratic Solvent Selection for Optimal Separation in Liquid Chromatography

Haleem J. Issaq / Program Resources, Incorporated, NCI-Frederick Cancer Research Facility, Frederick, Maryland

I. INTRODUCTION

The selection of a solvent system which will give optimum resolution in liquid chromatography is not a simple matter. The most important considerations are the properties of the material being separated and the stationary phase. The mobile phase can be selected only when these two factors have been defined. A trial-and-error procedure is generally used to find a mobile phase that will satisfactorily resolve all the components of the mixture. In high-performance liquid chromatography (HPLC) the mobile phase is generally a mixture of two or more pure solvents. This is especially true when the reversed phase mode is employed where the mobile phase is composed of water and an organic modifier. If the composition of the mobile phase remains the same during the entire procedure, we speak of isocratic elution, however, if the composition of the mobile phase is continuously changing, within a predetermined time period, we speak of gradient elution.

This review will deal mainly with the selection of an isocratic mobile phase using systematic statistical and graphical approaches. Although the discussion will concentrate on reversed-phase HPLC, the methods discussed can also be applied to other chromatographic modes, such as ion exchange or normal phases.

II. ROLE OF THE MOBILE PHASE

The mobile phase in liquid chromatography plays an important if not a major role in the separation process. The mobile phase determines not only the separation of the components in a mixture, with the assumption that other parameters have been optimized, but the degree of resolution (R_s), how far the peaks are from one another, and how wide are they. Solute retention on the column and selectivity, that is, the order of elution, are also affected by this factor. Mobile phase strength (polarity) determines retention times (R_t) and mobile phase composition determines selectivity and R_s.

Selection of the mobile phase is based on the properties of the stationary phase, which in turn is chosen after consideration of the properties of the solute mixture to be separated. This will determine the chromatographic process, that is, adsorption, partition, or ion exchange. When a single solvent is used, it is selected from the eluotropic series, in which hexane is the least polar and water the most polar. The eluotropic series is not used when selecting a mobile phase that consists of a mixture of pure solvents. Snyder [1] gave an equation for determining the polarity of the mixed mobile phase, which will be discussed later. In liquid chromatography the relation between the solute and the mobile and stationary phases may be represented by the peaks of a triangle. This means that there are three types of interacton: solute–mobile phase, mobile phase–stationary phase, and solute–statinary phase. Note that when a mixture of solvents is used solvent–solvent interaction and solvent demixing (in thin-layer chromatography) must also be considered. Separation of a mixture is achieved when an optimum balance is reached between these factors. If the interaction between one pair (e.g., solute-mobile phase) is much stroenger than between the others, poor separation or no separation at all results. In other words, to resolve the components of a mixture the chromatographer must create conditions under which the solutes are eache forced to favor both mobile and stationary phases differently. To achieve this the stationary phase or the mobile phase must be altered. When a mixture of compounds does not elute off the column, as a result of strong solute–stationary phase interaction or weak sol–ute mobile phase interaction, the mobile phase or the material should be changed to achieve stronger solute–mobile phase interaction and weaker solute–stationary phase interaction. If, on the other hand, the mixture elutes with the solvent peak, then the mobile phase or the column material should be changed so that solute–mobile phase interaction is weaker than before. It is, of course, cheaper to change the mobile phase than the column.

This chapter will deal with methods for selecting an optimum mobile phase for isocratic elution. These methods are based on sound statistical approaches which eliminate operator intuition or guessing and result in savings in time and material.

III. THEORETICAL CONSIDERATIONS

Selection of a pure solvent in liquid chromatography is, as mentioned earlier, based on the eluotropic series. However, the selection of a binary solvent mixture, ternary, and so on, and prediction of the resolution and elution times of the components of a mixture is more complex. Snyder [1] gave the following equation for calculating the polarity of a solvent mixture:

$$P' = \Sigma \Phi_i P_i \tag{1}$$

Φ_i and P_i are the volume fractions in the solvent mixture which has a polarity of P'. This relation and polarity of pure solvents does not apply to normal phases where the calculations are more complicated [1].

The relation between retention time and solvent strength is described by the following equation [1]:

$$\log(K_1/K_2) = \alpha A_s(\varepsilon_2^0 - \varepsilon_1^0) \tag{2}$$

where K' is the capacity factor, ε^0 is the solvent strength parameter of solvents 1 and 2, A_s is the molecular area of the adsorbed sample, α is the adsorbent surface activity function, and

$$K' = (R_t - R_{t0})R_{t0} \tag{3}$$

R_t and R_{t0} are the retention times of retained and unretained solutes. It was found that the relation in Eq. (2) does not always hold [2–4]. For example, when two mobile phases (acetonitrile–water and methanol–water) which have the same solvent strength [calculated according to Eq. (1)] were used to elute the same solutes (naphthalene and biphenyl) on the same solid phase, the retention times were not the same [4]. A term was later added to Eq. (2) to account for solvent–solute interactions [2,3]. Since different solvents give different selectivities [5–8], changing the solvent composition may result in a different elution order, depending on the properties of the sample mixture and the solvent chosen. For a mobile phase mixture, solvent strength (polarity) in general determines elution distance of the solutes (i.e., R_t), while mobile phase composition determines its selectivity. The composition of the mobile phase would determine the degree of separation α, between two adjacent peaks i and ii, where

$$\alpha = K'_{ii}/K'_{i} \tag{4}$$

Based on Snyder's theory [1], Saunders [7] presented a graphical representation based on ε^0 for selecting a solvent for adsorption liquid chromatography. The application of these graphs is rapid and provides a reasonable first approximation to a solvent mixture appropriate for a given sample. It must be stressed that the results are approximate, and in some cases the solvent mixture will not be ideal [7].

For a given sample and adsorbent, log K' varies linearly with ε^0. This is generally true for K' values between 1 and 10, which is an acceptable working range allowing separation of a component from a mixture, but which does not lead to dilution of the sample or long retention times. Solvent strength (polarity) gives a general indication of solute retention, but it may not predict the correct retention times [5,6,9]. Resolution between two adjacent peaks is defined by

$$R_s = (R_{t_2} - R_{t_1})/\tfrac{1}{2}(W_1 + W_2) \tag{5}$$

where R_t is the time of elution of peak maximum, and W is the baseline width of the peak in units of time.

Also, resolution in liquid chromatography has been defined [6] by the following equation:

$$R_s = \tfrac{1}{4}(\alpha - 1)N^{1/2}[K'/(1 + K')] \tag{6}$$

where the number of theoretical plates, N is defined as

$$N = 16(R_t/W)^2 \tag{7}$$

Note that all the above factors are a function of R_t [Eqs. (3)−(7)].

The three terms in Eq. (6) should be optimized to achieve maximum resolution; however, if the experimental conditions (flow rate, column dimensions, particle size, and properties of the sample) are kept constant, the only parameter affecting separation will be the mobile phase. The composition of this will determine not only the retention times of the solutes, but also their order of elution. It is important to have a solvent which will give reasonable retention times for all components of the mixture, R_t between 5 and 40 min, in HPLC, and an R_f value of 0.2 to 0.7 thin-layer chromatography.

IV. BINARY, TERNARY, QUATERNARY SOLVENT MOBILE PHASES

Two solvent mobile phases may consist of two binary mixtures (85% methanol−water and 70% acetonitrile−water) or of a binary mixture and a pure solvent (60% methanol–water and pure acetonitrile).

Multisolvent mobile phases may be selected to give optimal separation of a mixture based on a statistical approach or graphical representation. Both approaches are sound and can give optimum results. The statistical approach requires the use of a computer, while the graphical one may not.

A ternary solvent mixture may be three pure solvents, three binary mixtures or a mixture of both as mentioned earlier. In reversed-phase chromatography where methanol–water, acetonitrile–water, and tetrahydrofuran (THF)–water are thought of as three binary mixtures, they might also be considered as a quaternary solvent mixture, methanol–acetonitrile–THF–water.

V. STATISTICAL APPROACHES TO MOBILE PHASE OPTIMIZATION

Belinky [10] used a simplex statistical design strategy to achieve an isocratic mobile phase for the separation of a mixture of polycylic aromatic hydrocarbons (PAH). Using reversed-phase columns and methanol–acetonitrile–water, Belinky assumed that each solvent would occupy the vertex of a triangle (Fig. 1). Since the PAH will not elute from the column with water, a 65% methanol–water mixture was chosen as the weakest mobile phase along the AC axis and 55% acetonitrile–water as the weakest mobile phase along the AB axis. These points are labeled C' and B', respectively. This results in four vertices CBB'C', where C' and B' are pseudocomponents of methanol–water and acetonitrile–water, respectively. A simplex design for a three-component mixture requires a triangular coordinate system with three vertices.

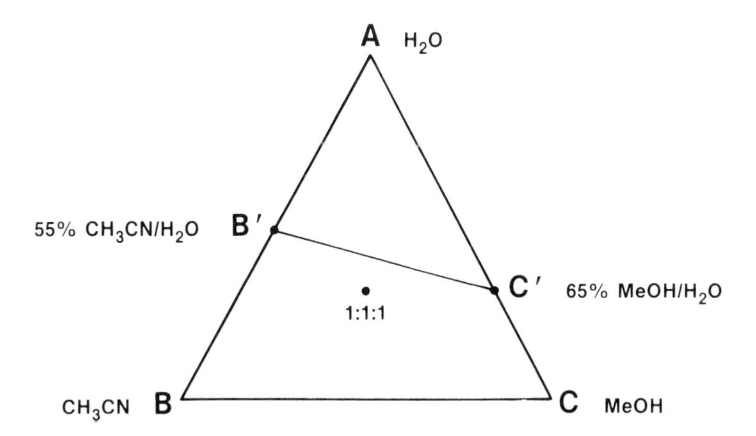

Figure 1 Ternary solvent system, using two organic modifiers.

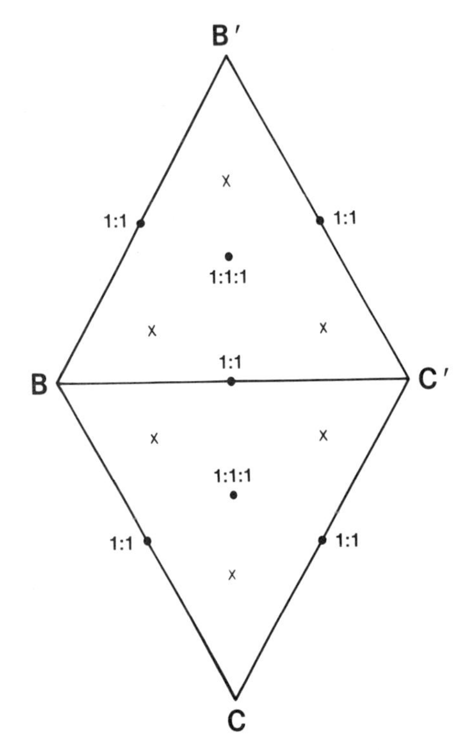

Figure 2 Orthogonal representation of two organic modifier solvent system as two ternary solvent systems BB'C' and BCC'.

In this case two triangular systems are obtained: C'CB and B'CB (Fig. 2), which may be represented by two simplex designs having three components each. Belinky's objective was to generate response surfaces which would allow the prediction of resolution at any point within C'CB and B'CB.
The response surface is described by the following equation:

$$Y = b_1X_1 + b_2X_2 + b_3X_3 + b_{12}X_1X_2 + b_{13}X_1X_3 + b_{23}X_2X_3 + b_{123}X_1X_2X_3 \qquad (8)$$

where Y is the capacity factor or resolution, which is the sum of the individual contributions of each component or pseudocomponent in the system X_1, X_2, and X_3. The b coefficients are determined experimentally; Y values are measured at each of the following concentrations for each of the simplex approaches:

C'	1.00	0	0.50	0.50	0	0.33
C	0	1.00	0.50	0	0.50	0.33
B	0	0	1.00	0.50	0.50	0.33

These seven generated values will be used to find the values of the b coefficients. Three extra points were used to measure the fit, these are:

C'	0.67	0.16	0.16
C	0.16	0.67	0.16
B	0.16	0.16	0.67

The calculation of the b coefficients is shown in the following equations:

$$b_1 = Y_A \tag{9}$$

$$b_2 = Y_B \tag{10}$$

$$b_3 = Y_C \tag{11}$$

$$b_{12} = 4(Y_{AB}) - 2(Y_A + Y_B) \tag{12}$$

$$b_{13} = 4(Y_{AC}) - 2(Y_A + Y_C) \tag{13}$$

$$b_{23} = 4(Y_{BC}) - 2(Y_B + Y_C) \tag{14}$$

$$b_{123} = 27(Y_{ABC}) - 12(Y_{AB} + Y_{AC} + Y_{BC} + 3(Y_A + Y_B + Y_C) \tag{15}$$

Substituting the b values in Eq. (8) will allow the prediction of Y for any solvent composition within the confines of the model, which is assumed to be an accurate representation of the response surface. A total of 17 different mobile phase compositions were required to generate the response surfaces for the two simplex designs. An optimum mobile phase was found which separated the PAH mixture.

One disadvantage of this approach, which is generally sound, is that 17 different mobile phase compositions are required, that is, 17 chromatographic runs, to find an optimum mobile phase. This is, needless to say, time-consuming, since the chromatographer has to identify the elution order of the peaks in each of the 17 runs.

Two statistical approaches similar to that of Belinky [10] have been published. Glajch et al. [11] and Issaq et al. [12] discussed a statistical design based on peak pair resolution and overlapping resolution mapping (ORM) for selecting a mobile phase that will give optimum resolution of the components of a mixture. Both works are based on a statistical design by Snee [13]. The reader is referred to Snee's work [13] for the equations and the procedure used for the ORM of each pair of peaks. The procedure used by Issaq et al. [12] will be described here in detail. For the computer programs used, see Appendixes 1 and 2 in Ref. 12.

A combination of the three initial solvents is devised according to Table 1. Other combinations may also be used. The initial solvents A, B, and C may be pure, a mixture of two organic solvents (normal phase), or a mixture of water and an organic modifier for reversed phase (see Fig. 3). After selecting the solvents and proportions to be used (Table 1), 10 data points, one for each solvent combination, are collected. These are used to calculate the resolutions of each pair of compounds in the mixture. If no peak crossover takes place, the resolution between each pair (1,2,2–3,3–4,. . .) is used. If peak crossover does occur, the resolution between all the peaks is calculated (1–2, 1–3, 1–4, 2–3, 2–4, 3–4 . . .) and used in determining the optimum mobile phase.

Two computer programs are used to predict optimum solvent composition. The first (Appendix 1 of Ref. 12) is a FORTRAN program (PEAKIN) which rearranges resolutions to correct for crossover, and, if necessary, prints a table similar to Table 2 and produces a data file suitable for use in the next program. The second program (Appen-

Table 1 Combination of Solvents A, B, and C Used in the Issaq et al. Study [12] to Predict Optimum Mobile Phase Compositions

Percentage of solvent A	Percentage of solvent B	Percentage of solvent C
100	0	0
0	100	0
50	50	0
50	0	50
0	50	50
33	33	33
67	16	16
16	67	16
16	16	67

Table 2 Sample of Mobile Phase Ratios and Resolutions as Arranged by the PEAKIN Program

Solvent	Mobile phase ratios									
10% acetone	1.00	0.0	0.0	0.0	0.5	0.50	0.33	0.16	0.16	0.67
5% MeOH	0.0	1.00	0.0	0.50	0.50	0.0	0.33	0.16	0.67	0.16
30% ethyl acetate	0.0	0.0	1.00	0.50	0.0	0.50	0.33	0.67	0.16	0.16
$R_s(B_1 - B_2)$	6.20	9.60	7.10	6.70	7.50	7.90	7.10	5.30	6.70	7.50
$R_s(B_2 - G_1)$	4.20	5.80	3.30	6.20	5.80	3.70	5.00	4.50	6.30	5.00
$R_s(G_1 - G_2)$	5.80	0.40	5.00	3.80	2.50	4.10	5.00	4.80	3.30	4.50

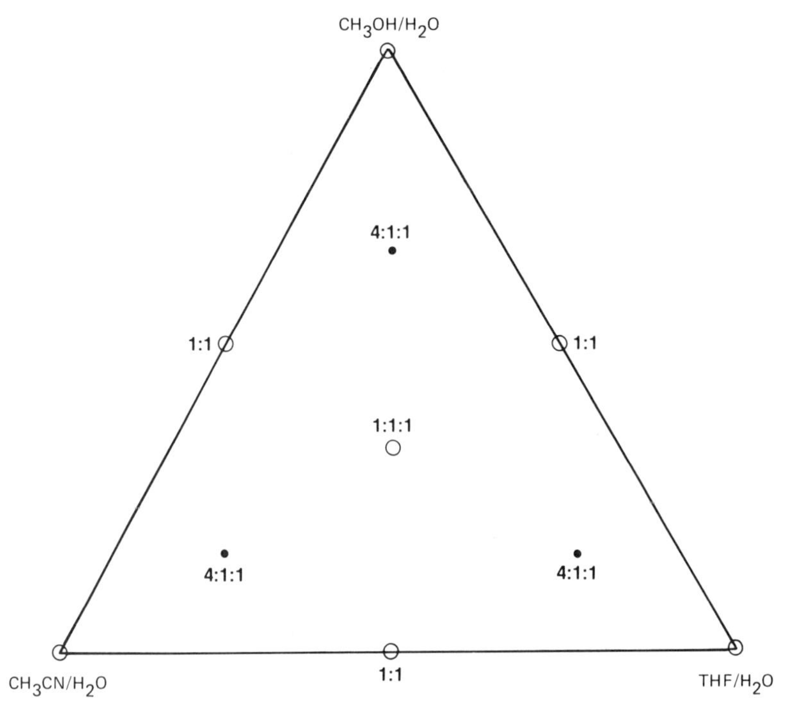

Figure 3 Combination of three organic modifiers for reversed-phase HPLC used in ORM calculations.

dix 2 of Ref. 12) is a Statistical Analysis System (SAS, version 79.5) route [14]. A DATA paragraph converts the three-dimensional solvent compositions to a two-dimensional triangle representation as used by Snee [13]. The data are fitted into a cubic model for a three-dimensional system. The parameters of the cubic equation for each set of peak resolutions are computed using the general linear model (GLM) procedure.

The PRINT procedure lists predicted resolutions of each peak pair for all solvent combinations, varying each solvent from 0 to 100% by 2% increments (see, e.g., Table 3). Contour plots of the region where the predicted resolution above a desired level determined by the analyst are produced (see, e.g., Fig. 4) using the PLOT procedure. The union of these plots showing the region where all resolutions are above this level and plots showing the area of maximum total resolution are also produced using PLOT. A flow chart of the procedure is shown in Fig. 5. The programs are run on an IBM model 370/168, and use 210 K of core.

Table 3 Tabulation of Peak Pair Resolutions as Predicted by the Computer Based on ORM Calculations

Obs.	A[a]	B	C	Peak 1	Peak 2	Predicted resolution[b]
2451	0.08	0.88	0.04	1	2	3.9000
2452	0.08	0.88	0.04	1	3	6.7315
2453	0.08	0.88	0.04	1	4	11.2939
2454	0.08	0.88	0.04	1	5	13.2736
2455	0.08	0.88	0.04	2	3	2.8315
2456	0.08	0.88	0.04	2	4	7.3938
2457	0.08	0.88	0.04	2	5	9.3731
2458	0.08	0.88	0.04	3	4	4.5623
2459	0.08	0.88	0.04	3	5	6.5417
2460	0.08	0.88	0.04	4	5	1.9793
2461	0.08	0.90	0.02	1	2	3.7630
2462	0.08	0.90	0.02	1	3	6.7713
2463	0.08	0.90	0.02	1	4	11.2940
2464	0.08	0.90	0.02	1	5	13.4224
2465	0.08	0.90	0.02	2	3	3.0083
2466	0.08	0.90	0.02	2	4	7.5310
2467	0.08	0.90	0.02	2	5	9.6608
2468	0.08	0.90	0.02	3	4	4.5227
2469	0.08	0.90	0.02	3	5	6.6521
2470	0.08	0.90	0.02	4	5	2.1298
2471	0.08	0.92	0.00	1	2	3.6190
2472	0.08	0.92	0.00	1	3	6.8080
2473	0.08	0.92	0.00	1	4	11.2887
2474	0.08	0.92	0.00	1	5	13.5692
2475	0.08	0.92	0.00	2	3	3.1891
2476	0.08	0.92	0.00	2	4	7.6697
2477	0.08	0.92	0.00	2	5	9.9536
2478	0.08	0.92	0.00	3	4	4.4807
2479	0.08	0.92	0.00	3	5	6.7635
2480	0.08	0.92	0.00	4	5	2.2839
2481	0.10	0.00	0.90	1	2	3.2316
2482	0.10	0.00	0.90	1	3	2.1402
2483	0.10	0.00	0.90	1	4	6.2222
2484	0.10	0.00	0.90	1	5	4.9304
2485	0.10	0.00	0.90	2	3	-1.0914
2486	0.10	0.00	0.90	2	4	2.9906
2487	0.10	0.00	0.90	2	5	1.7639
2488	0.10	0.00	0.90	3	4	4.0821
2389	0.10	0.00	0.90	3	5	2.8358
2390	0.10	0.00	0.90	4	5	-1.2267

Table 3 (continued)

Obs.	A[a]	B	C	Peak 1	Peak 2	Predicted resolution[b]
2491	0.10	0.02	0.88	1	2	3.3939
2492	0.10	0.02	0.88	1	3	2.2967
2493	0.10	0.02	0.88	1	4	6.4227
2494	0.10	0.02	0.88	1	5	5.1312
2495	0.10	0.02	0.88	2	3	-1.0972
2496	0.10	0.02	0.88	2	4	3.0288
2497	0.10	0.02	0.88	2	5	1.7978
2498	0.10	0.02	0.88	3	4	4.1260
2499	0.10	0.02	0.88	3	5	2.8769

[a]A = CH_3OH, B = CH_3CN, C = THF
[b]A negative value indicates a peak crossover.

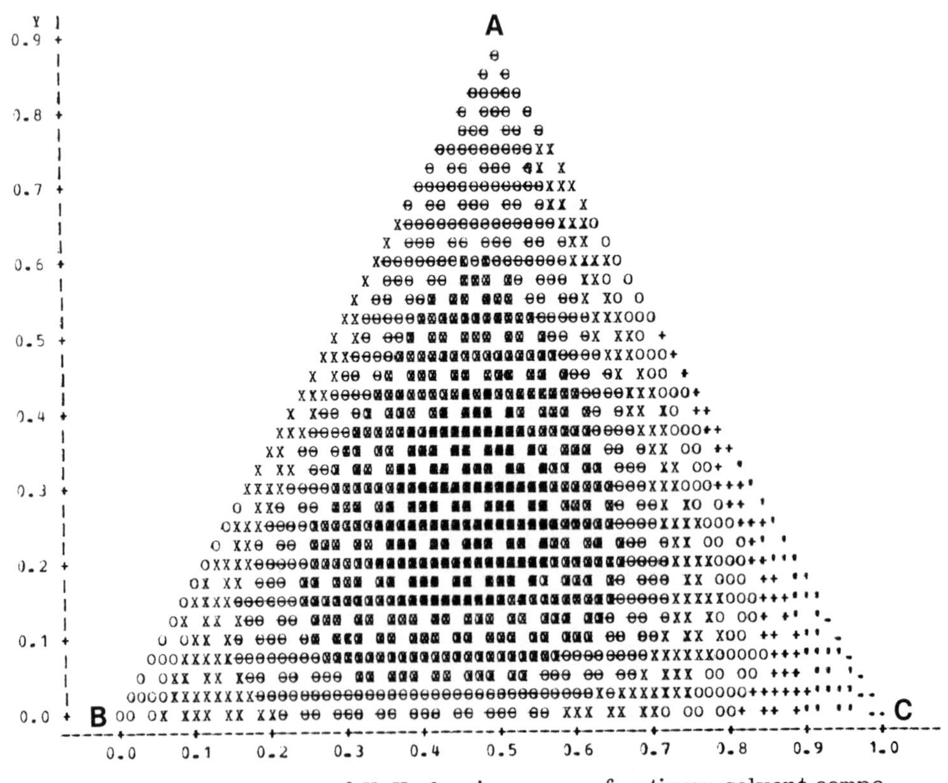

Figure 4 Contour plot of Y*X showing areas of optimum solvent composition.

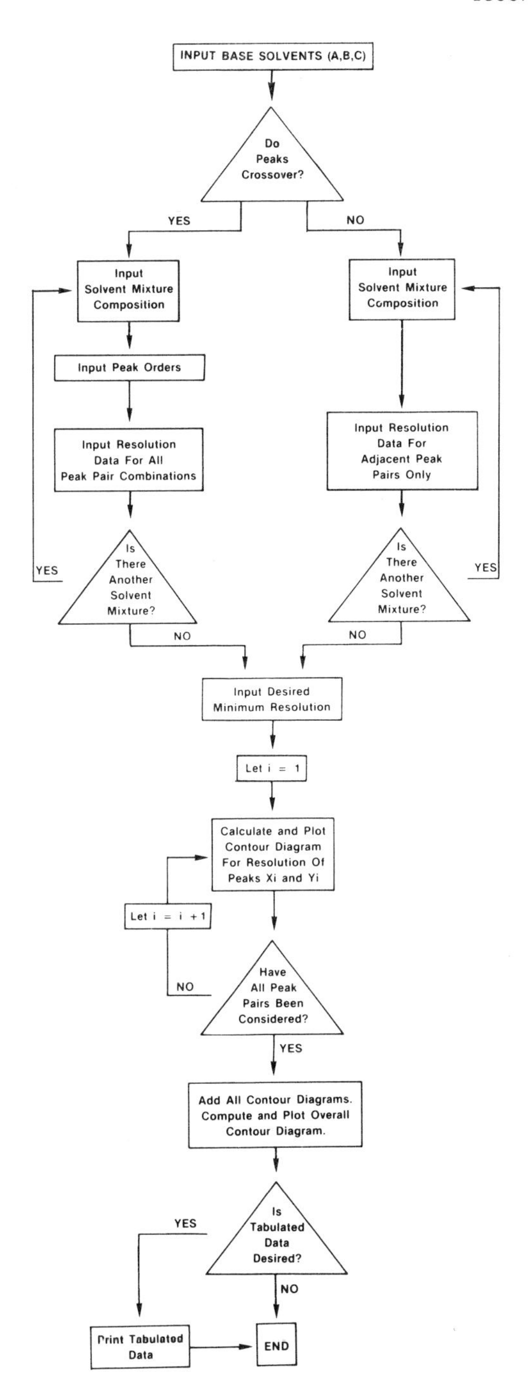

Figure 5 Flow chart of the procedure used for the ORM method.

Ideally, where a combination of three modifiers and a base solvent is used, the region of the optimum mobile phase mixture found from the ORM calculations will be in the center of the triangle (Fig. 4). If one of the solvents (e.g., C) is not ideal, the optimum mixture will be composed of the other two solvents (A and B) with only a small amount of C (see Fig. 6). Therefore the optimum region can indicate which of the three solvents is a poor choice. The base solvents are water for reversed phase and hexane for normal phase [11]. Other solvents for normal phase may also be used.

It is also possible to select one solvent (e.g., B) which gives better resolution of the components of a mixture than the other two solvents (A and C). The contour plot will show a bias toward solvent B. In this case, other solvents should be substituted for A and C. These examples show that the initial selection of the individual mobile phases is an important step which can lead to good resolution using the three organic modifiers.

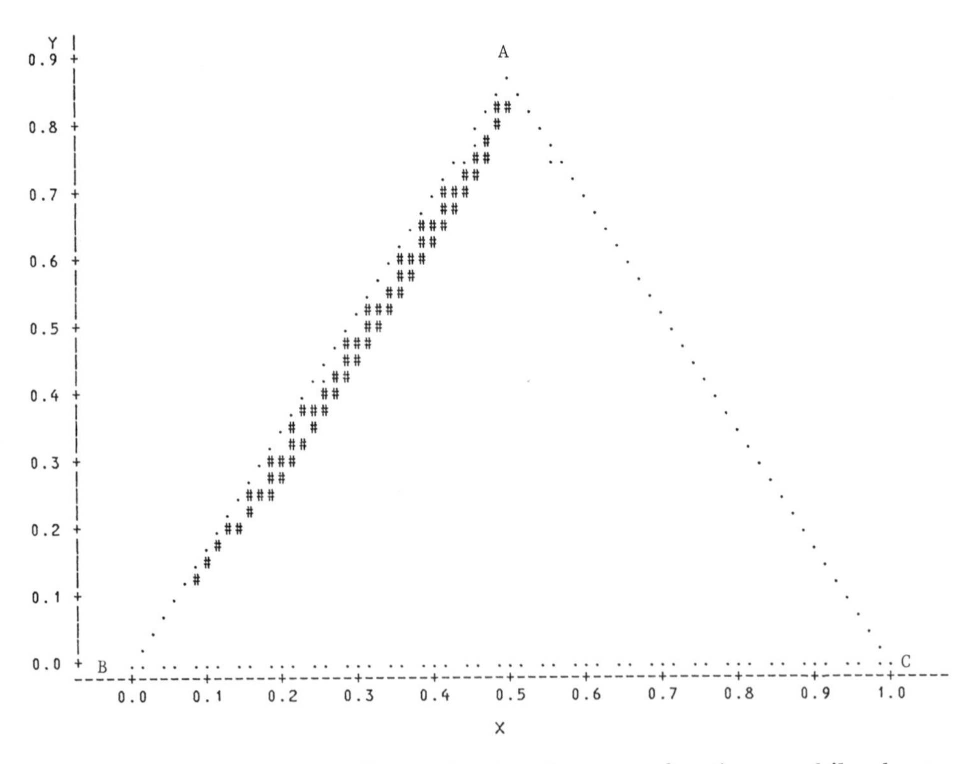

Figure 6 Contour plot of X*Y showing the area of optimum mobile phase where the resolution between two adjacent peaks is greater than 2.5 (A = CH_3OH, B = CH_3CN, C = THF; Peak 1 = 4, Peak 2 = 5).

The HLPC and TLC results indicate that this solvent selection system can be successfully applied to mobile phase optimization. Peak crossover due to different solvents can easily be handled by this method for both HPLC and TLC. Figure 7 shows the separation of a naphthalene–biphenyl–anthraquinone–methylanthraquinan–ethylanthraquinone mixture. Note the peak crossover in each of the solvents used. Figure 8 shows the separation using the predicted mobile phase mixture on reversed-phase C_8 columns.

Today manufacturers are equipping their instruments with pumps and microprocessers that will generate any mobile phase mixture from up to four different solvents, without operator supervision. This makes the chromatographer's job much easier. Also, instruments which are equipped with automatic injection will inject a set of standards so that the peak elution order can be determined, that is, which peak corresponds to what component. This will be discussed in detail later. The above procedure (ORM) was used for the separation of 26 fentanyl homologs and analogs [15] and for the separation of mixtures of selected steroids [16].

VI. GRAPHIC PRESENTATION OF MOBILE PHASE OPTIMIZATION

The ORM approach works extremely well when three organic modifiers and a base solvent are necessary to achieve optimum resolution of all components of a complex mixture.

Another approach to solvent optimization has been published based on the linear relationship between log K' and the log mole fraction of the solvent [17]. This approach is not as sound or as generally applicable as the statistical approaches discussed earlier. Recently a more practical approach to mobile phase optimization with two organic modifiers has been published [18]. Only five chromatographic runs were required for the data base, and the subsequent mathematical treatment of the data is much less involved. The method is based on the window diagram technique which was originally developed by Laub and Purnell [19–21] for the optimization of separations in gas–liquid chromatography. Recently a review of the window diagram application to gas chromatography, electrochemistry, and spectroscopy was published [22]. Contrary to the conclusion of Glajch et al., who dismissed the window diagram technique, the method is not limited to linear retention behavior or to two-component solvent systems. This will be discussed further later. Peak crossovers are also easily handled. The method was successfully applied to the optimization of separation of a five-component mixture in reversed-phase HPLC with two organic modifiers and water-based solvents (22).

For reversed phase, the three most widely used solvents are acetonitrile–water, methanol–water, and tetrahydrofuran–water. The initial ratios of organic–water selected are approximately 70 to 75%

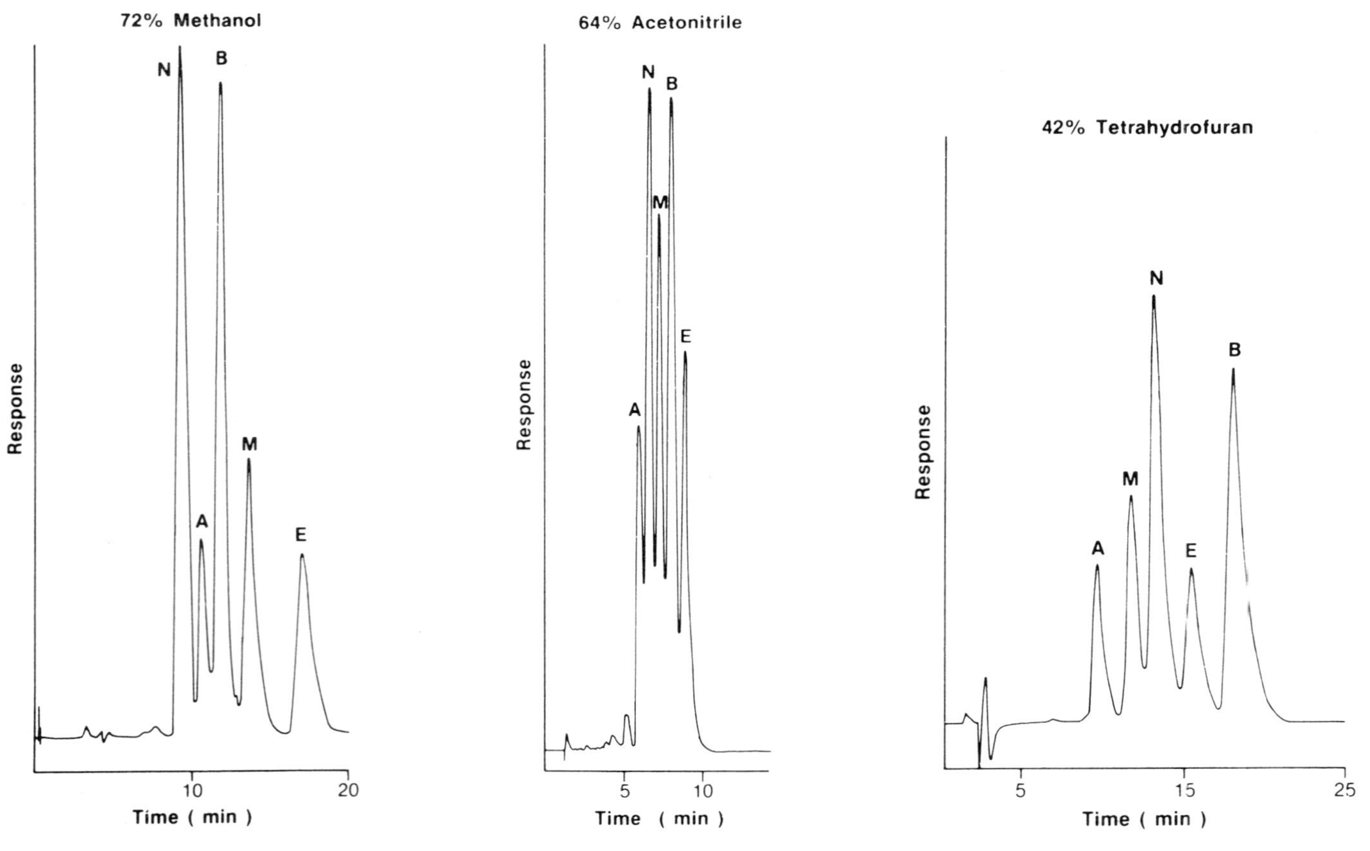

Figure 7 HPLC separation of naphthalene (N), anthraquinone (A), biphenyl (B), methylanthraquinone (M), and ethylanthraquinone (E) on reversed-phase C_8 column using 72% methanol–water (left), 64% acetonitrile–water (center), and 42% THF–water (right) at a mobile phase flow rate of 1.2 ml/min.

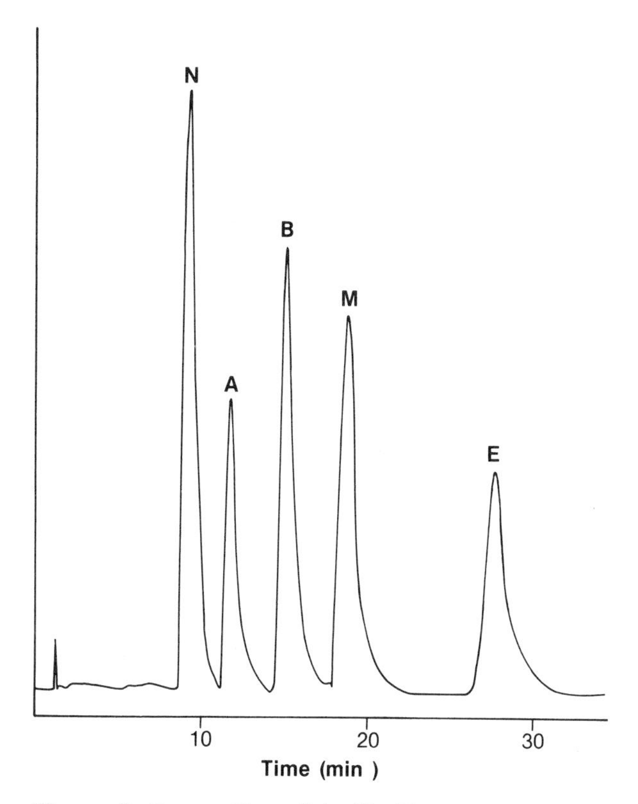

Figure 8 Separation of A, N, M, B, and E on reversed-phase C_8 column using a predicted mobile phase of 64% CH_3CN : 72% CH_3OH : 42% THF (10:67:23), at a flow rate of 1.2 ml/min.

methanol–water, 60 to 65% acetonitrile–water and 40 to 50% tetrahydrofuran–water. The strategy for selecting two of these three is illustrated below. For simplicity, assume that a five-component mixture is to be separated. The sample is first injected where 60% acetonitrile (AN)–water is the mobile phase. Should four peaks be obtained, standarads are used to identify the two coeluting peaks. Only these two are then reinjected and eluted using the two coeluting peaks. Only these two are then reinjected and eluted using a different solvent, for example, 50% tetrahydrofuran–water. Should two components be separated, the different mobile phase compositions are prepared using 60% acetonitrile as solvent A and 50% THF as solvent B. Should 50% THF–H_2O fail to separate the pair coeluting, the percentage of THF is adjusted or another mobile phase is selected. This approach is simple and time saving because the analyst has only to separate the pair not resolved; also, the identification of two components is simpler than identifying all components in a mixture.

This separation strategy was used to separate anthraquinone, 2-methylanthraquinone, 2-ethylanthraquinone, naphthalene, and biphenyl. The sample solution was chromatographed with 60% acetonitrile—H_2O. Only three peaks were observed. Anthraquinone and naphthalene coelued, as did 2-methylanthraquinone and biphenyl. However, both solute pairs were separated with 40% THF—H_2O. This demonstrates that each of the four pairs had been resolved in at least one of the initial solvents.

After selecting the initial solvents and the proportions of each in the three-solvent combination, the retention times data base is generated by recording the retention times of each solute in each of the different solvent combinations, 75% A, 50% A, and 25% A.

Table 4 shows the composition of solvents used and the retention times for each of the five solutes with each different mobile phase. The retention data as a function of mobile phase composition was fit to a polynomial of the fourth order by least-squares analysis.

Figure 9 shows plots of the calculated retention times for each solute as a function of mobile phase composition. The experimental points are also indicated. Note that, in contrast to gas chromatography, the plots are not linear. This is due to the complicated nature of solute—mobile phase, solute—stationary phase, and mobile phase—stationary phase interactions [8].

Figure 10 is a window diagram showing plots of retention time ratios versus mobile phase composition for all 10 pairs of the five solutes. The region of retention time ratio values that are higher than the minimum found at each mobile phase composition is shaded. Note that when the relative retention is calculated to be less than unity (peak crossover), the reciprocal is taken such that the ratio is always greater than or equal to unity. The tops of the windows represent the mobile phase composition giving the best separation for the least-separated pair. Two windows are seen in Fig. 10: one at 21% B with a minimum retention ratio = 1.1, and a considerably smaller window (poorer separation) at 100% B. Thus the optimum mobile phase composition for this particular separation is predicted to be 27% B (10.8% THF—43.8% AN—44.4% H_2O), which does, in practice give base-line separation of the components of the mixture (Fig. 11).

The theoretical measure of separation of a solute pair in chromatographic techniques is the relative retention α, which in HPLC is defined as the ratio of the capacity factor k' of the more retained solute to the less retained solute. We calculated k' for all solutes at all solvent compositoins by correcting for the column dead volume. The k' data were treated in a similar way to the retention data of Table 4. Figure 12 shows plots of the calculated k' for each solute as a function of mobile phase composition, and Fig. 13 gives the window diagram. Figures 9 and 11 and Figs. 10 and 12 are strikingly similar. The op-

Table 4 Retention Times for Each of the Solutes at Different Mobile Phase Compositions

Mobile phase composition			Actual component percentage			Retention time (min)				
Solvent	A[a](%)	B[b](%)	AN	THF	H_2O	Anthra-quinone	2-Methylanthra-quinone	Naphthalene	2-Ethylanthra-quinone	Biphenyl
1	100	0	60	0	40	7.23	8.98	7.23	10.94	8.98
2	75	25	45	10	45	8.26	10.30	9.16	13.27	11.77
3	50	50	30	20	50	9.57	11.98	11.30	15.82	15.13
4	25	75	15	30	55	11.65	14.75	14.67	19.98	20.38
5	0	100	0	40	60	16.44	21.20	22.41	30.19	34.03

[a]60% acetonitrile (AN)−water.
[b]40% tetrahydrofuran (THF)−water.

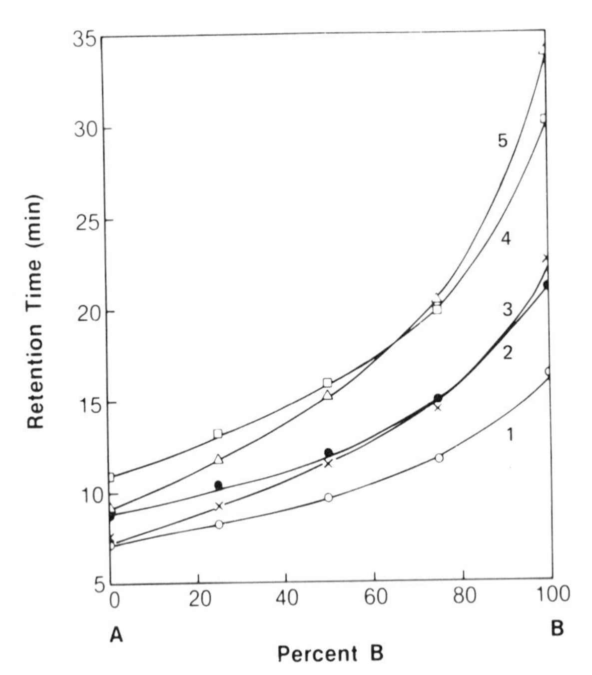

Figure 9 Retention time versus mobile phase composition for five solutes (solvents: A, 60% acetonitrile–water, and B, 40% tetrahydrofuran–water; solutes: 1, anthraquinone; 2, 2-methylanthraquinone; 3, naphthalene; 4, 2-ethylanthraquinone; 5, biphenyl).

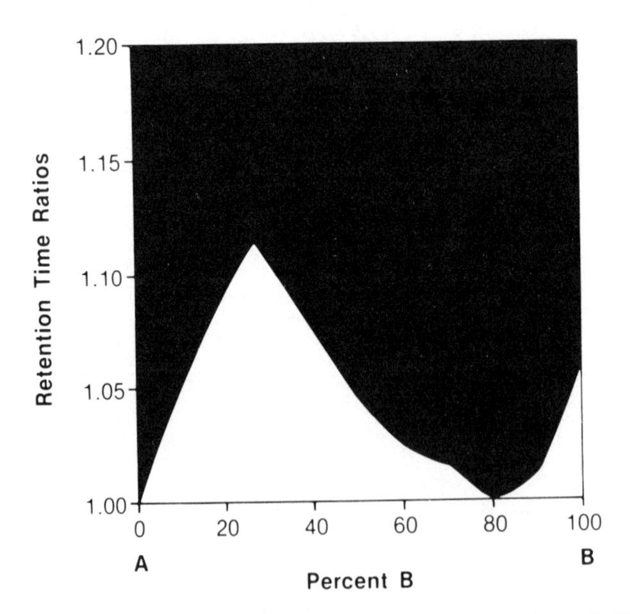

Figure 10 Window diagram for all 10 pairs of five solutes, based on retention data as in Fig. 9.

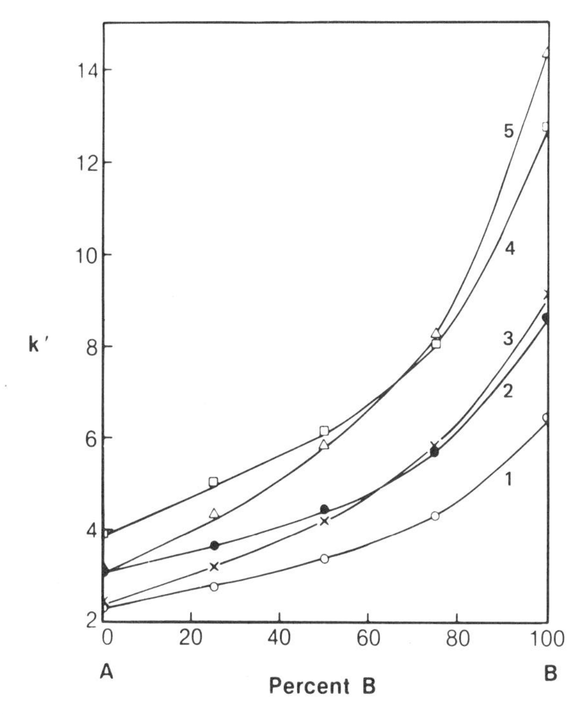

Figure 11 K' versus mobile phase composition for five solutes (symbols for solvents and solutes as in Fig. 9).

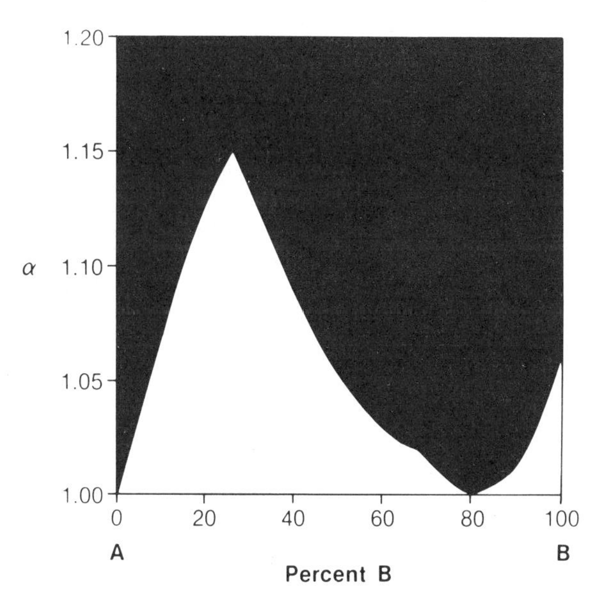

Figure 12 Window diagram for all pairs of five solutes, based on K' data as in Fig. 11.

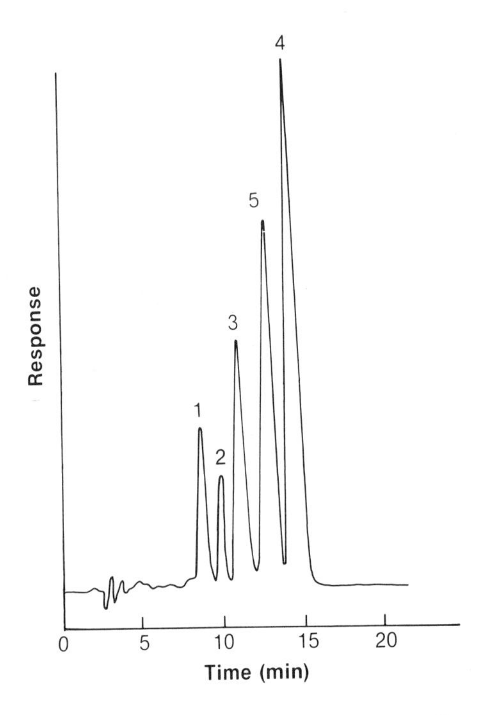

Figure 13 Chromatogram of the five solutes at optimum mobile phase composition (43.8% acetonitrile, 10.8% tetrahydrofuran, 44.4% H_2O) as determined from Fig. 9.

timum mobile phase composition obtained from the larger window of Fig. 13 is exactly the same as that obtained from Fig. 10. However, here we can obtain the minimum value of α (1.15 at 27% B). Using this value, we can calculate the minimum number of plates N_{req} for separation according to Purnell's equation [23]. In this instance, and assuming a capacity factor of five, N_{req} is calculated to be approximately 2350 plates. Note that accurate measurement of the column "dead volume" in HPLC is a difficult problem with no easy solution [24]. Any optimization techniques dependent on k' data suffer from the unavailability of accurate methods for the determination of column dead volume. The window diagram method presented here does not require the accurate determination of k'. As demonstrated earlier, the optimum solvent composition can be determined from raw retention data. When retention time is plotted against mobile phase composition, it can be seen that a total of five runs (Fig. 9) will give the mobile phase that will separate all the components of the mixture.

Deming and Turoff [25] used the window diagram technique to select an optimum pH of a buffer for the separation of a mixture of benzoic acids on reversed-phase HPLC. The procedure consisted of measuring the retention time of the weak acid of interest at pH 3 or higher, values in an appropriate buffer, after which models were fitted to the data and the model parameters were used to construct window diagrams from which an optimum pH could be selected. Deming and co-workers used the window diagram technique to study the effect of pH on other liquid-chromatographic separations [26,27]. Later they extended the simple factor window diagram technique to the multifactor case, where a two-factor study in which values of pH and the concentration of ion interaction reagent were chosen to give optimum separation [28,29]. Otto and Wegscheider [30] used a multifactor model to optimize the separation of dibasic substances on reversed-phase chromatography. Elution strength, ionic strength, and pH were studied in order to construct a three-dimensional semiempircal model window diagram for predicting retention times of dibasic compounds. The original article [30] should be consulted for details.

The optimization of temperature and solvent strength for the separation of phenylthiohydantoin amino acids on reversed-phase columns was achieved using the window diagram technique by plotting the logarithm of retention time versus solvent strength or temperature [31].

Lindberg et al. [32] discussed the application of factorial design and response surface calculations to the optimization of a reversed-phase ion-pair chromatographic separation of morphine, codeine, noscapine, and papaverine.

Kowalski [33] has reviewed the above statistical approaches in detail and therefore will not be discussed here. The reader can also refer to the original article of Lindberg et al. [32].

Schoenmakers et al. [34] devised a simple procedure for the optimization of reversed-phase separations with ternary mobile phase mixtures. The procedure is based on the use of isoeluotropic mixtures, that is, mixtures that are expected to yield the same capacity factor for a hypothetical average solute [35], and the "window diagram" technique. Their procedure is summarized here: A gradient run from pure water to pure methanol would show if all solutes in the mixture can be eluted isocratically within a given time, a k' value between 1 and 10 [36]. The sample is then run in two binary isoeluotropic mixtures (acetonitrile–water and tetrahydrofuran–water), and the peaks are identified. Then a linear plot is constructed of ln K' versus the solvent composition, and the ternary mixture that gives the best separation is selected. The sample is run using the optimum mobile phase. If the experimental results are not satisfactory, the calculations are repeated until there is no change in the optimal composition and the best separation has been achieved. It is assumed that in K' versus mobile phase composition is linear, which may or may not be true.

This will affect the quality of the results. Our results [18] indicate that K' versus mobile phase composition is not linear. In a later study Dronen et al. [37] utilized the chromatographic response function (CRF) of Morgan and Deming [38,39] to develop a more general approach to the selection of mixed mobile phases. The CRF procedure has been described in detail in Ref. 38. Berridge used the sequential simplex procedure, CRF, to optimize reversed-phase [40] and normal-phase [41] HPLC separations. He also tested the quality of the separation by examining the second derivative of the final chromatogram. If two peaks coeluted, the second derivative will reveal it, which will then require a rerun of the optimization procedure using different criteria and/or a different selection of solvents. The CRF is one that describes the separation quality in terms of the number of peaks, resolution, and time of analysis. The CRF used by Berridge [41] is defined as follows:

$$\mathrm{CRF} = \sum_{i=1}^{L} R_i + L^a - b(T_L - T_A) + c(T_1 - T_0) \tag{16}$$

where R is the resolution between adjacent peaks, L is the number of peaks detected, T_A is the specified analysis time (in minutes), T_L is the retention time of the last peak, T_1 is the retention time of the first peak, T_0 is the specified minimum retention time, and a,b, and c are operator-selectable weightings (set to 2). Using this above procedure, the separation of a five-component mixture required 23 experiments, in which two of the five peaks were not base-line separated. Although the procedure is automated, we feel that 23 experiments are a waste of time and material, unless the separation is an extremely difficult one.

Toon and Rowland [42] proposed a simple graphical method for the selection of a binary mobile phase that will give optimum separation. Six barbiturates were to be separated on a reversed-phase column using acetonitrile–water. The solution, as stated, lay in the empirical observation that for all barbiturates studied a linear relationship existed between $\log(1-R_0/R_t)$, designated as R_Q, and the mobile phase, where R_t and R_0 are retained and nonretained solutes. Optimum resolution is defined as complete separation of the solute peaks at their base in the shortest time. By plotting R_0 values, estimated for both the front and the back of solute peaks against percentage solvent composition, one can find the optimum mobile phase. Optimum resolution is defined as the point where the plots of the R_0 front of one solute intersect with the R_0 back of another solute. It is claimed that this procedure worked for multicomponent (n) mixtures after some trial and error, since there will be an n-1 points of intersection. It is much easier, in our opinion, to adopt the window diagram approach or a gradient run from pure water to pure acetonitrile to select an isocratic mobile phase. However, this procedure of Toon and Rowland [42] was criticized by Tomlinson [43], who found that R_0 is always negative and at variance with Figs. 1, 3, 5, and 8 in Ref. 42. Other points were also raised by Tomlinson [43] which will not be discussed here. Rafel [44] compared the resolution of nitroaromatics and flavones by three different methods: the simplex method (45), the extended Hooke–Jeeves direct search method [46], and the Box–Wilson steepest ascent path [47]. The results showed that the optimal resolutions obtained by any of these methods are similar, however, the Hooke–Jeeves method following a 2k-factorial design appeared to be a faster and easier approach.

VII. PEAK ELUTION ORDER (COMPONENT IDENTIFICATION)

One of the most difficult and time-consuming aspects of searching for a mobile phase that will give optimal separation is the identification of the eluted peaks. To establish such a mobile pahse, the analyst, depending on the method chosen, must run 5 to 23 experiments using different solvent combinations. Since these various methods require the use of solvents of different selectivities, this can lead to different elution orders of the components, that is, peak reversal. Thus the analyst must identify each eluted peak at the end of each chromatographic run, which is normally done by spiking the sample with a known standard or injecting each of the standards separately, a time-consuming procedure. Issaq and McNitt [48] developed a computer program whereby the eluted peaks are identified by the area ratio of each peak compared with the other eluted peaks in that experiment. The computer program takes into consideration peak reversal, peak coalescence, and peak splitting. The analyst can also use the ratio of the areas under the peaks generated using two different wavelegnths, for example, 254 and 280 nm. This computer program is also suited for use with radiolabeled compounds, since the number of counts is directly proportional to the amount of radioactivity.

A more accurate method for identifying the eluted components is one based on the molecular and physical properties of the compounds under study. The use of mass spectrometry (HPLC/MS) would give accurate results, but this is relatively expensive. Another method would be to use fourier transform infrared spectroscopy (FTIR), but this is not without problems such as background interference from the solvent and the sample. This may be overcome if the analyst can subtract interference with the aid of a computer. A reasonably priced FTIR spectrophotometer which may be suited for such studies is the model IR/32, made by IRM instruments Company, which can record an infrared spectrum every second. Barring interferences, this could be an excellent and accurate method for component identification.

A spectrophotometer that can record the spectra in 1/100 the time it takes the FTIR to record the spectra, and at half the price, is the high-speed spectrophotometric detector, recently introduced by Hewlett-Packard. The model 1040A detector records the ultraviolet—visible spectra of the eluting component in 10 msec without interrupting the chromatographic process, unlike stopped-flow ultraviolet detectors. This is based on the combination of a series of photodiodes which monitor all wavelengths (190 to 600 hm) simultaneously, with a data acquisition processor network, for fast analog-to-digital conversion and fast data storage. A detailed description of the detector has been published [49]. Other companies have introduced similar ultraviolet detectors based on the photodiode array principle.

VIII. CONCLUSIONS

This chapter has dealt with methods for selecting an optimum mobile phase for isocratic elution. It is clear that the window diagram technique is a very simple, noncomplicated approach to mobile phase optimization using two binary mobile phases. It is equally clear that the ORM approach is a very efficient method for selecting a ternary or quaternary mobile phase system. The ORM approach will require a powerful computer, while the window diagram will not in most cases. A graphic presentation is all that is needed, and in those instances where a computer is required, a microcomputer will suffice. These two approaches are systematic and require a defined set of conditions and a minimum number of experiments. Their use should be encouraged, since they will save the chromatographer time and money.

ACKNOWLEDGMENTS

By acceptance of this article, the publisher or recipient acknowledges the reight of the U.S. Government to retain a nonexclusive, royalty-free license in and to any copyright covering the article. This work was supported by contract no. NO1-CO-23910, with the National Cancer Institute, National Institutes of Health, Bethesda, Maryland 20205.

REFERENCES

1. L. R. Snyder, *Principles of Adsorption Chromatography*, Marcel Dekker, New York, 1968.
2. M. Jaroniec, J. K. Rozzylo, J. A. Jaroniec, and B. Oxcik-Mendyk, J. Chromatogr. *188*, 27 (1980).
3. D. E. Martire and R. E. Boehm, J. Liquid Chromatogr. *3*, 753 (1980).
4. H. J. Issaq, Pittsburgh Conference on Analytical Chemistry and Applied Spectroscopy, Atlantic City, 1981.
5. L. R. Snyder, J. Chromatogr. Sci. *92*, 223 (1974).

6. S. R. Baklayar, R. McIlwrick, and E. Roggendorf, J. Chromatogr. *14*, 343 (1977).
7. D. L. Saunders, Anal. Chem. *46*, 470 (1974).
8. L. R. Snyder and J. J. Kirkland, *Introduction to Modern Liquid Chromatography*, 2nd ed., Wiley, New York, 1979.
9. D. Rogers, Am. Lab. *12*, 49 (1980).
10. B. R. Belinky, *Analytical Technology and Occupational Health Chemistry*, ACS Symposium Series, Volume 220, American Chemical Society, Washington, D.C., 1980, pp. 149–168.
11. J. L. Glajch, J. J. Kirkland, and K. M. Squire, J. Chromatogr. *199* 57 (1980).
12. H. J. Issaq, J. R. Klose, K. L. McNitt, J. W. Kaky, and G. M. Muschik, J. Liquid Chromatogr. *4*, 2091 (1981).
13. R. D. Snee, Chem. Technol. *9*, 702 (1979).
14. SAS Institute, *SAS User's Guide*, Raleigh, N.C., 1979.
15. I. S. Lurie, A. C. Allen, and H. J. Issaq, J. Liquid Chromatogr., *7*, 463 (1984).
16. P. E. Antle, Chromatographia *15*, 277 (1982).
17. S. Hara, K. Kunihiro, H. Yamaguchi, and E. Doczewinski, J. Chromatogr. *239*, 687 (1982).
18. H. J. Issaq, G. M. Muschik, and G. M. Janini, J. Liquid Chromatogr. *6*, 259 (1983).
19. R. J. Laub and J. G. Purnell, Anal. Chem. *48*, 1720d (1976).
20. R. J. Laub, J. H. Purnell, and P. S. Williams, J. Chromatogr. *134*, 249 (1977).
21. R. J. Laub, A. Pelter, and J. H. Purnell, Anal. Chem. *51*, 1878 (1979).
22. R. J. Laub, Am. Lab. *13*, 47 (1981).
23. J. H. Purnell, J. Chem. Soc. 1268 (1960).
24. E. Grushka, H. Colin, and G. Guiochon, J. Liquid Chromatogr. *5*, 1297 (1982).
25. S. N. Deming and M. S. H. Turoff, Anal. Chem. *50*, 546 (1978).
26. W. P. Price, Jr., R. Edens, D. L. Hendrix, and S. N. Deming, Anal. Biochem. *93*, 233 (1979).
27. W. P. Price, Jr., and S. N. Deming, Anal. Chim. Acta. *108*, 227 (1979).
28. R. C. Kong, B. Sachok, and S. N. Deming, J. Chromatogr. *199*, 307 (1980).
29. B. Sachok, R. C. Kong, and S. N. Deming, J. Chromatogr. *199*, 317 (1980).
30. M. Otto and W. Wegscheider, J. Chromatogr. *258*, 11 (1983).
31. C. M. Noyes, J. Chromatogr. *266*, 451 (1983).
32. W. Lindberg, E. Johansson, and K. Johansson, J. Chromatogr. *211*, 201 (1981).
33. B. R. Kowalski, Anal. Chem. *52*, 112R (1980).
34. P. J. Schoenmakers, A. C. J. H. Dronen, H. A. H. Billiet, and L. de Galan, Chromatographia *15*, 688 (1928).

35. P. J. Schoenmakers, H. A. H. Billiet, and L. de Galan, J. Chromatogr. *218*, 261 (1981).
36. P. J. Schoenmakers, H. A. H. Billiet, and L. de Galan, J. Chromatogr. *205*, 13 (1980).
37. A. C. J. H. Drouen, H. A. H. Billiet, P. J. Schoenmakers, and L. de Galan, Chromatographia *16*, 48 (1983).
38. S. L. Morgan and S. N. Deming, J. Chromatogr. *112*, 267 (1975).
39. S. L. Morgan and S. N. Deming, Sep. Purif, Methods *5*, 333 (1976).
40. J. C. Berridge, J. Chromatography *244*, 1 (1982).
41. J. C. Berridge, Chromatographia *16*, 174, (1973).
42. S. Toon and M. Rowland, J. Chromatogr. *298*, 341 (1981).
43. E. Tomlinson, J. Chromatogr. *236*, 258 (1982).
44. J. Rafel, J. Chromatogr. *282*, 287 (1983).
45. W. Spendley, J. R. Hext, and F. R. Himsworth, Technometrics *4*, 441 (1962).
46. R. Hooke and T. A. Jeeves, J. Assoc. Comput. Mach. *8*, 221 (1961).
47. G. E. P. Box and K. B. Wilson, R. J. Stat. Soc. B *13*, *1* (1951).
48. H. J. Issaq and K. L. McNitt, J. Liquid Chromatogr. *5*, 1771 (1982).
49. J. M. Miller, S. A. George, and G. B. Willis, Science *218*, 241 (1982).

4

Electrochemical Detectors for Liquid Chromatography

Ante M. Krstulović*, Henri Colin and Georges A. Guiochon / Ecole Polytechnique, Palaiseau, France

I. INTRODUCTION

With the rapid ascent of contemporary liquid chromatography (LC) and recent trends toward miniaturization of LC equipment, all instrument components are now being critically examined. The use of small-bore columns appears to be particularly attractive in view of reduced solvent consumption, decreased sample dilution and the possibility of on-line

---

**Current affiliation*: L.E.R.S.-Synthelabo, Paris, France

coupling with mass spectrometry. The analyses of small amounts of samples demand highly sensitive detectors with low dead volume, which explains the current efforts in the optimization of the already existing devices and the development of new ones.

Flow-through detectors for LC are of two types: the *bulk property* detecting and the *solute property* detecting. The former, such as the refractive index, conductometric, and capacitance detectors, monitor the bulk properties of the mobile phase. While having the advantage of being nondestructive and relatively "universal," these detectors generally afford moderate sensitivities and have a limited dynamic range. Solute property detectors, such as ultraviolet absorbance, fluorimetric, amperometric, polarographic, and coulometric detectors, selectively characterize certain components in the mobile phase.

Regardless of the principle of detection, the desirable properties of an LC detector are the following:

1. It should possess high sensitivity and thus afford low detection limits.
2. It should have a wide dynamic range.
3. The signal should be reproducible and stable.
4. The time constant should be low enough in order not to distort the signal.
5. The cell should have the lowest possible volume and should be designed in such a way as to avoid remixing of the separated bands.
6. It should exhibit approximately the same sensitivity for all solutes detected, and the signal should not be temperature dependent.

No currently available LC detector satisfies all of these requirements, and a judicious choice is needed in selecting the optimal detection mode for a specific application.

The electrochemical or electrical properties of components of chromatographic effluenets have been utilized as a basis for detection long before the advent of high-performance liquid chromatography (HPLC). In 1940 Troitskii [1] detected the adsorption boundaries in LC by measuring the dielectric constants, while Müller [2] designed a continuous flow system with an amperometric cell for the measurement of low levels of hydroquinones. The first applications of combined use of polarographic detectors and liquid chromatography can be found in the work of Drake [3] and Kemula [4]. However, the full potential of the tandem operation of thin-layer electrochemistry and LC was not fully realized until the detection technique was further adapted and refined by Adams [5] and Kissinger et al. [6]. It became immediately evident that thin-layer electrochemistry, although not as widely applicable as optical detection, would become unrivaled in certain areas of biomedical and environmental research in terms of sensitivity and linear dynamic range. Since then, this technique has been experiencing a spectracular ascent

Table 1 Electrolytic Methods and Methods Based on Electric Properties of Solutions Used for the Continuous Monitoring of Substances in Flowing Liquids

Method	Quantity measured[a]	Quantity controlled
Electrolytic methods		
Equilibrium potentiometry	$E = f(a)$	—
Polarography and steady-state voltametry	$I_{lim} = k_I c$	E
Nonstationary voltametric techniques (ac, square wave, pulse, differential pulse)	$I_{ac} = f(t)$	E(t)
Constant-potential coulometry	$Q = \int_0^t I \, dt$	E
Methods based on electric properties of solutions		
Low-frequency conductometry		—
Capacitance	$Z = f(\kappa + \varepsilon)$	—

[a]Symbols: E, potential; E(t) potential as a periodic function of time; I, current; I_{lim}, limiting current; I_{ac}, ac component of the electrolytic current; Q, charge; Z, impendance; a, activity; c, concentration; k_I, diffusion current constant; κ, specific conductance; and ε, permittivity.
Source: Ref. 10.

and its principles [7–14] and main areas of application [15–22] have been reviewed in numerous publications.

This article will review the detector contribution to band broadening, the basic characteristics and the main areas of application of the detectors exploiting the electric properties of solutions and electrolytic methods. The basic principles of operation of both types of detectors are outlined in Table 1.

II. CONTRIBUTION TO BAND BROADENING

Any detector contributes to band broadening and sample dilution because of its cell volume (acting as a mixing chamber) and its time constant. These extracolumn effects can be characterized by their variances: $\sigma^2_{d.v.}$ for the cell volume and $\sigma^2_{d.t.}$ for that of the time constant. It has been shown [23,24] that these variances are given by the following equations:

$$\sigma^2_{d.v.} = V_d^2/F^2 \tag{1}$$

$$\sigma^2_{d.t.} = \tau^2 \tag{2}$$

where V_d, F, and τ are the detector cell volume, the mobile phase flow rate, and the detector time constant, respectively.

The extra column effects decrease the efficiency of the system. If N_0 is the time column plate number and N the observed plate number, the relative loss in efficiency is given by

$$(N_0 - N)/N_0 = \theta^2/(1 + \theta^2) \tag{3}$$

where θ^2 is the ratio of variances of the extra column effect and that of the column. The contributions of the cell volume and time constant will be discussed separately.

As far as the cell volume is concerned, it is necessary to express the column variance on a volume basis:

$$\sigma_0^2 = \left(\frac{\pi d_c^2}{4} \varepsilon(1 + k')hd_p\sqrt{N} \right)^2 \tag{4}$$

where d_c, ε, k, h, d_p, and N have their usual meaning.

The combination of Eqs. (1) and (4) gives the relationship between the cell volume and the characteristics of the column [25–28]:

$$V_d = \theta \frac{\pi d_c^2}{4} \varepsilon(1 + k')hd_p\sqrt{N} \tag{5}$$

It is clear that, for a given accepted loss of efficiency (a given value of θ), the smaller the capacity ratio and the better the column (smaller h), the smaller the cell volume has to be. At constant particle size and flow rate, decreasing the plate number (and thus the column length) and the column diameter results in a smaller required cell volume.

For a typical 150 X 4.6 mm column packed with 5-µm particles and operated at the optimum flow rate (h = 2.5), the cell volume has to be less than 7 µl if a less than 5% relative decrease in efficiency is expected for a moderately retained solute (k' = 1). In the case of 250- and 1000-mm-long, 4-mm inner diameter columns, the volumes are 1.9 and 3.9 μl, respectively.

In order to characterize the detector time constant, it is necessary to express the column variance in time units:

$$\sigma_0^2 = \left(\frac{h d_p^2 (1 + k')}{\nu D_m} \sqrt{N} \right)^2 \qquad (6)$$

where ν and D_m are the reduced velocity and solute diffusion coefficient, respectively. The combination of Eqs. (2) and (6) gives the expression for

$$\tau = \theta \frac{h d_p^2 (1 + k') \sqrt{N}}{\nu D_m} \qquad (7)$$

Under the same conditions as above, and assuming $\nu = 3$ and $D_m = 1.5 \times 10^{-5}$ cm^2/sec, it can be calculated that the values of the time constant are 1.2, 1.4, and 3.14 sec, respectively.

III. ELECTROLYTIC DETECTORS

The electrochemical (EC) detectors in this category, such as the polarographic, voltametric, coulometric, and potentiometric detectors, operate by generating an electrical current as a result of either the direct oxidation or reduction of the solute, or through a secondary reaction with the electrode material. The compounds which are amenable to EC detection must have an "electroactive" functionality and must accommodate the loss or gain of electrons upon undergoing the reaction. In order to find out whether a particular compound can be detected within the usable range of commonly employed EC detectors, one should consult the redox reaction data. A useful source of this information is the series *Encyclopedia of Electrochemistry of the Elements* [29], or any other texts on organic electrochemistry (e.g., see Ref. 30).

It the compound of interst is not amenable to direct electrolysis, indirect pre- or postcolumn reactions can be used [7]: Amino acids can thus be detected by reaction with copper (II) [31]; sugars can be oxidized by hexacyanoferrate (III) [32], and unsaturated compounds can be reacted with electrolytically generated Br_2 [33]. It should be pointed out that detections based on secondary reactions usually afford poorer detection limits, especially if the disappearance of an electroactive species is measured [33]. In addition, the linear range may also be slightly lower for these reactions. Certain compounds, particularly some biological species, may exhibit irreversible behavior resulting from slow heterogeneous electron transfer [34]. This results in adsorption of reaction products on the surface of the electrode (and possible formation of polymeric films), leading to electrode poisoning. These problems can be circumvented by using redox mediators to facilitate the electron transfer. An impressive survey of compounds with redox mediation properties can be found in Ref. 34. An ideal "electron shuttle" should exhibit a well-defined electron stoichiometry, should have a known formal potential ($E^{\circ\prime}$), and should not alter the redox potential of the solute under study. Furthermore, the oxidized and reduced mediator forms should be stable and soluble in the mobile phase.

It should be pointed out that an electrochemical detector can be considered as a postcolumn reactor in which a portion (in polarographic and voltametric detectors) or the entire amount of analyte in the cell (in potentiometric and coulometric detectors) is chemically converted. Thus the basis of detection is a chemical rather than a physical process. Therefore it is understandable that the selection and optimization of detection conditions is more critical with EC detectors than some other types, and the limitations posed on the mobile phase properties are more stringent. Thus the electrical conductivity, pH, and solvent composition must be considered. If these detector operational requirements are incompatible with the chromatographic separation (e.g., in normal-phase chromatography), postcolumn mixing can remedy the situation. Alternatively, expensive salts such as tetraalkylammonium hexafluorophosphate can be used [14]. However, most EC-HPLC applications are carried out using conducting mobile phases, and thus either the reversed-phase or ion-exchange mode of separation. A sufficiently high ionic strength of the mobile phase (average electrolyte molarities of approximately 0.05 M and above) affords a good control of electrode potential (low IR drop between the electrodes) and the signal-to-noise ratio. Insufficient conductivity results in potential losses, which limits the linear range. Since most redox reactions of organic compounds are accompanied by a gain or loss of protons, the pH of the supporting electrolyte may have a profound importance both on the half-wave potential ($E_{1/2}$) and the rate of electron transfer. From the Nernst equation, a 60-mV shift in $E_{1/2}$ can be expected for a unit pH change for rapid reactions in which the total number of electrons transferred equals the number of electrons gained or lost.

In addition, a change in pH can dramatically alter the nature of the chemical reaction and thus the signal. This is illustrated with catecholamines in Fig. 1. At lower pH values epinephrine is oxidized into the corresponding o-quinone accompanied by the loss of two protons and two electrons. This reaction constitutes the basis for the highly sucessful EC-HPLC assays for catecholamines. At elevated pH, however, the free base of the side chain attacks the aromatic ring to form indoline, which can be oxidized further if the rearrangement is fast enough compared to the residence time of the molecule in the cell. Thus the resulting signal is doubled compared to that at low pH. Con-

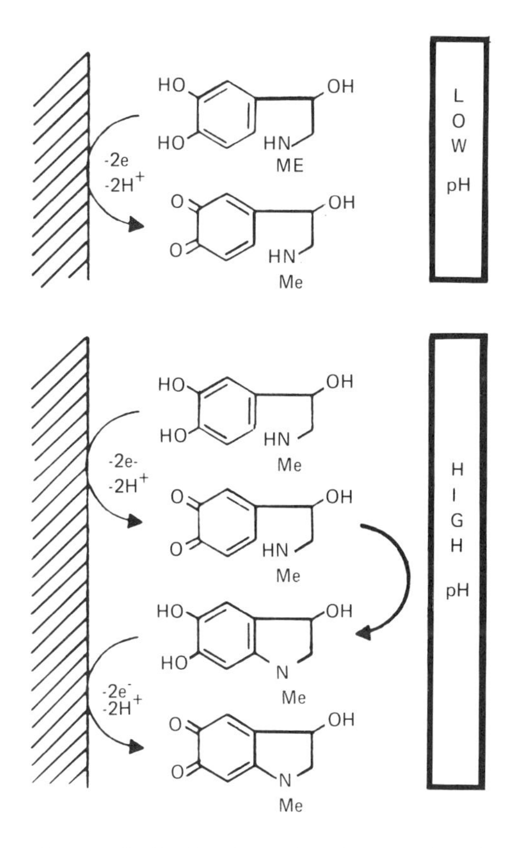

Figure 1 Mechanisms for the electrochemical oxidation of epinephrine. (From Ref. 7.)

versely, the responses for norepinephrine and dopamine do not follow the same trend, which is consistent with the differences in the rates of nucleophilic addition reactions for the three catecholamines. Generally, oxidations are carried out more easily at higher pH values. However, care must be taken to avoid the use of basic solutions which would lead to dissolution of the column packing material in the case of auto-oxidation catecholamines. If a basic medium is absolutely necessary for the EC detection, the solution pH can be raised by post-column addition of the reagent. Conversely, reductions are more conveniently carried out at lower pH.

The chemical nature of the buffer can also influence the detector performance if the solutes react with the buffer components to form complexes of different electroactivities than that of the original solute molecule. Buffer components can also interact with the electrode surface and thus modify the nature of the electron transfer. Contrary to the oxidative mode, lower buffer concentrations are desirable for the reductive mode (10^{-3} to 10^{-2} M) in order to minimize the concentrations of transition metals which can be deposited on the electrode surface. The trace metals can be removed by pre-electrolysis of the eluents [14].

A. Amperometric Detectors

These detectors are by far the most commonly employed among all EC detectors. This category comprises those detectors in which the working electrode is maintained at a selected potential versus the reference electrode and the signal (current) results from the oxidation or reduction of a small fraction (usually 5%) of the electroactive species. The current measured is proportional to the solute concentration in the mobile phase. The relationship between the signal and solute concentrations is usually linear over four to five orders of magnitude of concentration.

At the electrode—solution interface the conversion of electron current to ion current occurs in the presence of an electroactive species capable of receiving or donating electron(s) from or to the electrode. The overall sensitivity of detection is a function of the rate of mass transfer of solute molecules toward the electrode and the ease of charge or electron transfer at the surface of the working electrode. The flux of electroactive species results from the bulk motion of the solution past the electrode surface and the diffusion process. Migration of charged species can be ruled out in the presence of a sufficiently high concentration of supporting electrolyte.

The molar flux J of solute molecules is a function of the mass transfer coefficient T (which depends on detector geometry, size, solvent flow rate, viscosity, and temperature), and the concentration gradient C between the bulk of the solution and that at the electrode surface, C_0:

$$J\ (\text{mol/cm}^2\ \text{sec}) = T\ (\text{cm/sec})[C - C_0](\text{mol/cm}^3) \tag{8}$$

The resulting current I is proportional to the flux, electrode area A, and the number of electrons transferred per molecule n:

$$I\ (\text{c/sec}) = J\ (\text{mol/cm}^2\ \text{sec})\ n(\text{Eq/mol})F\ (\text{C/Eq})A(\text{cm}^2) \tag{9}$$

where F is Faraday's constant (96,500 C/sec).

Electron transfer reactions between the solute and the electrode surface depend on the thermodynamics and kinetics of the reaction. However, a reaction which is thermodynamically favorable can nevertheless be kinetically slow. In these cases, the reaction rate can be increased by applying a larger potential difference between the electrodes. The disadvantages of this will be discussed later. The mass flux will be controlled by kinetics if the concentration gradient is zero

($C - C_0 = 0$). Conversely, for fast reactions, the concentration of solute molecules in the immediate vicinity of the electrode will approach zero, and thus the current will be mass controlled. Most analytical applications of EC detection are carried out in this region where the overall sensitivity depends on the electrode, cell design, mobile phase flow, and temperature (a change of 1°C causes an approximate charge of 3% of the current). The thickness of the diffusion layer at the working electrode is of critical importance for the detection sensitivity; it depends on the cell geometry and solvent flow rate in the vicinity of the electrode surface (the thickness is a power function of the cell volume and the value of the exponent depends on the cell hydrodynamics). These factors will be discussed separately.

1. Detector Construction

Amperometric detectors operate almost exclusively on a three-electrode principle (working, auxiliary, and reference electrodes) in order to suppress the ohmic (IR) drop in the flow cell and achieve a good control of the operational potential and its constancy. During operation, large values of potential drop between the working and reference electrodes (which depends both on solution resistance and electrolysis current) may cause a shift in the potential of the working electrode. This is particularly dangerous if the operative voltage is on the rising portion of the voltametric wave (current versus potential). Thus it is very important that the reference and auxiliary electrodes be positioned as closely as possible to the working electrode.

Several cell designs, each one having different hydrodynamic properties, have been described in the literature. The three most commonly employed (Fig. 2) are the *tubular*, *wall jet*, and *channel* types. Some characteristics of these hydrodynamic systems are outlined in Table 2.

Recently Hanekamp and van Nieuwkerk [38] combined the three equations given in Table 2 to obtain a general equation for all types of cell geometries:

$$I = knFDC^* Sc^{\beta} w\ Re^{\alpha} \tag{10}$$

Where Re is the Reynolds number [$Re = \bar{v} l_x / \eta$], Sc is the Schmidt number ($Sc = \eta / D$), $\bar{v}$ is the average linear fluid velocity, k, l_x, α, and β are the numerical constants which depend on cell geometry, w is a characteristic for the electrode width, C^* is the bulk solute concentration, η is the kinematic viscosity, and D is the diffusion coefficient.

In the channel- or thin-layer-type cell [6], the column effluent is fed into a rectangular channel, and the electrode is imbedded in the channel wall. It has been shown that the highest signal-to-noise ratios can be obtained with very small electrode surface area and small cell height [39,40]. Generally, the noise in EC detection is proportional to the surface area of the detector. The signal-to-noise ratios for different hydrodynamic systems and electrode dimensions are given in Table 3.

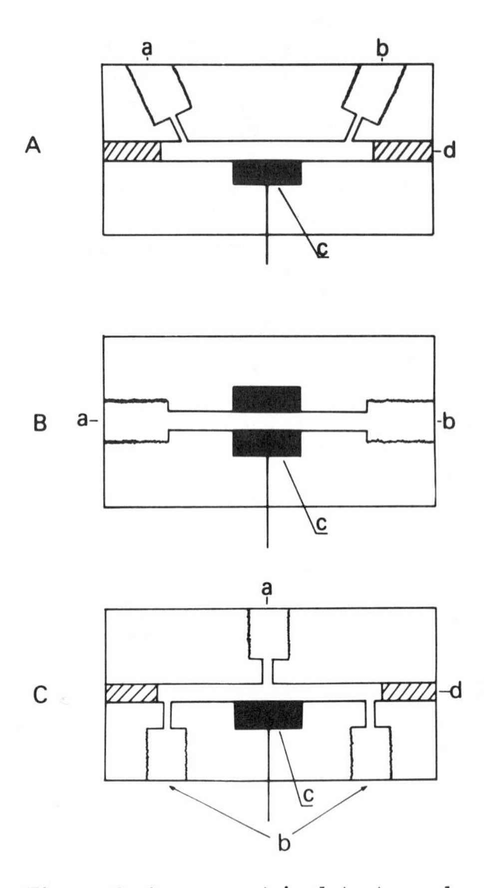

Figure 2 Amperometric detector schematic: (A) channel, (B) tubular, and (C) wall-jet (a, entrance; b, exit; c, working electrode; d, spacer). (From Ref. 12.)

With the polarographic detectors the best results were obtained when the liquid and mercury streams were at right angles to each other [41]. For detectors using solid-state electrodes dimensions are given in Table 3. The effective cell volume of this type of detector is below 1 μl [42]. Recently a flow through a two-electrode wall-jet cell with a platinum measuring electrode and a 20-nl cell volume has been described [43]. A schematic diagram is shown in Fig. 3. Because of the very small detector volume, its contribution to peak broadening is negligible and thus it can be used with short microbore columns.

Table 2 Some Characteristics of Hydrodynamic Systems Commonly Employed for Amperometric Detection

Cell type	Equation for limiting current[a]	Flow	Reference
Channel (thin layer)	$I = 0.68nFDCd(\eta/D)^{1/3}(Ul/\eta)^{1/2}$	Laminar	35
Tubular	$I = 2.035\pi nFD^{2/3}CU^{k}l^{2/3}r^{2/3}$	k = 0.33 for laminar flow k = 1 for turbulent flow	36
Wall jet	$I = 1.60KnFD^{2/3}CV^{3/4}\eta^{-5/12}d^{-1/2}r^{3/4}$	Turbulent	37

[a]Symbols: I, limiting current; D, diffusion coefficient; U, maximal linear fluid velocity in the parabolic Poiseuille flow velocity distribution; l, length of tubular electrode; K, constant; η, kinematic viscosity; d, nozzle diameter; r, radius of the disk electrode; V, the volume flow rate; F, Faraday's constant; n, the number of electrons transferred per molecule; and C, the solute concentration.

Table 3 Signal-to-Noise Ratios for Different Cell Geometries and Electrode Dimensions (y)

Detector type	Area	$(S/N)_{H,y}$	y
Solid state			
Tube	$2\pi rl$	$85.2\ y^{-1/3}$	rl
Thin layer	bl	$138.6\ y^{-1/2}$	ℓ
Wall jet	πr^2	$181.9\ y^{-1/2}$	r
Polarographic			
Opposite	$4\pi r^2$	$160.0\ y^{-1/2}$	r
Parallel	$4\pi r^2$	$39.8\ y^{-2/3}$	r
Normal	$4\pi r^2$	$185.2\ y^{-1/2}$	r

Source: Ref. 41.

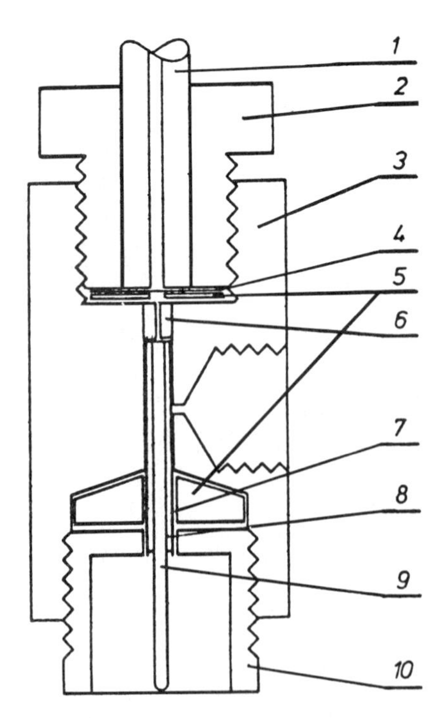

Figure 3 Schematic of the 20-ml detection cell: (1) glass separation column; (2) stainless steel column screw; (3) stainless-steel detector body, (4) filtering paper, (5) Teflon sealing, (6) glass capillary (inner diameter, 50 μm), (7) stainless steel capillary, (8) Teflon insulation, (9) platinum wire (0.5 mm in diameter), and (10) connector. (From Ref. 43.)

The theoretical principles of tubular detectors have been described by Blaedel and co-workers [44,45]. They employed tubular electrodes of various materials (platinum, cadmium, gold, carbon, etc.) and had limited performance owing to mechanical problems associated with the preparation, polishing, and cleaning of very smooth electrode surface inside the tube [9]. Thus it is not surprising that these detectors have not gained widespread use.

It should be pointed out that all EC detectors are like antennae which pick up low-frequency noise. Thus shielding is necessary if work is to be carried out at high sensitivities. At present most EC detectors are provided with a Faraday cage. Pump flow rate pulsations also have an adverse effect on the signal stability. Excessive oscillations should be dampened hydrodynamically [11]. Thermal instability and chemical pollution from trace impurities in the solvent system may also cause problems. Temperature changes affect the liquid junction potential and the background current. Thus, for trace analysis, both the column and the detector cell should be thermostated. The impurities in the solvent system, particularly the trace metals, may cause base-line problems. Some authors suggest rinsing the whole chromatographic system with a dilute (10 M) solution of ethylenediaminetetraacetic acid (EDTA) [11].

2. Amperometric Detectors with Multiple Working Electrodes

Although EC detection is considered to be selective (in view of the limited number of solutes which can undergo a chemical reaction under certain conditions), this property can be further enhanced by several methods. One simple approach is the use of a dual detector which employs two working electrodes held at different potentials and monitored simultaneously [46]. This is particularly advantageous for the detection of closely eluting peaks, provided that the two solutes have different redox properties [47,48]. Also, easily oxidizable or reducible substances can be detected simultaneously with those requiring higher potentials for their detection. In addition, the ratios of currents detected at the two chosen potentials are characteristic of the molecules undergoing EC reactions. This is completely analogous to the use of absorbance ratios obtained from dual or multiple wavelength detection [49]. Schematics of the dual detectors with the two working electrodes connected in parallel and in series are shown in Fig. 4.

With a serial dual-electrode arrangement the products generated at the upstream electrode can be detected at the downstream electrode. An application of the series dual-electrode reduction—oxidation detection is illustrated with chromatograms of a synthetic mixture of explosive compounds shown in Fig. 5. Chromatogram A illustrates the interference of oxygen in the reductive mode (upstream electrode). At the downstream electrode (chromatogram B) the product of the upstream oxygen reduction (H_2O_2) is not oxidized at the potential em-

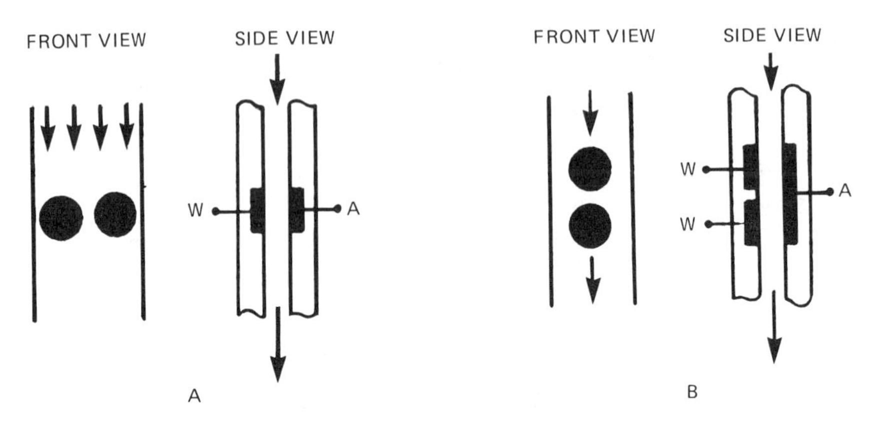

Figure 4 Thin-layer transducers with two working electrodes in (A) parallel and in (B) series. (From Ref. 14.)

ployed. Thus the chromatogram is free from O_2 interference. This is particularly advantageous if oxygen removal is impossible because of the limited volume of the sample available. Chromatograms C and D show the upstream and downstream detection of the same sample which was previously deoxygenated. This dual detection mode is useful for the elimination of other interferences resulting from irreversible electrochemical reactions of some components of the background medium.

The removal of dissolved oxygen (approximately 10^{-3} mol/liter is present in aqueous solutions in equilibrium with the atmosphere) is an important step prior to EC detection in the reductive mode. The following reactions take place in acidic and basic–neutral media:

$$\text{Step 1:}\quad O_2 + 2H^+ + 2e \longrightarrow H_2O_2 \qquad (11)$$

$$\text{Step 2:}\quad H_2O_2 + 2e + 2H^+ \longrightarrow 2H_2O \qquad (12)$$

$$\text{Step 1:}\quad O_2 + H_2O + 2e \longrightarrow HO_2^- + OH^- \qquad (13)$$

$$\text{Step 2:}\quad HO_2^- + 2e^- + H_2O \longrightarrow 3OH \qquad (14)$$

In practice, deoxygenation of LC solvents can be carried out using several physical methods: nitrogen sparging [50], vacuum degassing [50], ultrasonic agitation [50], nitrogen-activated nebulization [51], and refluxing [52].

Removal of O_2 from the samples can be carried out by sparging with nitrogen presaturated with the solvent in order to prevent solvent loss by evaporation [14].

In the dual detection approach used by Schieffer [48] the upstream electrode was replaced with a coulometric cell which enabled a complete removal of interferences prior to the downstream detection. The difference in selectivities obtained using a conventional amperometric

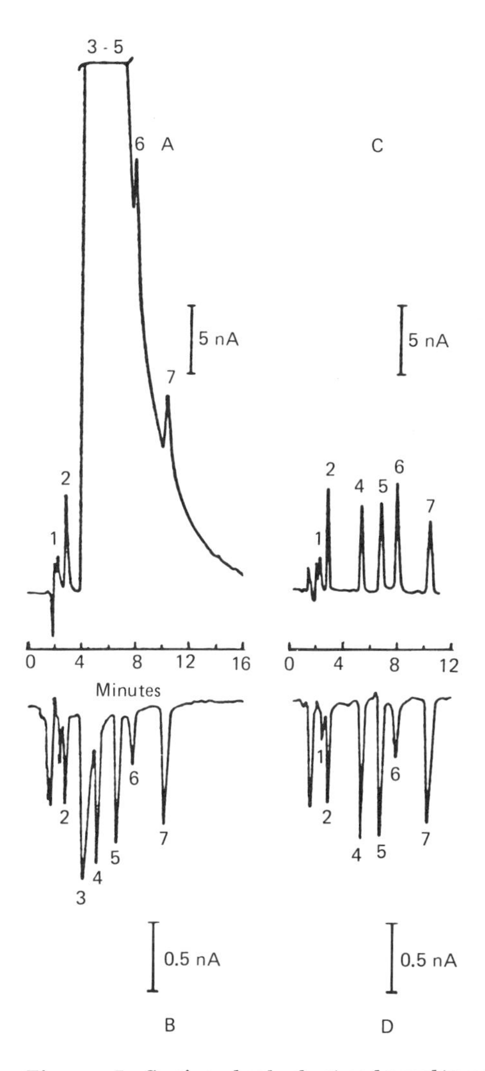

Figure 5 Series dual-electrode voltamograms of a synthetic mixture of explosive compounds (A,B) without and (C,D) with sample deoxygenation. Traces A and C are for the upstream electrode (reductive, E = −0.9 V) and traces B and D are for the downstream electrode (oxidation, E = +0.8 V). The mixture contained 11 pm of HMX (1), 5 pmol of picric acid (2), 18 pmol of 4-nitrophenol (4), 9 pmol of 2,4,6-trinitrotoluene (5), 30 pmol of nitroglycerin (6), and 14 pmol of 2,4-dinitrotoluene; peak 3 is oxygen [column, 25 × 0.46 cm; Biophase C_{18}; mobile phase—0.02 M monochloracetic acid, 0.015 M sodium acetate, 0.001 M EDTA in 17% (v/v) 1-propanol and 5% (v/v) ethanol, pH 3.5; upstream electrode (Au–Hg) set at 0.90 V and downstream electrode (glassy carbon) set at +0.80 V; flow rate 1.7 ml/min]. (From Ref. 14.)

detector (A), a dual coulometric–amperometric cell, and differential pulse detection is shown in Fig. 6. The use of dual-detection dc amperometry is an attractive alternative to differential pulse voltametry in terms of equipment simplicity. Is should be pointed out that it is not at all necessary to measure the output of the coulometric cell, since its sole purpose in this case is to completely eliminate the interferring substances [48].

An interesting development in amperometric detection is the use of regenerative flow cells for the analysis of reversible redox reactions. It is known that an increase in electrode area increases both the sensitivity and the noise. Thus the signal-to-noise ratio is usually not improved, and often diminished. Regenerative flow cells offer enhanced sensitivities while maintaining a small electrode surface [54]. The product of the electrochemical reaction at the working electrode diffuses away until it reaches the auxiliary electrode, at which point it is regenerated to its original form. If this process is repeated several times, considerable current amplification can be obtained [54].

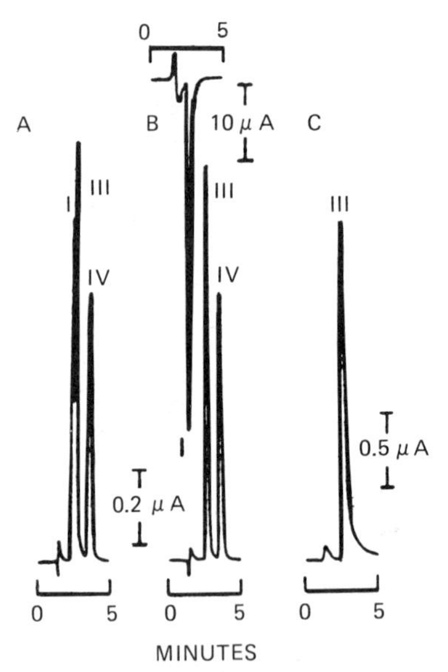

Figure 6 Chromatograms of (I) 6-hydroxydopa, (III) 1-dopa, and (IV) tyrosine: (A) conventional amperometry; (B) separation of I and III using dual coulometric–amperometric cell, coulometric chromatogram of I offset 1 min; and (C) differential pulse chromatogram [column, μBondapak C_{18} (250 X 4.6 mm inner diameter); solvent, 0.01 M NaH_2PO_4, pH 3.5; flow rate, 2.0 ml/min; detection potential, (A) +0.85 V versus Ag/AgCl; (B) coulometric cell, +0.25 V; (C) 50-mV modulation amplitude, 0.5-sec pulse frequency, +0.38 V initial potential]. (From Ref. 48.)

An alternative approach for increasing detection sensitivity is the use of a rotating disk electrode [55,56]. This system employs the wall-jet geometry, and the working electrode can be rotated at different speeds using a belt-driven pulley system (Fig. 7). Since the rotation of the working electrode decreases the thickness of the diffusion layer (as a function of speed), the rate of mass transport (Fick's law) and, consequently, the sensitivity are increased. In addition, the response of this type of detector is less dependent on flow rate than with the conventional amperometric detectors.

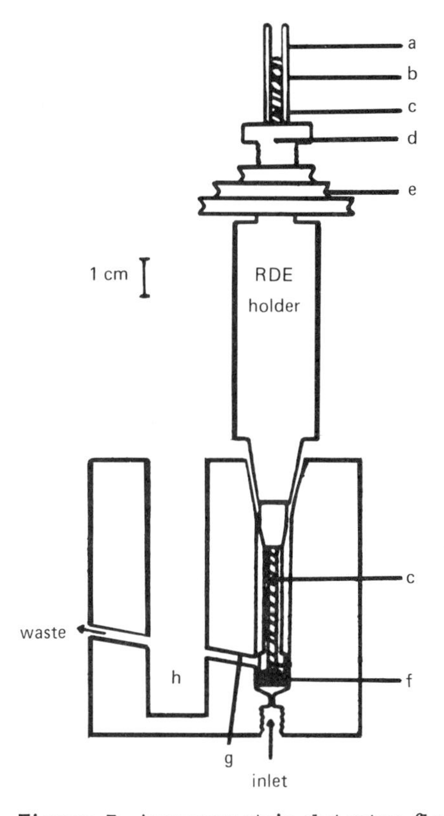

Figure 7 Amperometric detector flow cell with a rotating disk working electrode: (a) Kel-F tube, (b) mercury contact, (c) brass rod, (d) screw with rubber O-ring to fasten the electrode in the holder, (e) pulleys to rotate the electrode at different rotation speeds, (f) carbon paste electrode, (g) connecting channel, and (h) compartment for the reference and auxiliary electrodes. (From Ref. 55.)

3. Amperometric Detectors for Nonconducting Mobile Phases

In chromatographic separations carried out in normal-phase or nonaqueous reversed-phase modes, the nonconducting mobile phases employed do not satisfy the fundamental requirement for EC detection. Thus special amperometric detectors have been designed for these purposes [53]. A schematic representation of this detector is shown in Fig. 8. The detector cell is composed of two compartments. In the first one, which can be considered as a postcolumn open tubular reactor, a certain amount of an electrolyte is added to the nonconducting column effluent. The detection is carried out in the second compartment, which is a three-electrode amperometric flow cell.

4. Electrode materials

The choice of electrode material is dictated by the range of operating potential needed for monitoring an analyte in a given solvent system. The main reactions which limit the useful range are the evolution of the components of the mobile phase and/or various trace impurities which may be present therein. The useful potential ranges for some currently employed electrode materials are given in Table 4.

The most commonly employed electrode materials are mercury (hanging or stagnant drops), various forms of carbon, platinum, gold, and amalgams. The advantages and disadvantages of mercury will be

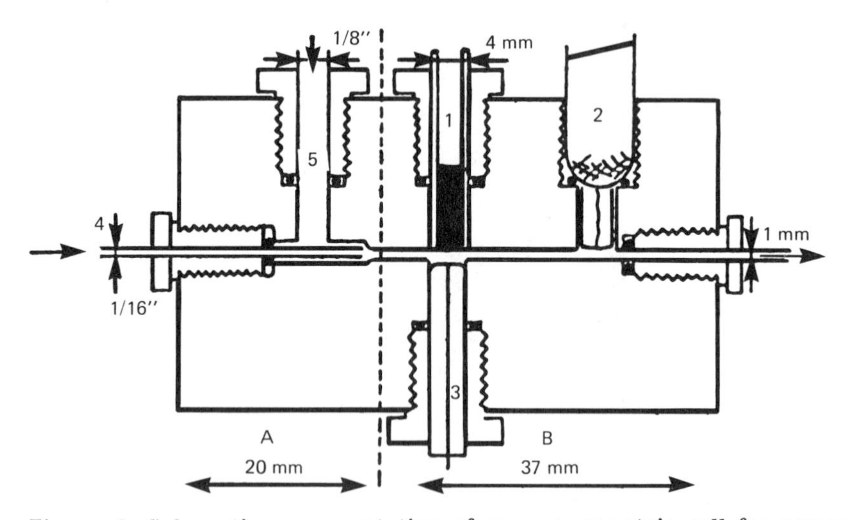

Figure 8 Schematic representation of an amperometric cell for nonconducting media: (A) mixing chamber and (B) measuring chamber [1, working electrode (glassy carbon); 2, reference electrode (saturated calomel); 3, auxiliary electrode (platinum disk, 4 mm); 4, eluent inlet; and 5, supporting electrolyte inlet]. (From Ref. 53.)

Table 4 Useful Potentential Ranges from Some Currently Employed Electrode Materials

Material	Potential limits				
	Aqueous buffer, pH 4.5 (V/SCE)		Nonaqueous media $[(V/NHE)_{aq}]$		
Mercury	-2.0	+0.4			
Mercury film	-1.0	+0.4			
Glassy carbon	-0.8	_1.2			
Pyrolytic graphite	-1.5	+1.2			
Impregnated graphite	-0.6	+1.2			
Graphite paste	-1.6	+1.1			
Platinum	-0.5	+1.2	Acetonitrile/$LiC\ell O_4$	-3.2	+3.2
			Nitromethane/$LiC\ell O_4$	-2.7	+4.2
			Dimethylsulfoxide/$LiC\ell O_4$	-3.5	+1.3
			Dimethylformamide/$LiC\ell O_4$	-3.4	+1.6
			Dichloromethane/tetraalkylammonium salt	-2.4	+2.6
			Tetrahydrofuran/tetraalkylammonium salt	-3.3	+1.8

Source: Ref. 16.
Symbols: V, volts; SCE, standard calomel electrode; and NHE, normal hydrogen electrode.

discussed in another section (polarographic detectors). Pyrolytic graphite, glassy (or vitreous) carbon, and carbon paste have found common usage as electrode materials for voltametric applications. The main advantages of carbon electrodes are good electrocatalytic properties, lower susceptibility to surface poisoning than those of other materials such as platinum and gold, inertness to chemical attack, and small pore size. A schematic of the structural model of glassy carbon is shown in Fig. 9. It is produced by carefully controlled heating (up to 1200°C) of a resin (e.g., phenol–formaldehyde) in an inert atmosphere. The resulting material is made up from aromatic ribbons of carbon atoms stacked together in a form of twisted and intertwined microfibrils [57]. Prior to their use, glassy carbon electrodes must be pretreated in order to obtain reproducible results. Microscopic pitting resulting from excessively fast heating on evolution of gases during the carbonization process must be eliminated by abrasion and subsequent polishing with suspensions of alumina or chromium (III) oxide (particle size 0.3 μm) [58]. This pretreatment is supposed to eliminate the functional groups which may be present on the surface (carbon–oxygen functionalities). The physicochemical properties of

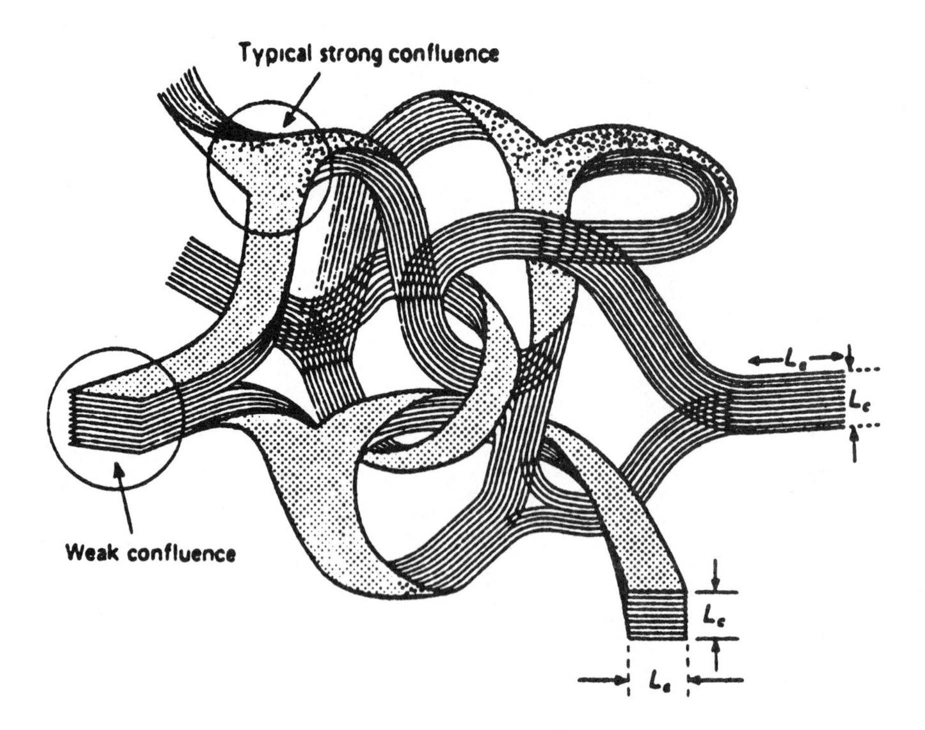

Figure 9 Schematic structural model for glassy carbon. (From Ref. 57.)

glassy carbon vary to a certain extent from manufacturer to manufacturer [59]. The difficulties in obtaining highly reproducible surfaces have led to contradictory results concerning certain electrode properties. Thus some authors [60,61] have obtained higher current densities for the oxidation of hexacyanoferrate [11] with glassy carbon than with platinum or gold. These findings were contradicted by others [62].

The use of basal plane pyrolytic graphite [63] and reticulated vitreous carbon [64], low-temperature isotropic carbon [65], highly orientated pyrolytic graphite [63], and carbon cloth [32,66] has been described in the literature, but their application has not become widespread. Solid carbon electrodes can be fouled by adsorption of reactants or reaction products: The surface can be cleaned chemically by rinsing with ethanol, chloroform, or nonoxidizing mineral acids [58]. Deactivated surfaces can be reactivated electrochemically by means of a voltage pulse train [58, 67] or mechanically [58]. It has been reported that the roughening of the electrode surface resulting from the voltage pulse train did not have any influence on the background current or the noise level [67].

Numerous electrodes based on carbon or graphite particles mixed with liquid or solid diluents such as Nujol or other mineral oils [6], ceresin wax [68], silicone rubber [69], polypropylene [70], Kel-F [71], and Teflon [72] have been described in the literature. Classic carbon paste electrodes prepared with liquid diluents such as mineral oil (Nujol) or other dielectric materials cannot be used with mobile phases containing more than approximately 20 to 30% organic modifier (methanol, acetonitrile) owing to their rapid dissolution [10,14]. Under normal conditions they can be used for several days or weeks without resurfacing [14]. It has been reported [73] that reactions on these electrodes are slower than on well-polished solid electrodes. Detection limits are usually low, owing to lower noise (10). There is disagreement concerning the background currents which arise from surface adsorption of traces of oxygen on the surface [10] and its absorption in the paste materal [10,14]: Some authors report low residual currents [14], and others high [10].

More satisfactory seems to be carbon paste materials with a solid matrix, since they have longer lifetime (at least 1 month) and they are more resistant to organic modifiers [10].

With all carbon-based electrodes, 30 min to 1 hr of conditioning is mandatory at a potential a few hundred millivolts higher than the protocol conditions, followed by a 30 min of equilibration at the operating potential.

Platinum and gold electrodes are useful for detection in nonaqueous solvents; in aqueous eluents they suffer from formation of oxide layers and/or filming [14]. Their major use has been in the detection of inorganic species [74,75]. Amalgamated gold and platinum have been used for many electrochemical reductions [14,76–78].

B. Coulometric Detectors

Coulometric detectors are destructive, since the analyte is electrolyzed completely (100% efficiency). The resulting current I and the quantity of solute undergoing the conversion are given by the following equations:

$$I = nFC\bar{U} \tag{15}$$

and (Faraday's law)

$$Q = nFm \tag{16}$$

where $\bar{U}$ is the average volume flow rate, Q is the number of coulombs, and m is the number of moles of analyte electrolyzed. All other symbols have their usual meaning.

If the requirement for 100% conversion is satisfied, the detector response is independent of the flow rate, cell geometry, eluent viscosity, temperature, and so on.

In order to achieve 100% electrolytic efficiency, the effective surface of the working electrode must be large and the flow rate small. At the same time, it is desirable to maintain the cell volume as small as possible, since the chromatographic performance would otherwise be degreaded. Unfortunately, the increase in electrode surface entails some serious problems. Since the current in the coulometric cell is at least in the microampere range (nanoampere for amperometric detectors), low cell resistance is mandatory for maintaining a reasonably wide linear range, otherwise the voltage drop will cause pronounced deviations from linearity at higher analyte concentrations. Since the noise and residual current increase in proportion to the electrode surface, contrary to what one might expect, coulometric detectors afford lower sensitivities than amperometric ones [11,13].

Numerous designs of flow through coulometric cells have been described in the literature [32,48–68,79–81]. Generally speaking, the cell designs fall into two categories: those in which the column effluent is fed through the electrode in an open path [39,68,82] and those in which the solution passes through the electrode in a reticulated path [32,83,84]. The schematic diagrams of the two types of design are shown in Fig. 10.

C. Polarographic Detectors

Since its introduction by Heyrovsky in the 1920s, polarography has become popular for the analyses of trace compounds. Miniaturization of the standard polarographic cell and its application to HPLC was therefore a logical development. A polarographic detector is an amperometric detector which employs the dropping mercury electrode (DME) as a working electrode. The advantages of the DME over other types of electrodes are well documented in many textbooks for instrumental analysis; however, some major points are worth mentioning. The DME ex-

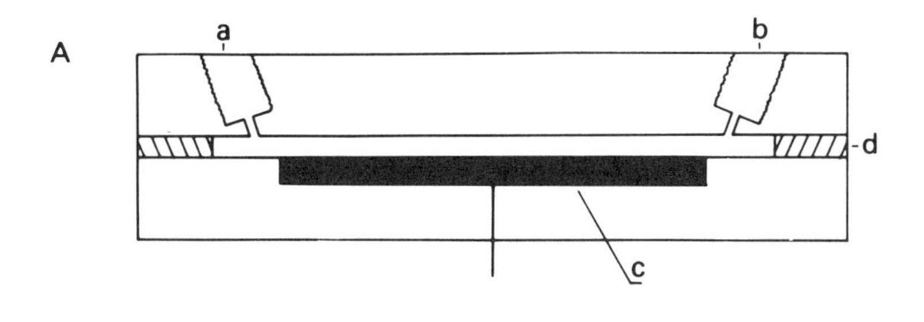

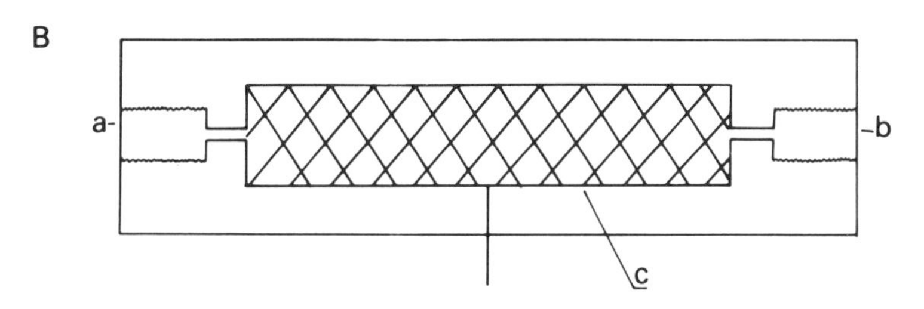

Figure 10 Schematic representation of coulometric detectors: (A) open type and (B) reticulated type (a, solvent inlet; b), solvent outlet; c), working electrode, and d, spacer). (From Ref. 11.)

hibits a very high hydrogen overvoltage. Thus its cathodic range extends up to -1.5 V versus the standard calomel electrode (SCE) in acidic solutions, and it is even higher in neutral or basic media. Therefore the DME is ideally suited for the reductive mode. In addition, the electrode surface is continuously renewed, which diminishes the problem of electrode contamination. Since oxygen will be reduced at the DME, deaeration of solutions is mandatory.

The anodic range is very limited with polarographic detectors, since at approximately +0.4V versus the SCE, mercury is oxidized, giving rise to a high background current.

The DME also possesses some shortcomings, such as the unwanted charging current. The surface of the drop and the electrical double layer in its vicinity act as a capacitor. Since the capacitance is a function of electrode surface, it will increase during the growth of the drop, giving rise to a charging current. The current oscillations must be suppressed either by using an RC filter (however, this will increase the detector time constant, which is indesirable) or by using a short drop time. The latter can be achieved by using a horizontal capillary [85,86] which gives a drop time of the order of 10 nsec and controlling the drop time with a movable pin [52] or short voltage pulses [87]. A schematic diagram of a polarographic cell using a horizontally placed mercury capillary is shown in Fig. 11. Since the signal of a polarographic detector is also flow sensitive, the hydrodynamics of the cell are important. Two fluid flows must be taken into consideration: the mercury flow and the eluent flow. There are three different cell designs based on flow considerations: the mercury flow can be parallel, opposite, or perpendicular to the column effluent flow. It has been demonstrated that the limiting current is a function of the square root of linear solvent velocity for the opposite and normal flow [88]; for the parallel arrangement the limiting current was found to vary as the cubic root of the linear flow velocity [89]. From the above relationships it was concluded that the normal mercury flow with respect to the solution would give the largest currents [38].

There also exist detectors which employ a stationary rather than a dropping mercury electrode [90]. Although their construction is simpler, the main advantage of the DME, the continuously renewed surface, is thus lost.

D. Methods of Measurement

In addition to the classic dc amperometry where the instantaneous dc current is measured as a function of the time at constant potential,

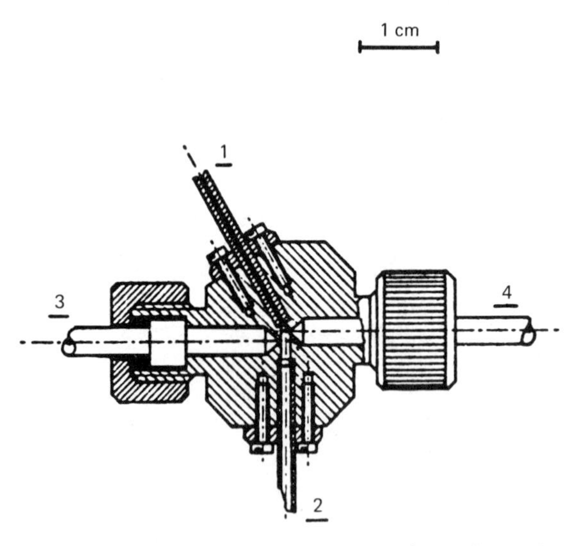

Figure 11 Schematic representation of a polarographic detector with a horizontal dropping mercury electrode; (1) eluent inlet, (2) eluent outlet and auxiliary electrode, (3) reference electrode, and (4) the DME; capillaries and reference electrode are conically ground in order to minimize the cell volume. (From Ref. 86.)

other derivative techniques are also used. Their main characteristics are illustrated schematically in Fig. 12.

Significant improvements in detection selectivity can be achieved by means of the differential pulse (DP) technique. In DP amperometry the potential of the working electrode is pulsed around $E_{1/2}$ (pulse heights are usually 25 to 50 nV), and the current is sampled at the beginning and at the end of the pulse. The difference $\Delta I = I_1 - I_2$ is displayed as a function of potential. Although larger pulses increase the signal, the noise is also increased. Thus lower detection limits are obtained with larger pulse heights. By fine tubing of the base potential around which the pulse is applied, a considerable control of selectivity can be achieved. This detection mode reportedly minimizes the adsorption of reaction products on the surface of the electrode [91].

In square-wave voltametric detection the potential is scanned within seconds, and the current is sampled before and after each pulse step. The signal measured is $I_1 - I_2$. This detection mode is "multidimensional," since it enables simultaneous multipotential monitoring. While the selectivity is better than with dc amperometry, the detection limits are usually lower [14]. This technique still needs further refinements prior to becoming a routine tool.

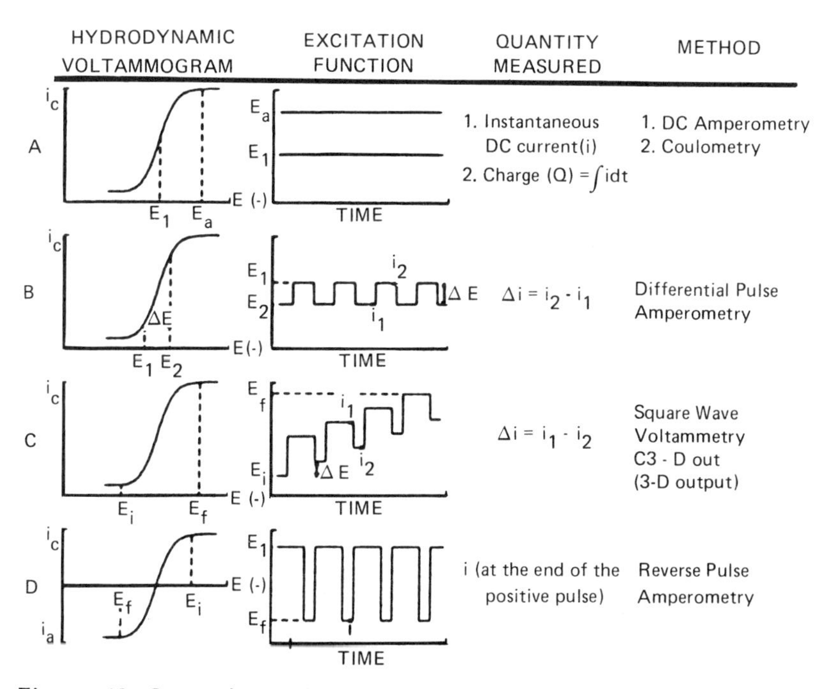

Figure 12 Operational characteristics of several methods of EC measurement. (From Ref. 14.)

Reversed-pulse amperometry is a relatively new technique which has not yet been fully explored [14,92]. It involves application of a positive pulse at the end of which the current is sampled. This technique holds promise for the analysis of amalgam-soluble metals without oxygen interference and the reduction of organic substances which can undergo a follow-up oxidation and can either adsorb onto the electrode surface or react with it [14].

E. Potentiometric Detectors

Potentiometric detectors measure the equilibrium potential difference between two electrodes at constant current conditions. Practically all detectors of this type operate at zero current conditions, which offers some advantages. Since no net electrolysis takes place at the working electrode, the voltage drop is eliminated, as well as the contamination of electrode surface by adsorption of reaction products.

These detectors usually employ ion-sensitive electrodes or redox electrodes. Their response is usually slow, which adversely affects the chromatographic separation. The ion-selective electrodes exhibit high selectivity, which limits their applicability. More general detection can be achieved by using less selective electrodes, such as the liquid membrane electrodes [93], by employing preliminary chemical reactions in a reaction placed before the detector, or by titrating the solutes with an agent monitored by the ion-sensitive electrode.

This type of detection is limited mostly to inorganic analyses, and its coupling with HPLC has not yet found extensive use.

IV. DETECTORS BASED ON ELECTRIC PROPERTIES OF SOLUTIONS

Although very often the conductometric and capacitance detectors are not included in similar reviews, they nevertheless exploit the electrical properties of solutions, and thus in a broader sense they belong to the category of electrochemical detectors.

A. Conductometric Detectors

The conductance of a solution is the reciprocal of the electrical resistance:

$$R = \rho(l/A) \tag{17}$$

where R is the resistance, ρ the specific resistance (resistance of 1 cm^3 of solution), and l and A are the solution length and electrode area, respectively. The specific conductance κ, which is the reciprocal of ρ, is given by

$$\kappa = 1/\rho = l/RA \tag{18}$$

Since, in practice, the behavior of ions in solution is nonlinear, the conductance is a nonlinear function of ion concentration over a wider concentration range.

Conductometric detectors employ a fixed l/A ratio, and the measurements are usually carried out using a Wheatstone bridge. Since a dc voltage would cause polarization of the cell electrodes, it cannot be applied, and ac voltage is used instead. This type of detector is useful for ionized solutes. It should be pointed out that mobile phases with very high ion concentrations diminish the detection sensitivity. The signal of this type of detector exhibits a strong temperature dependence; thermostating within a 10^{-3} °C is mandatory [10].

Several cell designs have been described in the literature [94–96]. Most of them have volumes in the microliter range and employ inert electrodes (e.g., platinum or stainless steel). An example of the conductometric cell design is shown in Fig. 13. The internal cell volume was reported to be less than 0.5 μl, the minimum detection limit was 1.4×10^{-8} g/ml KCl, and the linear dynamic range was 10^6.

B. Capacitance Detectors

Capacitance detectors measure changes in the permittivity of the mobil phase as a function of its composition:

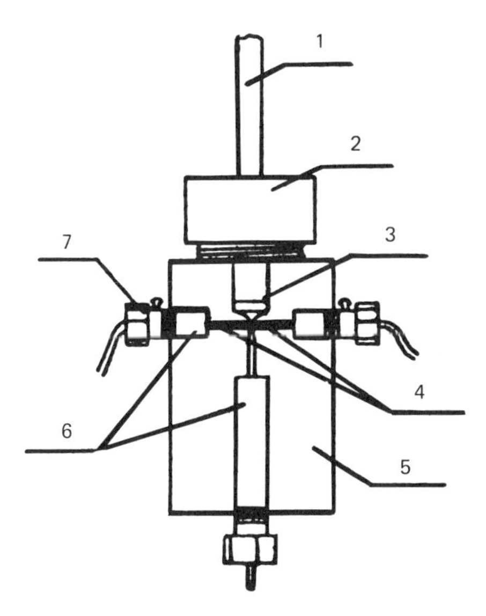

Figure 13 Schematic representation of a conductometric detector cell: (1) column, (2) connector, (3) silicon rubber seal, (4) platinum electrodes, (5) organic glass, (6) silicon rubber seal, and (7) electrode contacts. (From Ref. 94.)

$$C = \varepsilon_0 \varepsilon_T G_c \tag{19}$$

where C is the capacitance, ε_0 the permittivity of vacuum, ε_T the relative dielectric constant of the mobile phase in the flow cell, and G_c the constant dependent upon the cell geometry (parallel-plate versus cylindrical capacitor). Several cell designs have been proposed [97–104]. The commonly empolyed frequencies range from 1 to 5 KHz [102] to 18 MHz [101].

With high frequencies, low capacitance cells of relatively low volumes (<10 μl) can be used while still maintaining sufficient sensitivity. The signal is highly dependent on temperature (mainly because of the changes in permittivity and cell dimensions), and this is circumvented either by high-precision thermostating [99], feedback circuits (104), or the use of differential cells [100]. An example of the capacitance flow cell design is illustrated in Fig. 14. This design employed a common electrode (nickel-plated copper block) for the reference and measuring cells. In order to minimize the temperature effects, the effective cell volume was 7 μl. The minimal detection limit corresponded to a permittivity change of approximately 0.5×10^{-6}. Because of the moderate sensitivity and pronounced temperature dependence high-frequency impedance detectors are similar to refractometric detectors. Thus they are more suited for semipreparative and preparative chromatography than for trace analysis.

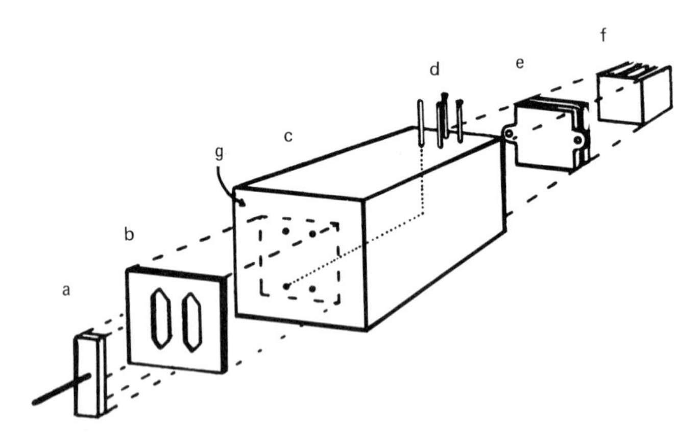

Figure 14 Schematic diagram of a capacitance flow cell design: (a) electrode and connecting wire, (b) spacer, (c) copper block, (d) connecting capillaries (0.25 mm inner diameter), (e) Peltier element, (f) heat sink, and (g) nickel-plated side of the copper block showing end of connecting capillaries. (From Ref. 99.)

V. SELECTED APPLICATIONS

A complete survey of all areas of applications of EC detection is beyond the scope of this review. The utility of this technique will be illustrated with some of its applications to biochemical–biomedical research. Examples of these and other applications are briefly outlined in Table 5.

The tandem operation of electrochemistry and HPLC has become a preeminent tool for the analysis of neurochemical substances, such as catacholamines, in limited amounts of biological samples. Measurements of the parent amines, their precursors and metabolites, as well as the activities of related biosynthetic enzymes provide very important determinants for the diagnosis and prognosis of many disease states, such as hypertension [105], Parkinson's disease [106], affective disorders [107], certain heart conditions [108], tumors of the neutral crest [109–111], hyper- and hypothyroidism [112], familial dysautonomia [113], and muscular distrophy [114].

The catecholamines (A), 5-hydroxyindoles (B), and O-methylated catechol derivatives (C) are detected amperometrically according to the following reactions:

A HO, HO, NH_2 $\longrightarrow$ O, O, NH_2 $+ 2H^+ + 2e^-$ (20)

B $_2$HN, NH, HO $\longrightarrow$ O, N, NH_2 $+ 2H^+ + 2e^-$ (21)

C CH_3O, HO, NH_2 $\longrightarrow$ O, O, NH_2 $+ H^+ + CH_3OH + 2e^-$ (22)

The detection of O-methylated compounds (C) demands a slightly higher activation energy, and thus a higher oxidation potential. The exact mechanism of the oxidation of tryptophan metabolites is not known exactly; however, it is believed that the reaction products are polymeric [17].

The commonly employed oxidation potentials for these compounds, obtained from cyclic voltametry, are listed in Table 6. By fine tuning of the oxidation potential, the detection selectivity can be increased considerably. It has been stated before that enhanced signals are ob-

Table 5 Recent Examples of EC-HPLC Applications

Compounds investigated[a]	Sample	LC mode	EC mode	EC method	Reference
NE,DA,AA,DHPG, MHPG,DBA, NMN,DOPAC	Rat brain	RP-IP	Oxidative	DC amperometric	128
NE,E,DA	Plasma	RP-IP	Oxidative	DC amperometric	129–135
NE,E,DA,D	Urine	RP-IP	Oxidative	DC amperometric	134–136
MHPG[1], 5-HTP[1],DA[1] NE[1],DOPAC[1], HVA[1],5-HT[2], 5-HIAA[2]	Cerebro-spinal fluid	RP RP-IP	Oxidative Oxidative	DC amperometric DC amperometric	116[1] 117[2]
TH	Rat brain	RP-IP RP	Oxidative Oxidative	DC amperometric DC amperometric	118 119
COMT	Tissues and red blood cells	RP-IP	Oxidative	DC amperometric	120
DβH	Cerebro-spinal fluid	RP-IP	Oxidative	DC amperometric	121

GABA	Rat brain	RP	Reductive	DC amperometric	122
Vitamin K_3	Standard	RP	Reductive	DC amperometric	123
Benzodia-zepines	Plasma	RP	Reductive	DC amperometric DP amperometric	91
Pesticides	Plants	NP	Reductive	Polarographic	124
Organometals	Foods	RP	Reductive	DP amperometric	125
Nitrosamines	Standard	RP	Reductive	Normal-pulse amperometric, DP amperometric, square-wave amperometric	126
Testosteroids	Standard	RP	Reductive	Polarographic	127

[a]NE, norepinephrine; E, epinephrine; DA, dopamine; D, dopa; AA, anthranilic acid; DHPG, dihydroxyphenylethyleneglycol; MHPG, 3-methoxy-4-hydroxyphenylethyleneglycol; DBA, dihydroxybenzylamine; NMN, normetanephrine; DOPAC, 3,4-dihydroxyphenylacetic acid; HVA, homovanillic acid; 5-HTP, 5-hydroxytryptophan; 5-HT, serotonin; 5-HIAA, 5-hydroxyindole acetic acid; TH, tyrosine hydroxylase, COMT; catechol-0-methyl transferase; GANA, γ-aminobutyric acid; and DβH, dopamine-β-hydrolase. Rp, reversed phase; RP-IP, reversed-phase ion pair; NP, normal phase: DP, differential pulse.

Table 6 Oxidation Potentials as Determined by Cyclic Voltametry Using a Carbon Paste Electrode and Ag—AgCl Reference at a Sweep RAte of 100 mV/sec[a]

Compound	Oxidation potential V
3,4-Dihydroxyphenylacetic acid (DOPAC)	+0.56
Homovanillic acid (HVA)	+0.76
3-Methoxy-4-hydroxyphenylglycol (MHPG)	+0.72
3-Methoxy-4-hydroxyphenethyl alcohol (MOPET)	+0.70
3,4-Dihydroxybenzylamine (DHBA)	+0.56
3-Methoxy-4-hydroxyphenylethylamine (3-methoxytyramine)	+0.74
3-Methoxy-4-hydroxyphenylethanolamine (normetanephrine)	+0.77
3-Methoxy-4-hydroxyphenethylamine (metanephrine)	+0.77
3,4-Dihydroxyphenylglycol (DHPG)	+0.57
Dopamine (DA)	+0.50
Norepinephrine (NE)	+0.55
Epinephrine (E)	+0.54
5-Hydroxytryptamine (5-HT,serotonin)	+0.57
5-Hydroxyindole acetic acid (5-HIAA)	+0.58

[a]An electrolyle solution of 1M sodium acetate, 0.1 M citric acid, pH 4.1, was used throughout.
Source: Ref. 22.

tained at higher oxidation potentials. Figure 15 illustrates the effect of increasing oxidation potential on the detection of some catecholamine metabolites. However, increased signal is achieved at the expense of detection selectivity. Thus the use of unnecessarily high detection potentials is not recommendable.

The reversed-phase ion-pair separation of catecholamines detected amperometrically is illustrated with the analyses of plasma samples from a patient with pheochromocytoma before (Fig. 16A and B) and after surgery (Fig. 16C). This tumor belongs to the neutral crest lesions, which are difficult to diagnose because of the absence of characteristic symptoms which would differentiate them from other etiologically different disorders (such as hypertension). Thus a reliable diagnosis is impossible without the assay of catecholamines and/or their

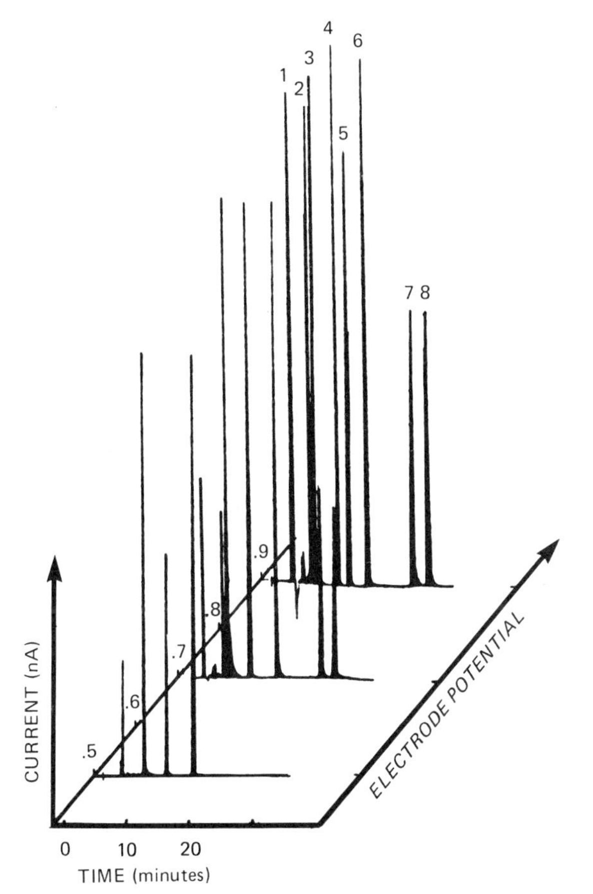

Figure 15 Effect of detector electrode potential on the detection of some catecholamine metabolites [column; Ultrasphere ODS (250 X 4.6 mm; inner diameter, 5 μm); eluent, 0.1 M citric acid, 0.1 M sodium acetate, 20% methanol; peak identities—(1) solvent, (2) MHPG (16.6 ng), (3) 5-HT (22 ng), (4) DOPAC (16.2 ng), (5) TRP (17.1 ng), (6) 5-HIAA (17 ng), (7) MOPET (15.5 ng), and (8) HVA (16 ng)]. (From Ref. 22.)

major metabolites. Figure 16A and B illustrates the enormously elevated levels of norepinephrine, indicating that the tumor was predominantly norepinephrine secreeting. It is interesting to note that after the surgical removal of the tumor, the catecholamine levels returned to normal (Fig. 16C). The peak identities can be confirmed by means of hydrodynamic voltamograms, as shown in Fig. 17. From repeated injections at several oxidation potentials, the current ratios ϕ can be

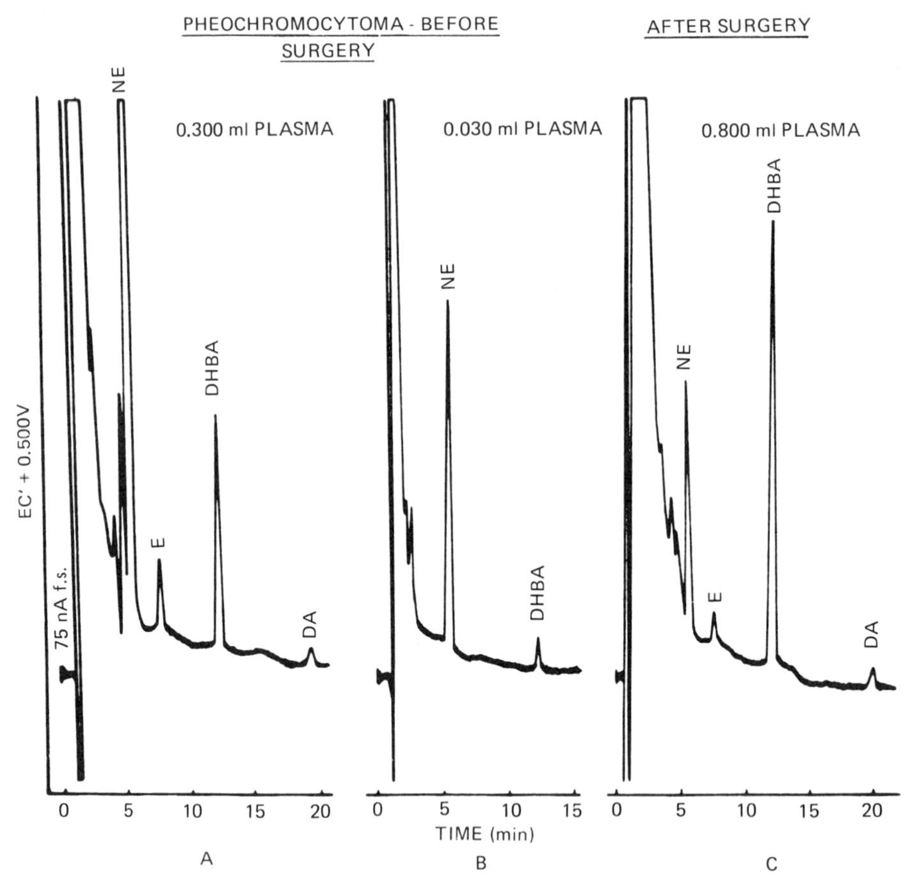

Figure 16 Chromatograms of the alumina extracts of plasma samples from a patient with pheochromocytoma (A,B) before and (C) after surgical removal of the tumor. The volumes indicated in the figure represent the volumes of original plasma corresponding to the quantities of extracts injected [column, Ultrasphere ODS (150 X 4.6 mm inner diameter, 5 μm); eluent, 0.0347 M KH_2PO_4, 3 mM sodium octylsulfate, 0.03 M citric acid, 14% methanol (v/v), pH 4.85; flow rate, 1.2 ml/min; temperature, ambient; detection, amperometric at +0.500V versus Ag—AgCl; concentrations—(A,B) NE, 145 ng/ml' E, 180 pg/ml; DA, 85 pg/ml; (C) NE, 350 pg/ml; E, 25 pg/ml; DA, 40 pg/ml]. (From Ref. 130.)

calculated and plotted versus potential applied. The result is a sigmoid curve which is also traced for the reference compound(s). The agreement between the curve for the sample peak and the corresponding reference compound is an important aid in peak characterization.

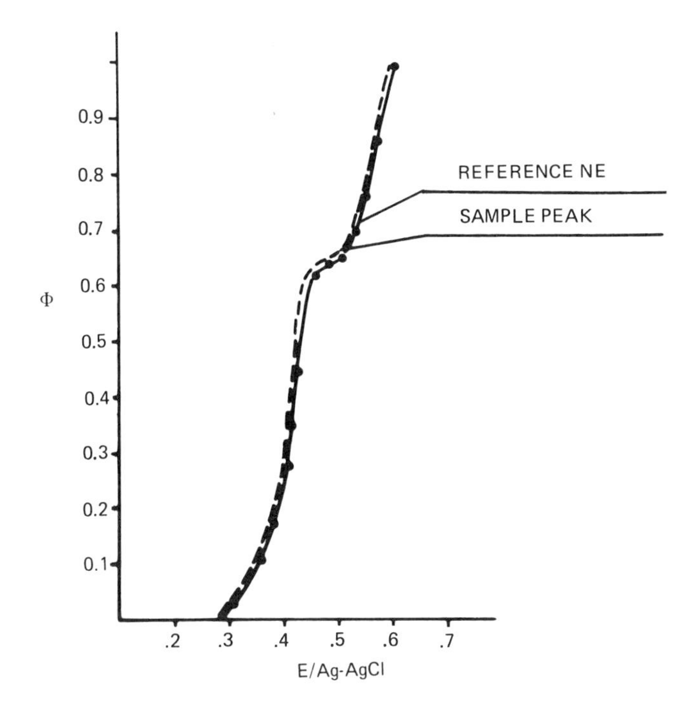

Figure 17 Hydrodynamic voltamograms for the NE reference compound and the peak with the corresponding retention time in the sample shown in Fig. 16A. The x axis represents oxidation potential versus Ag–AgCl; the y axis shows the ratio of the response (current) at a particular potential to the maximal response. (From Ref. 130.)

While the detection of picogram quantities of plasma catecholamines is the utmost proof of the sensitivity of EC detection and thus requires a certain amount of experimental skill, the analysis of their metabolites is considerably less complex. 3-Methoxy-4-hydroxyphenylethyleneglycol (MHPG) is a principal norepinephrine metabolite in mammals and its levels in various body fluids reflect the norepinephrine turnover; thus there is considerable interest in assessing its quantitative levels. Figure 18A and B illustrate the analysis of MHPG in human lumbar cerebrospinal fluid, detected at two oxidation potentials. It is evident that the detection at +1.00 V versus Ag–AgCl is considerably less selective than at +0.700 V, while the sensitivities for MHPG are not significantly different.

High-performance LC with EC detection is also ideally suited to the determination of enzyme activities and kinetics. Some examples of enzyme assays are listed in Table 5. The determination of blood levels

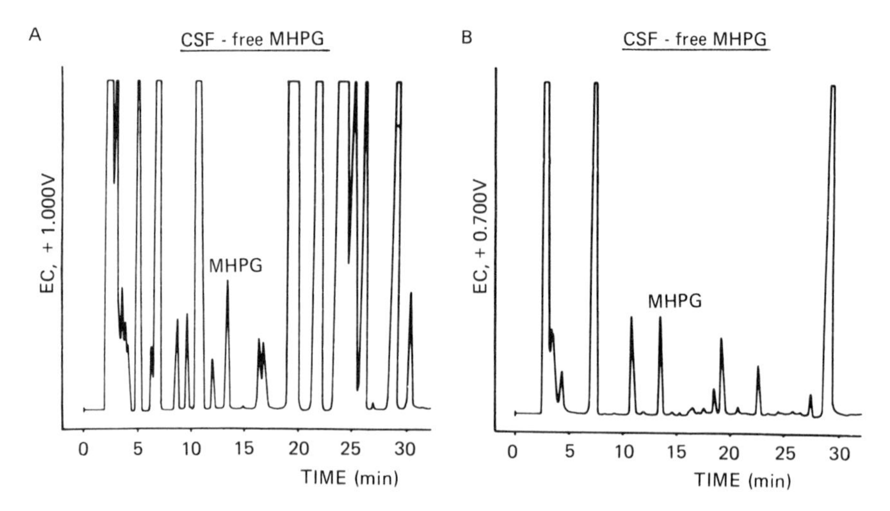

Figure 18 Chromatograms of the ethyl acetate extract of human lumbar cerebrospinal fluid, detected at (A) +1.00 V versus Ag—AgCl and (B) +0.700 V versus Ag—AgCl [column, μBondapak/C_{18} (300 mm X 4.6 m inner diameter, 10 μm); eluents: (low-strength), 0.1 M KH_2PO_4 pH 2.50; (high strength) CH_3OH-H_2O (3:2, v/v); gradient, linear from 0% to 60% of the high-strength eluent in 45 min; flow rate, 1.2 ml/min; temperature, ambient]. (From Ref. 116.)

of dopamine-β-hydroxylase is shown in Fig. 19. The dopamine-β-hydroxylase levels in blood or cerebrospinal fluid are important, since they offer an index of the sympathetic nervous system response to certain diseases (such as schizophrenia) or psychological invasions. Figure 19B illustrates the normal dopamine-β-hydroxylase activity in cerebrospinal fluid. The advantage of these assays is that, contrary to the nonspecific spectrophotometric methods, both the substrate and the reaction product(s) can be measured simultaneously, as well as the activities of other competing enzymes.

VI. CONCLUSIONS

High-performance liquid chromatography coupled with electrochemical detection is becoming an increasingly popular tool for trace analysis. It provides high sensitivity and the current flow-through cell designs have sufficiently low cell volumes to warrant their use with miniaturized equipment. New electrode materials of improved stability and greater batch-to-batch reproducibility, extended application of the multielectrode detection systems, and full exploitation of different measuring techniques will increase further the sucess of EC detec-

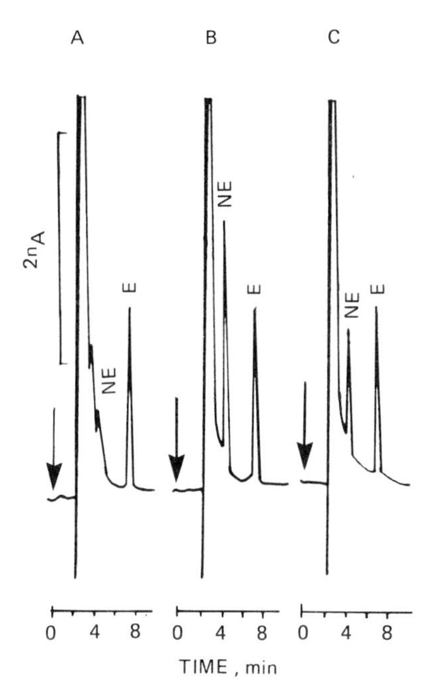

Figure 19 Elution pattern of dopamine-β-hydroxylase (DβH) incubation mixture with human cerebrospinal fluid as the source of enzyme (reaction: DA $\xrightarrow{D\beta H}$ NE): (A) fusaric acid blank incubation; (B) cerebrospinal fluid incubation; (C) 100 pmol of NE added to the furasic acid blank incubation; 500 pmol of E was added to each sample after incubation as an internal standard [column, Yanapak ODS (250 mm X 5.0 mm; inner diameter 10 μm); eluent, 0.1 M potassium phosphate buffer, pH 3.0, containing pentasulfonic acid (20 mg/100ml buffer); flow rate, 0.6 ml/min; detection, EC, +0.80V versus Ag–AgCl]. (From Ref. 121.)

tion. While at present it is possible to detect routinely picogram quantities of electroactive compounds, with future improvements the sensitivity of EC detectors should be extended to the femtogram range.

REFERENCES

1. G. V. Troitsii, Biokhimiya, *5,* 375 (1940).
2. O. H. Müller, J. Am. Chem. Soc. *69,* 2992 (1947).
3. B. Drake, Acta Chim. Scand. *4,* 554 (1950).
4. W. Kemula, Rocz. Chem. *26,* 281 (1952).
5. R. N. Adams, *Electrochemistry of Solid Electrodes,* Marcel Dekker, New York, 1969.

6. P. T. Kissinger, C. J. Refeshauge, R. Dreiling, and R. N. Adams, Anal. Lett. *6*, 465 (1973).
7. P. T. Kissinger, K. Bratin, G. Davis, and L. A. Pachla, J. Chromatogr. Sci. *17*, 137 (1979).
8. P. T. Kissinger, Anal. Chem. *49*, 447A (1977).
9. K. Brunt, in *Trace Analysis*, Vol. 1, Academic Press, New York, 1981.
10. K. Stulik and Y. Pacakova, J. Electroanal. Chem. *129*, 1 (1981); J. Chromatogr. *208*, 269 (1981).
11. S. G. Weber and W. C. Purdy, J. Electroanal. Chem. *115*, 175 (1980).
12. S. G. Weber and W. C. Purdy, Ind. Eng. Chem. Prod. Res. Dev. *20*, 593 (1981).
13. A. M. Krstulović and H. Colin, Analysis, *11*, 111 (1983).
14. K. Bratin and P. T. Kissinger, J. Liquid Chromatogr., *4*, (Supple. 2), 321(1981).
15. A. M. Krstulović, J. Chromatogr. Biomed. Appl. *229*, 1 (1982).
16. S. Allenmark, J. Liquid Chromatogr., 5, 1, (1982).
17. G. C. Davis, D. D. Koch, P. Kissinger, C. S. Bruntlett, and R.E. Shoup, in *Liquid Chromatography in Clinical Analysis* P. M. Kabra and L. J. Marton, eds.), Humana Press, Clifton, N. J., 1981.
18. W. A. MacCremhan, R. A. Durst, and J. M. Bellama, Anal. Lett. *10*, 1175 (1977).
19. D. A. Richards, Surface Technol. *15*, 113 (1982).
20. T. H. Müller and K. Unsicker, J. Neurosci. Methods *4*, 39 (1981).
21. P. T. Kissinger, C. S. Bruntlett, and R. E. Shoup, Life Sci. *28*, 455 (1981).
22. I. N. Mefford, J. Neurosci. Methods *3*, 207 (1981).
23. J. C. Sternberg, in *Advances in Chromatography*, Vol. 2 (J. C. Giddings and R. A. Keller, eds.), Marcel Dekker, New York, 1966, p. 205.
24. G. Guiochon, in *High-Performance Liquid Chromatography*, Vol. 2 (C. Horváth, ed.), Academic Press, New York, 1980, p. 46.
25. J. H. Knox and M. T. Gilbert, J. Chromatogr. *186*, 405 (1979).
26. J. H. Knox, J. Chromatogr. Sci. *18*, 453 (1980).
27. M. Martin, C. Eon, and G. Guiochon, J. Chromatogr. *108*, 229 (1975).
28. P. Kucera, J. Chromatogr. *198*, 93 (1980).
29. A. J. Bard and W. Lund, *Encyclopedia of Electrochemistry of the Elements*, Marcel Dekker, New York, 1969.
30. A. J. Bard and L. R. Faulkner, *Electrochemical Methods*, Wiley, New York, 1980, p. 38.
31. Y. Takata and G. Muto, Anal. Chem. *45*, 1864 (1973).
32. R. S. Deelder, M. G. F. Kroll, A. J. B. Beeren, and J. H. M. Van den Berg, J. Chromatogr. *149*, 669 (1978).
33. W. P. King and P. T. Kissinger, Clin. Chem. *26*, 1484 (1980).

34. M. L. Fultz and R. A. Durst, Anal. Chim. Acta *140,* 1 (1982).
35. K. Brunt and C. H. P. Bruins, J. Chromatogr. *172,* 37 (1979).
36. W. J. Blaedel and L. N. Klatt, Anal. Chem. *38,* 879 (1966).
37. J. Yamada and H. Matsuda, J. Electroanal. Chem. *44,* 189 (1973).
38. H. B. Hanekamp and J. J. van Nieuwkerk, Anal. Chim. Acta *121,* 13 (1980).
39. J. Lankelma and H. Poppe, J. Chromatogr. *172,* 375 (1976).
40. S. G. Webber and W. C. Purdy, Anal. Chim. Acta *99,* 77 (1978).
41. H. B. Hanekamp and H. G. de Jong, Anal. Chim. Acta *135,* 351 (1982).
42. B. Fleet and C. J. Little, J. Chromatogr. Sci. *12,* 747 (1974).
43. K. Slais and D. Kourilova, Chromatographia *16,* 265 (1982).
44. W. J. Blaedel and R. C. Engstrom, Anal. Chem. *50,* 476 (1978).
45. W. J. Blaedel and G. W. Schieffer, J. Electroanal. Chem. *80,* 259 (1977).
46. C. L. Blank, J. Chromatogr. *117,* 35 (1976).
47. R. E. Shoup (ed.), Bibliography of Recent Reports on Electrochemical Detection, BAS Press, W. Lafayette, Ind., 1981.
48. G. W. Schieffer, Anal. Chem. *52,* 1994 (1980).
49. A. M. Krstulović and H. Colin, TrAC, *3*(2), 43(1984).
50. J. N. Brown, M. Hewins, J. H. M. van der Linden, and R. J. Lunch, J. Chromatogr. *204,* 115 (1981).
51. C. Yarnitzky and E. Duziel, Anal. Chem. *48,* 2024 (1976).
52. L. Michel and A. Zlatka, Anal. Chim Acta *105,* 109 (1979).
53. M. Lemar and M. Porthault, J. Chromatogr. *130,* 372 (1977).
54. S. G. Weber and W. C. Purdy, Anal. Chem. *54,* 1757 (1982).
55. B. Dosterhuis, K. Brunt, B. H. C. Westerink, and D. A. Doornbos, Anal. Chem. *52,* 203 (1980).
56. W. J. Blaedel and J. Wang, Anal. Chim. Acta *116,* 315 (1980).
57. G. M. Jenkins and K. Kawamura, Nature *231,* 175 (1971).
58. W. E. van der Linden and J. W. Dieker, Anal. Chim. Acta *119,* 1 (1980).
59. H. Gunasingham and B. Fleet, Analyst *107,* 896 (1982).
60. J. F. Alder, P. O. Kane, and B. Fleet, J. Electroanal. Chem. *30,* 427 (1971).
61. M. Stulikova and F. Vydra, J. Electroanal. Chem. *38,* 349 (1972).
62. R. J. Taylor and A. A. Humfray, J. Electroanal. Chem. *42,* 347 (1973).
63. R. M. Wightman, E. C. Paik, S. Borman, and M. A. Dayton, Anal. Chem. *50,* 1410 (1978).
64. A. N. Strohl and D. J. Curran, Anal. Chem. *51,* 1050 (1979).
65. B. R. Hepler, S. G. Weber, and W. C. Purdy, Anal. Chim. Acta *102,* 41 (1978).
66. J. E. Girard, Anal. Chem. *51,* 836 (1979).
67. H. W. van Rooijen and H. Poppe, Anal. Chim. Acta *130,* 9 (1981).
68. R. J. Fenn, S. Siggia, and D. J. Curran, Anal. Chem. *50,* 1067 (1978).

69. E. Pungor and E. Szepesvary, Anal. Chim. Acta *43,* 289 (1968).
70. S. G. Weber and W. C. Purdy, Anal. Chim. Acta *100,* 531 (1978).
71. J. L. Andersen and D. J. Chesney, Anal. Chem. *52,*,2156 (1980).
72. D. N. Armentrout, J. D. McLean, and M. W. Long, Anal. Chem. *51,* 1039 (1979).
73. M. Stulikova and K. Stulik, Chem. Listy, *68,* 800 (1974).
74. J. A. Lown, R. Koile, and D. C. Johnson, Anal. Chim. Acta *116,* 33 (1980).
75. A. McDonals and P. D. Duke, J. Chromatogr. *83,* 331 (1973).
76. W. A. McCrehan, Anal. Chem. *53,* 74 (1981).
77. P. T. Kissinger, K. Bratin, W. P. King, and J. R. Rice, Am. Chem. Soc. Symp. Ser. *136,* 57 (1980).
78. W. A. McCrehan and R. A. Durst, Anal. Chem. *14,* 2108 (1978).
79. D. K. Roe, Anal. Chem. *36,* 2371 (1964).
80. Y. Takata and K. Fujita, J. Chromatogr. *108,* 255 (1975).
81. J. Devynick, A. Pique, and G. Delarue, Analusis 3, 417 (1975).
82. V. R. Tjaden, J. Lankelma, H. Poppe, and G. Muusze, J. Chromatogr. *125,* 275 (1976).
83. L. R. Taylor and D. C. Joanson, Anal. Chem. *50,* 240 (1978).
84. W. J. Blaedel and J. Wane, Anal. Chem. *51,* 799 (1979).
85. T. Wasa and S. Musha, Bull. Chem. Soc. Jpn. *48,* 2176 (1975).
86. H. B. Hanekamp, P. Bos, and R. W. Frei, J. Chromatogr. *186,* 489 (1979).
87. H. B. Hanekamp, W. H. Voogt, and P. Bos, Anal. Chim. Acta *105,* 109 (1979).
88. Y. Okinaka and I. M. Kolthoff, J. Electroanal. Chem. *73,* 3326 (1957).
89. A. Kimla and F. Strafelda, Coll. Czech. Chem. Commun. *29,* 2913 (1964).
90. D. L. Rabenstein and R. Saetre, Anal. Chem. *49,* 1036 (1977).
91. W. A. McCrehan, Anal. Chem. *53,* 74 (1981).
92. P. Maitoza and D. C. Hohnson, Anal. Chim. Acta *118,* 223 (1980).
93. F. A. Schultz and D. E. Mathis, Anal. Chem. *46,* 2253 (1974).
94. K. Tesarik and P. Kalab J. Chromatogr. *78,* 357 (1973).
95. V. Svoboda and J. Marsal, J. Chromatogr. *148,* 111 (1978).
96. S. Stankovianski, P. Cicmanec, and D. Kaniansky, J. Chromatogr. *106,* 131 (1975).
97. R. Vespalec and K. Hana, J. Chromatogr. *65,* 53 (1972).
98. V. Slavik, J. Chromatogr. *148,* 117 (1978).
99. H. Poppe and J. Kuysten, J. Chromatogr. *132,* 369 (1977).
100. S. Haderka, J. Chromatogr. *91,* 167 (1974).
101. R. Vespalec, J. Chromatogr. *108,* 243 (1975).

102. W. F. Erbelding, Anal. Chem. *47,* 1983 (1975).
103. Y. Hashimoto and M. Moriyassu, Microchim. Acta *2,* 159 (1978).
104. J. F. Alder and A. Thoer, J. Chromatogr. *178,* 15 (1979).
105. J. DeChamplin, L. Farley, D. Cousineau, and M-. R. van Amerigen, Circ. Res. *38,* 109 (1976).
106. O. Hornykiewicz Br. Med. Bull. *29,* 172 (1973).
107. J. J. Schildkraut, P. J. Orsulak, A. F. Schtzberg, J. E. Gudeman, J. O. Cole, W. A. Rhode, and R. A. Labrie, Arch. Gen. Psychiatry *35,* 1424 (1978).
108. O. Teodorescu, R. Cosmatchi, and S. Capalna, Physiologie *15,* 113 (1978).
109. S. E. Gitlow, D. Pertsemlidis, and L. M. Bertani, Am. Heart J. *82,* 557 (1971).
110. S. E. Gitlow, L. B. Dziedzic, L. Strauss, S. M. Greenwood, and S. W. Dziedzic, Cancer *32,* 898 (1973).
111. L. M. Bertaini-Dziedzic, A. M. Krstulović, and S. E. Gitlow, J. Chromatogr. *164,* 345 (1979).
112. P. Coulombe, J. H. Dussault, and P. Waker, Metabolism *25,* 973 (1976).
113. I. J. Kopin, R. C. Lake, and M. Ziegler, Ann. Intern. Med. *88,* 671 (1978).
114. J. J. Kabara, R. M. Riggin, and P. T. Kissinger, Proc. Soc. Exp. Biol. Med. *151,* 168 (1976).
115. R. C. Causon and M. E. Carruthers, J. Chromatogr. Biomed. Appl. *229,* 301 (1982).
116. A. M. Krstulović, L. Bertani-Dziedzic, S. W. Dziedzic, and S. E. Gitlow, J. Chromatogr. *223,* 305 (1981).
117. J. Wagner, P. Vitali, M. G. Palfreyman, M. Zraika, and S. Huot, J. Neurochem. *38,* 1241 (1982).
118. M. Messripour and J. B. Clark, J. Neurochem. *38,* 1139 (1982).
119. A. Togari, T. Kato, and T. Nagatsu, Biochem. Pharmacol. *31,* 1729 (1982).
120. R. E. Shoup, G. C. Davis, and P. T. Kissinger, Anal. Chem. *52,* 483 (1980).
121. H. Matsui, T. Kato, C. Yamamoto, K. Jujita, and T. Nagatsu, J. Neurochem. *37,* 289 (1981).
122. W. L. Caudill, G. P. Houck, and R. M. Wightman, J. Chromatogr. Biomed. Appl. *227,* 331 (1982).
123. P. T. Kissinger, K. Bratin, G. C. Davis, and L. A. Pachla, J. Chromatogr. Sci. *17,* 137 (1979).
124. J. G. Koen and J. F. K. Huber, Anal. Chim. Acta *51,* 303 (1970).
125. W. A. McCrehan, R. A. Durst, and J. M. Bellama, Nat. Bur. Spec. Publ. *519,* 57 (1979).
126. R. Samuelson and R. A. Osteryoung, Anal. Chim. Acta *123,* 97 (1981).

127. W. Kutner, J. Debowski, and W. Kemula, J. Chromatogr. *191*, 47 (1980).
128. L. R. Hegstrand and B. Eichman, J. Chromatogr. Biomed. Appl. *222*, 107 (1981).
129. I. N. Mefford, M. M. Ward, L. Miles, B. Taylor, M. A.Chesney, D. L. Keegan, and J. D. Barchas, Life Sci. *28*, 477 (1981).
130. A. M. Krstulović, S. W. Dziedzic, L. Bertani-Dziedzic, and D. E. DiRico, J. Chromatogr. *217*, 523 (1981).
131. G. Lachatre, G. Nicot, C. Hagnet, J. L. Rocca, L. Merle, and J. P. Valette, J. Liquid Chromatogr. 5, 1947 (1982).
132. F. Smedes, J. C. Kraak and H. Poppe, J. Chromatogr. Biomed. Appl., *231*, 25 (1982).
133. D. S. Goldstein, G. Feuerstin, J. L. Izzo, Jr., E. J. Kopin, and H. R. Keiser, Life Sci. *28*, 467 (1981).
134. R. C. Causon, M. E. Carruthers, and R. Rodnight, Anal. Biochem. *116*, 223 (1981).
135. C. L. Davies and S. G. Molyneux, J. Chromatogr. Biomed. Appl. *231*, 41 (1982).
136. M. Goto, T. Nakamura, and D. Ishii, J. Chromatogr. Biomed. Appl. *226*, 33 (1981).

5

Reversed-Flow Gas Chromatography Applied to Physicochemical Measurements

Nicholas A. Katsanos and George Karaiskakis / University of Patras, Patras, Greece

I. INTRODUCTION

Gas chromatography offers many possibilities for physicochemical measurements in addition to chemical analysis. Some of these methods lead to very precise and accurate results with relatively cheap instrumentation and a very simple experimental setup. They are widely used today, a fact which is emphasized by the edition of two books [1,2] dealing only with such physicochemical measurements. All are based on the traditional techniques of gas chromatography, namely, elution, frontal analysis, and displacement development, under constant gas flow rate. Recently, however, two methods have been developed which are based on perturbations imposed on the carrier gas flow.

The first of these methods is the stopped-flow technique, introduced in 1967 by Phillips and co-workers [3], for studying the kinetics of surface-catalyzed reactions. It consists in introducing a small amount of reactant as a vapor onto a chromatographic column acting both as a catalytic reactor and as a separation column, and then repeatedly *stopping* the carrier gas flow for short time intervals. Sharp symmetrical peaks (the "stop-peaks") follow each restoration of the gas flow, and these peaks have definite retention times and "sit" on the continuous chromatographic signal.

We have employed this technique for various studies, namely, the following: (1) In its simple original form the method was used for studying dehydrohalogenation reactions [4–6]; (2) after deriving the various equations describing the elution curve and the area under the curve of the stop peaks for a fairly general mechanism of surface reactions [7], we applied them to kinetic studies of deamination reactions [8–11], n-butene isomerization [12], and alcohol dehydrations on irradiated aluminium oxide [13]; (3) after a preliminary treatment and experimental verification of the results [14], the chromatographic theory for nonequilibrium stopped-flow gas chromatography was developed [15], and then applied to determine rate constants for desorption and equilibrium constants for adsorption of heptane on active surfaces, together with the thermodynamic parameters associated with them [16]; and (4) finally, the stopped-flow technique was used to measure gas diffusion coefficients, after having made the necessary theoretical analysis for that purpose [17].

When this method is employed in catalytic studies, it can yield valuable informations. It permits a direct determination of reaction rates for the surface reaction, not only for small conversions to products or for reaction times around zero, as most differential methods, but in the whole range of conversions covering an extended period of time. Moreover, *true* activation energies are calculated from the measured rates, since heats of adsorption are not involved in the calculations. Conversions calculated by the stopped-flow method bear on the fraction of the total surface being catalytically active [11].

The most severe limitations of the method are the following: (1) The reactant must have a relatively long retention time on the catalytic column, and (2) it does not seem to apply to reactions other than those with one reactant and first-order steps. Probably the main drawback of the stopped-flow technique is that it continuously switches the system under study from a flow dynamic one to a static system and vice versa, by repeatedly closing and opening the carrier gas flow. Longitudinal diffusion and other related phenomena, which are usually negligible during the gas flow, may become important when the flow is stopped.

The second flow perturbation method is the reversed-flow technique developed in this laboratory. A preliminary form of this [18] was used to measure the rate constants for the cracking of cumene, the dehydration of pronan-1-ol, and the deamination of 1-aminopropane, over a 13X molecular sieve. This was followed by a detailed theoretical analysis and application of the method to the dehydration of alcohols and deamination of primary amines [19–22].

The method was then extended to more complicated reactions with two reactants [23], and to other physicochemical measurements, like those of gas diffusion coefficients [24–27], adsorption equilibrium constants [28], and the kinetics of drying of catalysts [29]. It is the object of this article to review the application of the method in the above areas, with a unification approach to the underlying theory.

II. THEORETICAL ANALYSIS

Reversed-flow gas chromatography is really a sampling technique and consists in *reversing* the direction of flow of the carrier gas from time to time. If this carrier gas contains other gases at certain concentrations, recorded by the detector system of the chromatograph, each flow reversal creates perturbations on the chromatographic elution curve, having the form of extra peaks ("sample peaks"). If the concentration of a constituent in the flowing gas depends on a *rate process*, taking place within the chromatographic column, then, by reversing the flow, one performs a sampling of this process, for example, of a sufficiently slow chemical reaction. It is preferable that the rate process influencing the concentration c of a component be confined to a very short section near the middle of the chromatographic column, so that it does not mix with the chromatographic process itself, influencing also c. A basic sampling equation is now derived by reference to Fig. 1, under the following assumptions:

1. The adsorption isotherms in the chromatographic column are linear.
2. Axial diffusion of the gases along coordinate x or x' and other phenomena leading to nonideality (mass transfer resistances in the stationary phase, nonequivalent flow paths in the packed bed, etc.) are negligible. This seems reasonable for high enough flow rates.

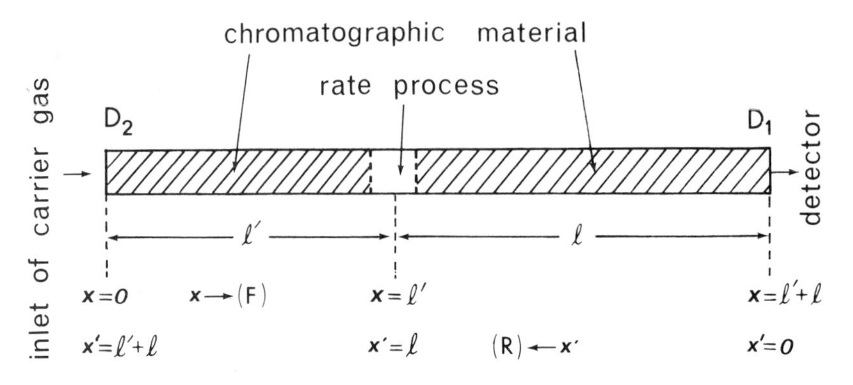

Figure 1 Representation of the chromatographic column in the reversed-flow method.

3. The rate process is taking place in a sufficiently short section of the total column length, so that its x distribution can be described approximately by a Dirac delta function, $\delta(x - l')$.

A. Notation

a	volume of gas phase per unit length of column, or cross-sectional area of void space (cm^2)
c,c'	concentrations of a solute vapor in the chromatographic column with the carrier gas flowing in direction F or R, respectively (mol cm^{-3})
c_z	concentration of the solute vapor in the diffusion column L (mol cm^{-3})
$\underline{C}$, $\underline{C}'$, C_z	Laplace transforms of c, c', and c_z with respect to t_0
$\overline{C}$, $\overline{C}'$	double Laplace transforms of c and c' with respect to t_0 and t'
$\overline{\overline{C}}$	triple Laplace transform of c with respect to t_0, t', and t
D	mutual diffusion coefficient of two gases (cm^2 sec^{-1})
f	area under the curve of R and F peaks (mol)
g	fraction of reactant adsorbed on catalytically active surface sites
h	height above the base line defined by Eqs. (24)–(27) (mol cm^{-3})
k, k_1, k_2	rate constants (sec^{-1})
k_A	partition ratio (dimensionless)
K_A, K_D	adsorption equilibrium constants of A and D (dimensionless)
l,l'	lengths of the two sections of the chromatographic column (Fig. 1) (cm)
L, L_1, L_2	lengths of the diffusion column (Figs. 4 and 12) (cm)
L_{eff}	effective length defined by Eq. (47)

m mass of a solute or reactant injected (mol)
n exponent defined in Eq. (44)
M catalyst mass (g)
N, N' constants defined by Eqs. (38) and (46), respectively
p_0, p', p transform parameters with respect to t_0, t', and t, respectively
P pressure (atm)
q parameter defined by Eq. (30)
q_0 initial concentration of solvent on a solid (mol cm^{-3}) con-
Q concentration parameter defined by Eq. (66)
$r(t_0)$ reaction rate at time t_0 (mol cm^{-2} sec^{-1} or mol g^{-1} sec^{-1})
S response of the detecting system (cm sec mol^{-1})
t_0 time measured from the injection of solute or reactant to the last backward reversal of gas flow (sec)
t' time interval of backward flow of carrier gas (sec)
t time measured from the last restoration to the forward direction of the gas flow (sec)
t_{tot} sum of times t_0 and t' (sec)
t_M, t'_M gas holdup time of the empty column section l or l', respectively (sec)
t_R, t'_R ideal retention time on the filled column section l or l', respectively (sec)
v linear velocity of carrier gas in interparticle space of the chromatographic column (cm sec^{-1})
v_A velocity defined by Eq. (18)
$\dot{V}$ volume flow rate of carrier gas (cm^3 sec^{-1})
V_N net retention volume (cm^3)
x, x', z distance coordinates defined in Fig. 4 (cm)
ε void fraction of column section L_2 (dimensionless)
θ, θ' time parameter defined by Eqs. (3) and (9), respectively (sec)
ρ volume ratio of solid and gas phases
τ, τ' time defined by Eqs. (20) and (14), respectively (sec)

B. Chromatographic Sampling Equation

The problem will be considered separately for various time intervals, in which the concentration of a representative gaseous component A (as a function of time and distance x or x') is determined by certain differential equations with given initial and boundary conditions.

1. Initial F Interval

When the carrier gas B flows in the direction F, that is, from D_2 toward D_1 (see Fig. 1), the concentration of component A at $x = l'$, $c(l', t_0)$, owing to the rate process taking place there, spreads out in the column section l, thus becoming a function of time and distance, $c(x, t_0)$. This is described by the following mass balance equation, under assumptions (1) and (2):

$$(1 + k_A)\frac{\partial c}{\partial t_0} = -v\frac{\partial c}{\partial x} + vc(l',t_0)\delta(x - l') \tag{1}$$

Taking the t_0 Laplace transform of this equation, under the initial condition $c(x,0) = 0$, we find an ordinary differential equation for C as a function of x. This can be integrated by means of its x Laplace transform, with the result

$$C = C(l',p_0)\exp(-p_0\theta)\ u(x - l') \tag{2}$$

where

$$\theta = (1 + k_A)(x - l')/v \tag{3}$$

and $u(x - l')$ is the Heaviside unit step function, which equals 0 for $x < l'$ and 1 for $x \geqslant l'$. At the detector, that is, at $x = l' + l$, $u(x - l')$ becomes $u(l) = 1$ for $l > 0$, and θ becomes $(1 + k_A)l/v = t_R$, that is, the ideal retention time of component A in the column section l. Thus Eq. (2) gives C at the detector as

$$C = C(l',p_0)\exp(-p_0 t_R) \tag{4}$$

and, according to the property "translation" of Laplace transformations, the inverse transform of Eq. (4) for $t_R \geqslant 0$ is

$$c = c(l',t_0 - t_R)\ u(t_0 - t_R) \tag{5}$$

This equation describes the breakthrough curve of component A at position D_1 (Fig. 1).

2. R. Interval

While the carrier gas is flowing in direction F, giving the curve described by Eq. (5), its flow is reversed to the direction R at a time $t_0 > t_R$. The time measured from the moment of reversal is called t'. The distance coordinate x is now changed to x', according to the obvious relation

$$x' = l' + l - x \tag{6}$$

and the concentration of A in this time interval, c', becomes a function of x' and t', $c' = c'(x',t')$. It is given by an equation analogous to Eq. (1):

$$(1 + k_A)\frac{\partial c'}{\partial t'} = -v\frac{\partial c'}{\partial x'} + vc'(l,t_0,t')\delta(x' - l) \tag{7}$$

$c'(l,t_0,t')$ now being the concentration due to the rate process at $x' = l$, which is a function of both t_0 and t'.

As with Eq. (1), we proceed by taking Laplace transforms of this equation with respect to time, but now the t_0 transform is taken first, and then the t' transform with initial condition

$$C'(x',p_0,0) = C'(1,p_0,0)\exp(p_0\theta')[1 - u(x' - 1)] \tag{8}$$

This is obtained from Eq. (2) by substituting $1 - u(x' - 1)$ for $x(x - l')$ and replacing $-\theta$, as defined by Eq. (3), by its equivalent:

$$\theta' = (1 + k_A)(x' - 1)/v = -\theta \tag{9}$$

The result of the above double Laplace transformation is

$$\frac{d\bar{C}'}{dx'} + \frac{(1 + k_A)p'}{v}\bar{C}' = \frac{1 + k_A}{v} C'(1,p_0,0)\exp(p_o\theta')[1 - u(x' - 1)] + \bar{C}'(1,p_0,p')\delta(x' - 1) \tag{10}$$

This ordinary differential equation is easily integrated by using x' Laplace transforms, giving $\bar{C}'(x',p_0,p')$:

$$\bar{C}' = \frac{C'(1,p_0,0)}{p' + p_0}\left[\exp(p_0\theta')[1 - u(x' - 1)] + \exp(-p'\theta')u(x' - 1) - \exp\left(-(p_0 l + p'x')\frac{(1 + k_A)}{v}\right)\right] + \bar{C}'(1,p_0,p')\exp(-p'\theta')u(x' - 1) \tag{11}$$

At the detector, that is, for $x' = l' + 1$, $u(x' - 1)$ becomes $u(l') = 1$, for $l' > 0$, and θ' becomes $(1 + k_A)l'/v = t'_R$, that is, the ideal retention time of component A in the column section l'. Then Eq. (11) becomes

$$\bar{C}' = \bar{C}'(1,p_0,p')\exp(-p',t'_R) + \frac{C'(1,p_0,0)}{p' + p_0}\{\exp(-p't'_R) - \exp[-p'(t'_R + t_R)]\exp(-p_0 t_R)\} \tag{12}$$

The breakthrough curve of A at the end D_2 of the column is now obtained by taking the inverse transforms, first with respect to p' and then with respect to p_0. The result is

$$c' = c'_1(1,t_0 + \tau')u(\tau') + c'_2(1,t_0 - \tau')[u(\tau') - u(\tau' - t_R)]u(t_0 - \tau') \tag{13}$$

where

$$\tau' = t' - t'_R \tag{14}$$

and $c'_1(1,t_0,t' - t'_R)$ is written $c'_1(1,t_0 + \tau')$, since the time t' is a continuation of t_0.

Equation (13) is the simplest chromatographic sampling equation. Its behavior for various values of τ' is interesting and is illustrated diagrammatically in Fig. 2. For $\tau' < 0$, that is, for $t' < t'_R$, $c' = 0$

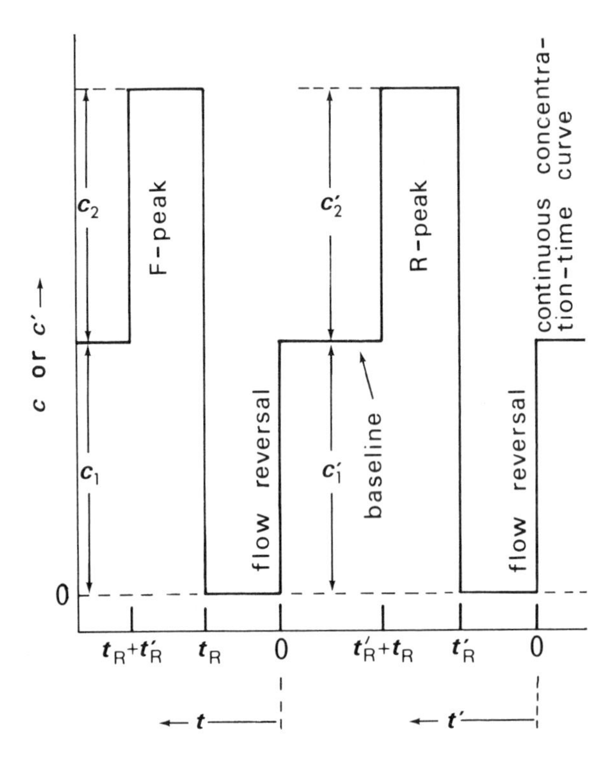

Figure 2 Elution curves predicted by Eqs. (13) and (21).

and no signal is recorded by the detector until the retention time t'_R is reached. Then $u(\tau') = 1$, and the chromatographic signal rises abruptly to $c'_1 + c'_2$. It falls again to c'_1 when $\tau' \geqslant t_R$ because the square function in brackets becomes zero. The $u(t_0 - \tau')$ factor remains at unity in the above interval, since the flow was reversed at $t_0 > t_R$. Thus the concentration of A at $x = l'$, $c(l',t_0)$, or at $x' = l$, $c'(l,t_0)$, due to the rate process at this position is shifted in time on reversing the flow direction. This time shift takes place in two opposite directions. One, which occurs "forward" to $c'_1(l,t_0, + \tau')$, starts at $t' = t'_R$ and continues uninterrupted. It is nothing else than the continuation of Eq. (5) at the other end of the column. The second shift $c'_2(l,t_0 - \tau')$ occurs "backward" and is barred in the interval $0 \leqslant \tau' \leqslant t_R$ or $t'_R \leqslant t' \leqslant t'_R + t_R$. Therefore it starts with the concentration $c'_2(l,t_0)$ and ends with that of a preceding time, namely, $c'_2(l,t_0 - t_R)$. This extra signal (R peak; see Fig. 2) adds to the forward shift and constitutes the simplest form of a "sample peak" for the rate process under investigation.

3. Second F Interval

After a certain time t' of backward flow, the carrier gas is again turned to the direction F, the time from this moment being denoted by t. The distance coordinate is changed from x' back to x, according to Eq. (6), and the concentration c(x,t) is again described by Eq. (1), with t substituted for t_0:

$$(1 + k_A)\frac{\partial c}{\partial t} = -v\frac{\partial c}{\partial x} + vc(l',t)\,\delta(x - l') \tag{15}$$

This is integrated by taking successive Laplace transformations with respect to t_0, t', and t. The initial condition at t = 0 (in the form of its t_0 and t' double transform) is obtained from Eq. (11) by changing x' to l' + l − x, θ' to −θ, and u(x' − l) to 1 − u(x − l'), according to Eqs. (6) and (9), with the result

$$\begin{aligned}\bar{C}(x,p_0,p',0) = {} & \frac{C(l',p_0,0)}{p' + p_0}\{\exp(-p_0\theta)u(x - l') \\ & + \exp(p'\theta)[1 - u(x - l')] \\ & -\exp[-p_0 + p')(1 + k_A)l/v + p'\theta]\} \\ & + C(l',p_0,p')\exp(p'\theta)[1 - u(x - l')]\end{aligned} \tag{16}$$

The result of the procedure outlined above is an ordinary differential equation in x, which can easily be integrated by using x Laplace transforms, giving $\bar{\bar{C}}(x,p_0,p',p)$:

$$\begin{aligned}\bar{\bar{C}} = {} & \frac{C(l',p_0,0)}{p' + p_0}\left\{\frac{\exp(-p_0\theta) - \exp(-p\theta)}{p - p_0}u(x - l')\right. \\ & - \frac{\exp(p'\theta) - \exp(-p\theta)}{p + p'}u(x - l') + \frac{\exp(p'x/v_A) - \exp(-px/v_A)}{p + p'} \\ & \left.\exp\left(\frac{-p'l'}{v_A}\right)\left[1 - \exp\left(\frac{-p_0 l}{v_A}\right)\exp\left(\frac{-p'l}{v_A}\right)\right]\right\} \\ & + \frac{\bar{C}(l',p_0 p')}{p + p'}\left[\exp(p'\theta) - \exp\left(-\frac{(p'l' + px)}{v_A}\right)\right. \\ & \left. - [\exp(p'\theta) - \exp(-p\theta)]u(x - l')\right] \\ & + \bar{\bar{C}}(l',p_0,p',p)\exp(-p\theta)u(x - l')\end{aligned} \tag{17}$$

where

$$v_A = \frac{v}{1 + k_A} \tag{18}$$

At the detector Eq. (17) is considerably simplified, since there $x = l' + l$, $u(x - l') = 1$, and $\theta = (1 + k_A)l/v = t_R$. Then, performing inverse Laplace transformations with respect to p, p', and p_0 in succession, one finds the final equation giving the value of c at the detector:

$$\begin{aligned} c = {} & c_1(l', t_0 + t' + \tau)u(\tau) \\ & + c_2(l', t_0 + t' - \tau)[1 - u(\tau - t')][u(\tau) - u(\tau - t'_R)] \\ & + c_3(l', t_0 - t' + \tau)u(t_0 + \tau - t')\{u(t - t')[1 - u(\tau - t'_R)] \\ & - u(\tau - t')[u(\tau) - u(\tau - t'_R)]\} \end{aligned} \tag{19}$$

where

$$\tau = t - t_R \tag{20}$$

Equation (19) is the desired general chromatographic sampling equation. It consists, on the right-hand side, of three concentration terms, denoted by c_1, c_2, and c_3. They all refer to $x = l'$, but to different values of the time variable, namely, $t_0 + t' + \tau$, $t_0 + t' - \tau$, and $t_0 - t' + \tau$, respectively. Each of the concentration terms is multiplied by a combination of unit step functions, so that it appears in a certain time interval and vanishes in all others. There are various possibilities, depending on the relative values of t_0, t', t, t_R, and t'_R.

Case Where $t' > t_R + t'_R$. This condition reduces Eq. (19) to

$$c = c_1(l', t_0 + t' + \tau)u(\tau) + c_2(l', t_0 + t' - \tau)[u(\tau) - u(\tau - t'_R)] \tag{21}$$

since the term c_3 becomes zero for all values of t, and the factor $1 - u(\tau - t')$ of the c_2 term remains at unity in the interval defined by the other factor $[u(\tau) - u(\tau - t'_R)]$. Equation (21) is the F-direction equivalent of Eq. (13) and its behavior is analogous to that. It is also depicted in Fig. 2 and predicts that $c = 0$ for $\tau < 0$, that is, $t < t_R$, and at $\tau \geqslant 0$ or $t \geqslant t_R$ two functions are recorded as a sum $c_1 + c_2$. The first is $c(l', t_0 + t')$ shifted forward by τ. This continues uninterrupted, as the first term of Eq. (21) shows. In the other function c_2 the total time $t_0 + t'$ is shifted backward by τ, and this function vanishes when $\tau \geqslant t'_R$ or $t \geqslant t_R + t'_R$. Thus a sample peak (F peak; see Fig. 2) is predicted in the interval $t_R \leqslant t \leqslant t_R + t'_R$, positioned on top of the otherwise continuing chromatographic curve.

If one repeats the reversal of the carrier gas flow in the direction R for a second time, then in the direction F for a third time, and so on, keeping the time between any two successive reversals greater than the total retention time $t'_R + t_R$, two series of peaks are produced. The R peaks are described by Eq. (13), while the F peaks are given by Eq. (21). Since the definitions "forward" and "reverse" are arbitrary, and the two above equations have the same form, one of them, say, Eq. (13), suffices to describe both kinds of peaks. In that case t_0 is taken to represent the total time passed from the beginning to the *last* reversal of gas flow, τ' the time passed *from* the last reversal of the flow diminished by the retention time of A in the new flow direction, and t_R the retention time in the opposite direction.

Case where $t' < t_R + t'_R$. This condition retains all three terms of the sampling equation, Eq. (19). The first term c_1 appears at $t = t_R$ ($\tau = 0$) and continues uninterrupted. The second term c_2 appears again at $t = t_R$ but it is cut down either at $t = t_R + t'$ ($\tau = t'$) or at $t = t_R + t'_R$ ($\tau = t'_R$), whichever comes first. This, of course, depends on whether $t' < t'_R$ or $t' > t'_R$.

The third term c_3 appears at $t = t'$ and vanishes at the same t value as the second term. Thus for $\tau \geqslant t'$ or $\tau \geqslant t'_R$ only c_1 remains as an "ending base line." Before that there are two possibilitied: (1) $t' < t_R$ and (2) $t' > t_R$. In case (1) the c_3 term appears first, and then "sitting" on it the sum $c_1 + c_2$. This is shown in Fig. 3A. In case (2) the sum $c_1 + c_2$ appears first and then on it c_3 (Fig. 3B).

If t' is small enough, the c_1 and c_3 terms will differ little in their time argument, and the situation depicted in Fig. 3A will have the appearance of a relatively narrow peak of two terms, with base line the remaining term (c_3 as the starting base line and c_1 as the ending base line). The smaller t' is, the narrower this *sample peak* becomes, as its width is clearly equal to t'. The situation in Fig. 3B is less favorable for measurements, because a single term appears to "sit" on a starting base line consisted of two terms ($c_1 + c_2$) and on an ending base line of only c_1.

In summary, the chromatographic sampling of a rate process, taking place at a certain point within a chromatographic column, is effected by reversing the flow direction of the carrier gas. This creates sample peaks in the elution curve. When the time elapsing between any two successive reversals of the flow is greater than the total retention time $t_R + t'_R$ of the component examined on column $l + l'$, a sample peak follows *every* reversal, as shown diagrammatically in Fig. 2. This is described mathematically by Eq. (13) or (21). But when the time passing between two successive reversals is less than $t_R + t'_R$, only one sample peak appears after the *two* successive reversals, as predicted by the general Eq. (19). This peak is either symmetrical when $t' < t_R$ (Fig. 3A) or unsymmetrical when $t' > t_R$ (Fig. 3B).

C. Calculation of Rate Coefficients from the Sample Peaks

If the concentrations c_1, c_2, and c_3 of Eq. (19), which pertain to $x = l'$, are due to a rate process taking place there, the rate coefficient of that process can be determined from the sample peaks created by applying the reversed-flow technique. For this purpose, either the area under the sample peaks f or their height h can be used.

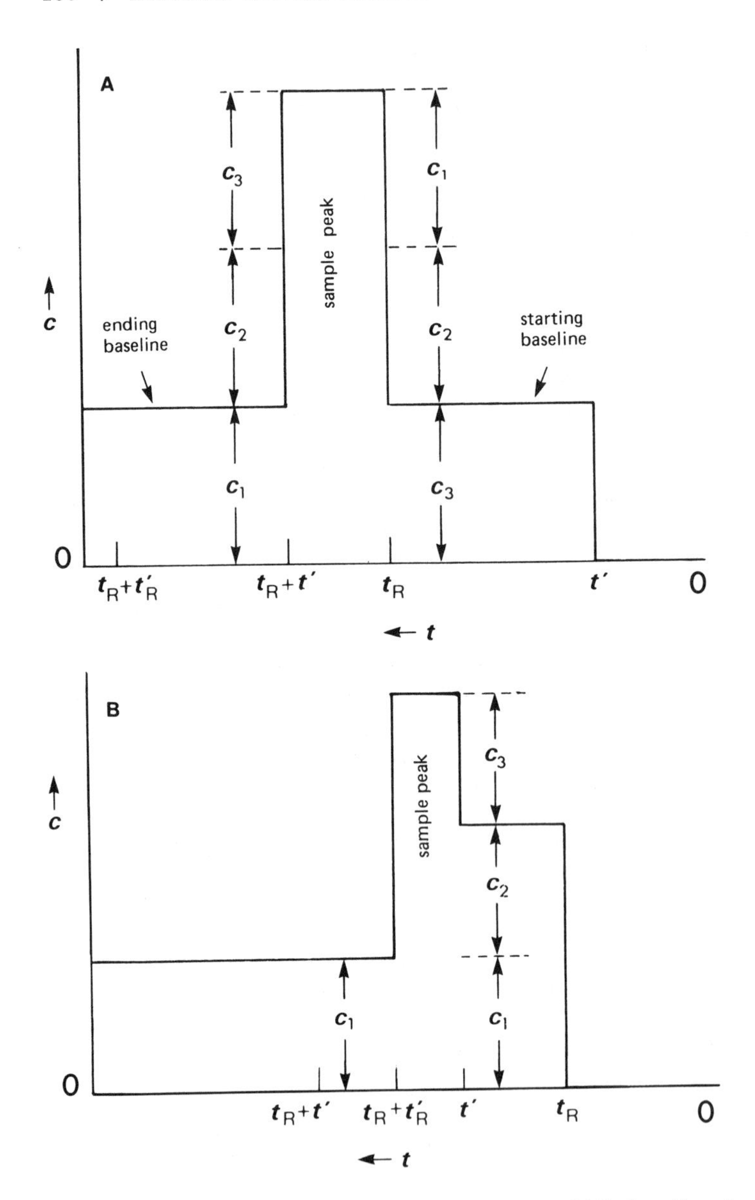

Figure 3 Elution curves predicted by Eq. (19) for $t' < t_R + t'_R$: (A) when $t' < t_R$ and $t' < t'_R$ and (B) when $t' > t_R$ and $t' > t'_R$.

Theoretically, the area f can most easily be calculated from the equations preceding Eq. (19) and its limiting case, Eq. (13), namely, Eq. (17) and (12), respectively. As an example, let us calculate the area under the R peak of Fig. 2. According to a well-known relation,

$$\mathcal{L}_{t_0} f = \int_0^\infty C' \, dV = \dot{V} \int_0^\infty C' \, dt' = \dot{V} \lim_{p' \to 0} (\bar{C}') \tag{22}$$

where in place of $\bar{C}'$ we use the second term of Eq. (12), since it is this term giving rise to the sample R peak, the first term being responsible for the ending base line. The result is

$$\mathcal{L}_{t_0} f = \frac{\dot{V}C'(1,p_0)}{p_0} [1 - \exp(-p_0 t_R)] \tag{23}$$

The area under the peak is found by simply taking the p_0 inverse transform of this expression.

The height h of the sample peaks above the continuous chromatographic signal (base line) is proportional to either one or two concentration terms of Eq. (19) (see Figs. 2 and 3). For the case of the sample peaks of Fig. 2 (R and F peaks) this height is proportional to the term c_2' or c_2 of Eq. (13) or (21), respectively. It can be measured form the base line to either of the discontinuities of the peak corresponding to $t' = t_R'$ ($\tau' = 0$) or $t' = t_R' + t_R$ ($\tau' = t_R$) for the R peak, and to $\tau = 0$ or $\tau = t_R'$ for the F peak. Thus for the R peak, using Eq. (13),

$$h_{\tau'=t_R} = c_2'(1, t_0 - t_R) \tag{24}$$

and for the F peak, using Eq. (21),

$$h_{\tau=0} = c_2(1', t_0 + t') \tag{25}$$

$$h_{\tau=t_R'} = c_2(1', t_0 + t' - t_R') \tag{26}$$

For the case of the sample peak of Fig. 3A, its maximum is taken to correspond to its middle time, that is, to $t = t_R + t'/2$ or $\tau = t'/2$. The height h of this maximum, measured from the ending base line, gives the sum $c_2 + c_3$. From Eq. (19), by putting $\tau = t'/2$ in these two terms, one obtains

$$h = c_2(1', t_0 + t'/2) + c_3(1', t_0 - t'/2)$$

The times in the two terms on the right-hand side differ only by t', and if this is small, both terms can be taken at a mean time t_0. Then the above relation becomes simply

$$h \approx 2c(1', t_0) \tag{27}$$

Thus sample peaks like that of Fig. 3A are approximately double those of Fig. 2 in height, and consequently the use of such peaks increases the precision of the method.

By following either the area f under the sample peaks or their height h as a function of the corresponding time, as indicated in Eqs. (23)–(27), the rate coefficient of the process responsible for the various concentrations is determined, as exemplified in what follows.

III. DETERMINATION OF GAS DIFFUSION COEFFICIENTS

The simplest conceivable case of the reversed-flow technique [24] is to have the sections l' and l shown in Fig. 1 entirely empty of any chromatographic material, the rate process at x = l' being the slow diffusion of a gaseous substance A into a carrier gas B, from a "diffusion" column L connected perpendicularly to column l' + l at x = l' (see Fig. 4). It permits the measurement of the mutual diffusion coefficient of A and B, this being the appropriate rate coefficient of the process. The component B enters at D_2 and meets the detector at D_1, or vice versa, flowing continuously through the chromatographic column l' +l, and filling also the diffusion column L. At the closed end of the latter the solute A is introduced, as a gas or vapor, in the form of a pulse (by means of a syringe).

The diffusion column L is relatively short (about 1 m) and can be straight or coiled.

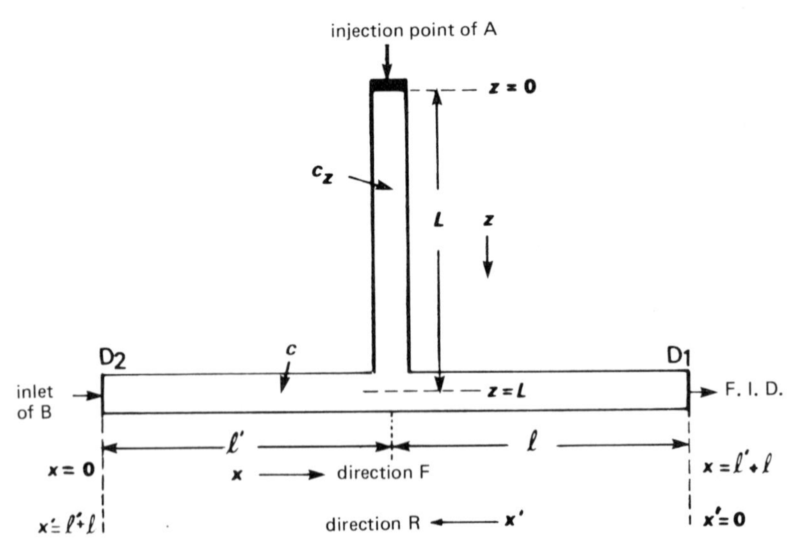

Figure 4 Arrangement of a diffusion column L and a chromatographic column l' + l for measuring gas diffusion coefficients. (From Ref. 24.)

A. Experimental

1. Apparatus and Materials

The experimental setup for the application of the reversed-flow method in this case is very simple and needs only a modification of a conventional gas chromatograph as shown diagrammatically in Fig. 5. Any kind of gas-chromatographic detector can be used, although a high-sensitivity device, like a flame ionization detector, is to be preferred. The reversing of the flow is effected by means of valve S (four-port or six-port with two alternate ports connected through a small piece of 1/16-in. tube). A restrictor can be placed at H to in-

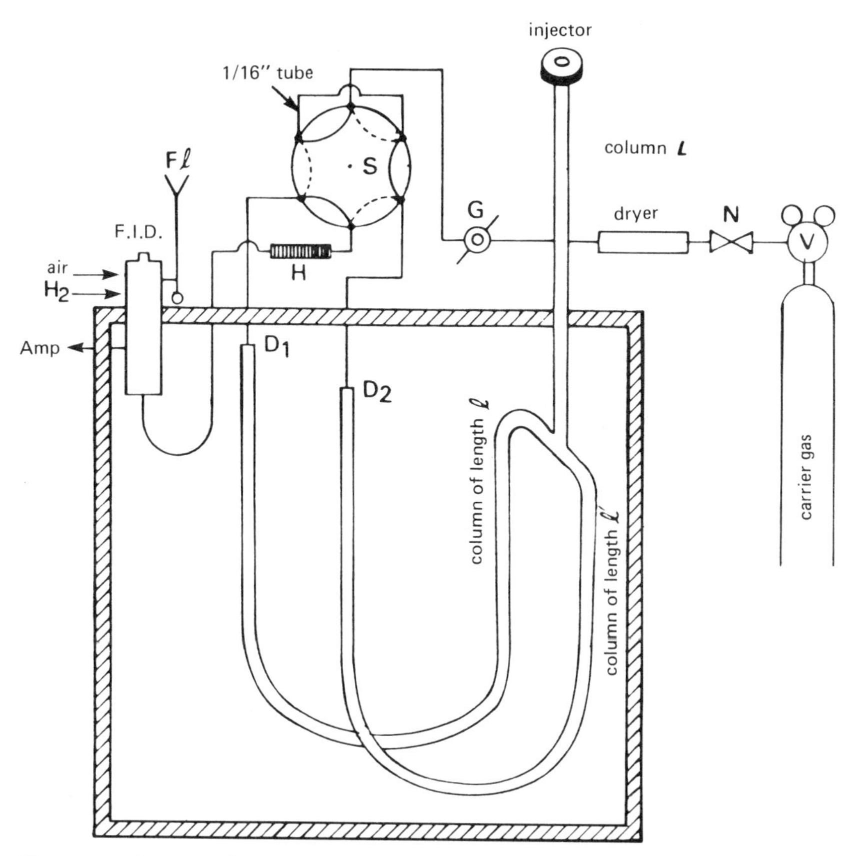

Figure 5 Schematic arrangement of the experimental setup for measuring gas diffusion coefficients: V, two-stage reducing valve and pressure regulator; N, needle valve; G, gas flow controller for minimizing variations in the gas flow rate; S, six-port gas sampling valve; H, restrictor; Fl, bubble flowmeter; and Amp, signal to amplifier. (From Ref. 24.)

crease the pressure within the whole system and/or to prevent the flame of a flame ionization detector from being extinguished when the valve is turned from one position to the other. Pure carrier gases (99.99%) and solutes (99.7 to 99.98%) were used.

2. Procedure

While carrier gas B is flowing in direction F (Fig. 5, valve S in position indicated by the solid lines), a small amount of solute A (usually 0.5 cm^3 of gas at atmospheric pressure) is injected in the diffusion column L. After a certain time, during which no signal is noted, an asymmetrical concentration–time curve of A is recorded, rising slowly and then decaying even more slowly. At a certain known time (from the moment of injection) the direction of the carrier gas is reversed by switching valve S to the other position (dotted lines). This reversal is repeated several times, the interval between successive reversals depending on whether the condition $t' > t_R + t'_R$ or $t' < t_R + t'_R$ is adopted. A whole series of sample peaks is thus obtained.

The pressure drop along column l' + l is negligible, and the diffusion coefficients are considered to have been determined at the pressure measured near the injection point, by means of a suitable manometer.

The temperature of the diffusion column L can be regulated, if necessary, by conventional methods. The simplest way is to coil this column and place it inside the chromatographic oven, with its closed end at the injector position of the chromatograph. This arrangement is particularly suitable for studying the temperature variation of gas diffusion coefficients [25].

B. Complementary Theory

In order to apply the chromatographic sampling equation, Eq. (19), and all other relations derived from it, the concentration of A at x = l' as a function of time, $c(l', t_0)$, is required. This is determined by the diffusion equation (Fick's second law):

$$\frac{\partial c_z}{\partial t_0} = D \frac{\partial^2 c_z}{\partial z^2} \tag{28}$$

where $c_z = c_z(z, t_0)$ is the concentration of the solute vapor A in the diffusion column L. Laplace transformation with respect to t_0 of this equation, under the initial condition $c_z(z,0) = (m/a)\,\delta(z)$, where $\delta(z)$ is the Dirac delta function, gives the linear second-order equation

$$\frac{d^2 C_z}{dz^2} - q^2 C_z = -\frac{m}{aD}\delta(z) \tag{29}$$

where

$$q^2 = p_0/D \tag{30}$$

Equation (30) can be solved by using z Laplace transformation, the result being

$$C_z = C_z(0)\cosh qz + \left(C_z'(0) - \frac{m}{aD}\right)\frac{1}{q}\sinh qz \tag{31}$$

where $C_z(0)$ and $C_z'(0)$ is the t_0 transform of the concentration c_z and its first z derivative, respectively, at z = 0.

From this point on, a slightly different derivation from that in the original article [24] is adopted, leading, however, to the same final result. Thus, since there is no flux across the boundary z = 0, the derivative $C_z'(0) = (\partial C_z/\partial z)_{z=0}$ is zero. At the other boundary z = L, or x = l', we have the conditions

$$C_z(L) = C(l') \tag{32}$$

and

$$-D\left(\frac{\partial C_z}{\partial z}\right)_{z=L} = vC(l') \tag{33}$$

By setting $C_z'(0) = 0$ in Eq. (31) and combining it with the boundary conditions (32) and (33), one finds

$$C(l') = \frac{m}{aDq}\,\frac{1}{\sinh qL + (v/Dq)\cosh qL} \tag{34}$$

The inverse Laplace transformation of this equation to find $c(l', t_0)$ is difficult. It can be achieved by using certain approximations, the first of which is to replace both sinh qL and cosh qL by exp(qL)/2. For not too long times and not too short diffusion columns, this is a good approximation, as for qL = 1.5 the error introduced is about 5% and for qL = 3 it is only 0.25%. These qL values fall within the realm of experimental practice, and thus Eq. (34) simplifies to

$$C(l') = \frac{2m}{aDq}\,\frac{\exp(-qL)}{1 + v/Dq} \tag{35}$$

A second approximation is based on the fact that for high enough flow rates, $1 + v/Dq \approx v/Dq$ in the denominator, and Eq. (35) becomes

$$C(l') = \frac{2m}{\dot{V}}\exp(-qL) \tag{36}$$

Inverse Laplace transformation of this equation with respect to p_0 gives

$$c(l',t_0) = \frac{N \exp(-L^2/4Dt_0)}{t_0^{3/2}} \tag{37}$$

where

$$N = mL/\dot{V}(\pi D)^{1/2} \tag{38}$$

Now using Eq. (37) in place of the functions c_1, c_2, and c_3 of Eq. (19), we can obtain c as an explicit function of time, thus describing analytically the concentration–time curve as recorded by the detector system. For example, c_1 will be given by the expression

$$c_1 = \frac{N \exp[-L^2/4D(t_0 + t' + \tau)]}{(t_0 + t' + \tau)^{3/2}} \tag{39}$$

where

$$\tau = t - t_M \tag{40}$$

according to Eq. (20), since column l is empty and the gas holdup time in it, t_M, must be substituted for t_R.

Also, Eq. (24) becomes

$$h_{\tau'=t_M} = \frac{N \exp[-L^2/4D(t_0 - t_M)]}{(t_0 - t_M)^{3/2}} \tag{41}$$

whereas Eq. (27) reads

$$h = \frac{2N \exp(-L^2/4Dt_0)}{t_0^{3/2}} \tag{42}$$

C. Results and Discussion

Examples of sample peaks obtained experimentally with gas diffusion as the rate process are shown in Figs. 6 and 7. These should be compared with Figs. 2 and 3, respectively. The comparison shows that the elution curves predicted by Eqs. (13), (21), and (19) are confirmed experimentally, apart from the fact that the peaks actually found are not square, owing to nonideality (axial diffusion in columns l and l', etc.).

The experimental verification of Eq. (41) is made by plotting $\ln[h(t_0 - t_M)^{3/2}]$ versus $1/(t_0 - t_M)$, where h is the height of the peaks like those in Fig. 6, this height being measured from the con-

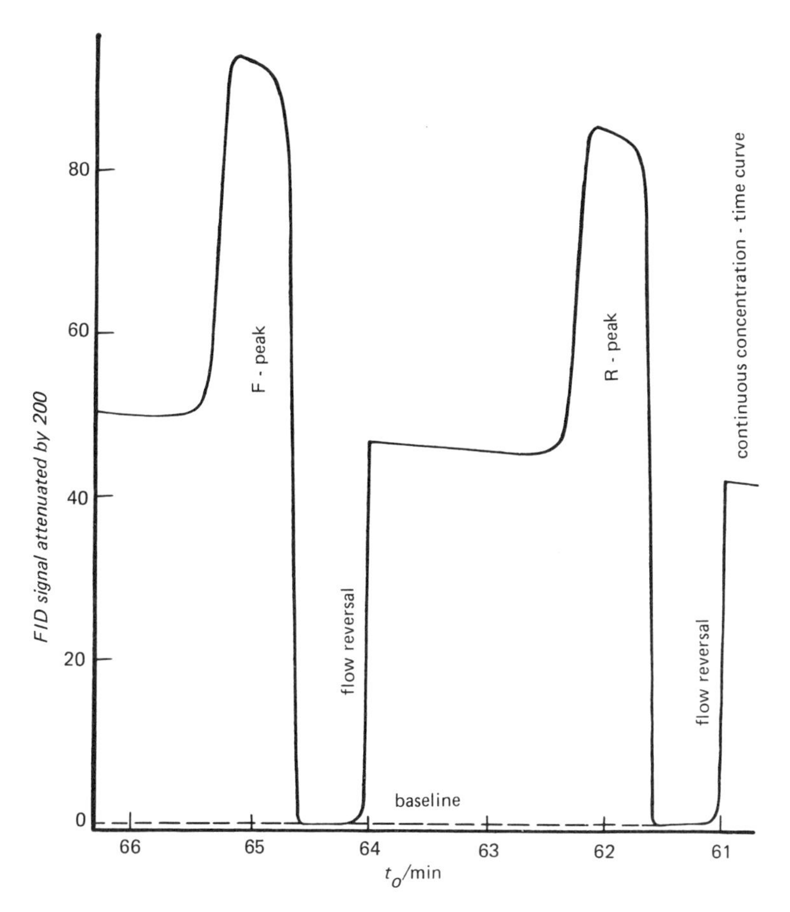

Figure 6 Sample peaks obtained with $t' > t_M + t'_M$ for the diffusion of C_2H_6 (0.5 cm^3) into N_2 ($\dot{V}$ = 0.27 $cm^3\ sec^{-1}$), at 293 K and 1.99 atm. A straight diffusion column L (61 cm X 4 mm inner diameter) and a chromatographic column l' + l = 40 + 40 cm X 4 mm inner diameter were used. (From Ref. 24.)

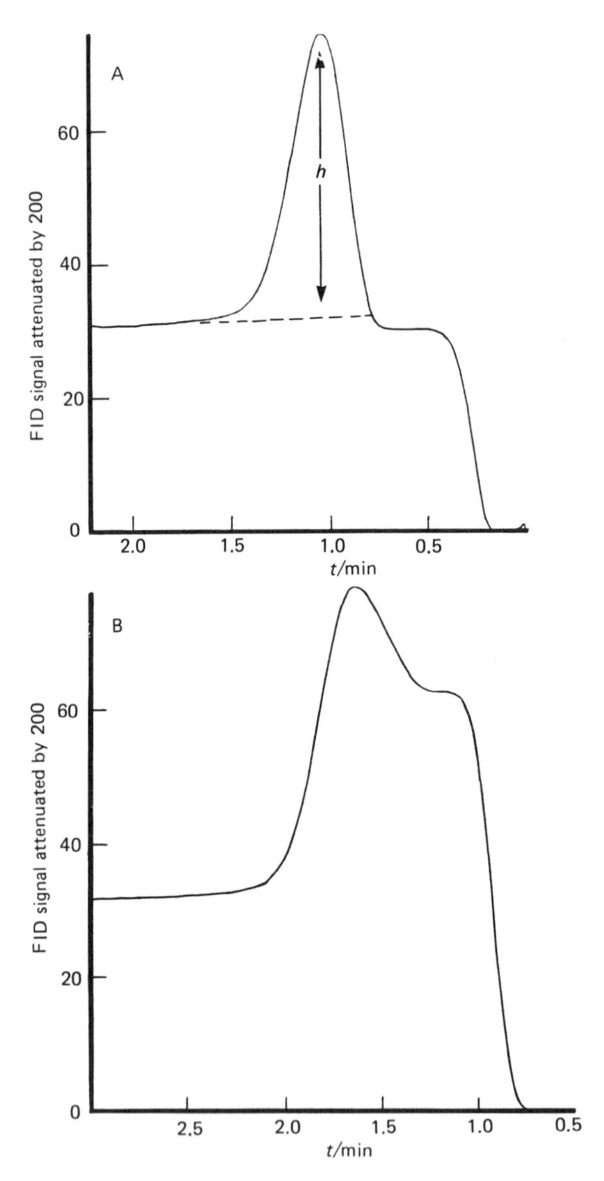

Figure 7 Sample peaks obtained with $t' < t_M + t'_M$ for the diffusion of C_2H_4 (0.5 cm^3) into He at 373.9 K and $\dot{V}$ = 0.34 cm^3 sec^{-1}. The diffusion column L (111.4 cm X 4 mm inner diameter) was coiled and placed inside the chromatographic oven. The two lengths l' + l of the empty chromatographic column (99.4 and 99.7 cm X 4 mm inner diameter) were also inside the oven: (A) $t' < t_M$ and $t' < t'_M$ and (B) $t' > t_M$ and $t' > t'_M$. (From Ref. 25.)

tinuous chromatographic signal to the maximum of the peak before returning to the base line. An example is given in Fig. 8.

Similarly, Eq. (42) is verified by plotting $\ln(ht_0^{3/2})$ against $1/t_0$, the height h being measured as shown in Fig. 7A. In Fig. 9 an example of such a plot is depicted.

In both cases above, the diffusion coefficient is computed from the slope of the straight lines and the known values of L. Table 1 lists some values of D found at ambient temperatures, while Table 2 shows the variation of D with temperature.

The results in Table 1 are reduced to 1 atm after multiplication by the pressure P of the experiment, given in the same table. The actual values of the diffusion coefficient determined experimentally are therefore given by D/P. From the values reported for C_2H_4–N_2 at three different pressures, it is seen that the variation of the results with small changes in pressure (and in $\dot{V}$) is small.

The precision of the present method, defined either as the relative standard deviation (%) or as the relative standard error (%) associated with each value, is as follows. From the values quoted for CH_4–He in Table 1 a relative standard deviation of 1% is calculated. From the

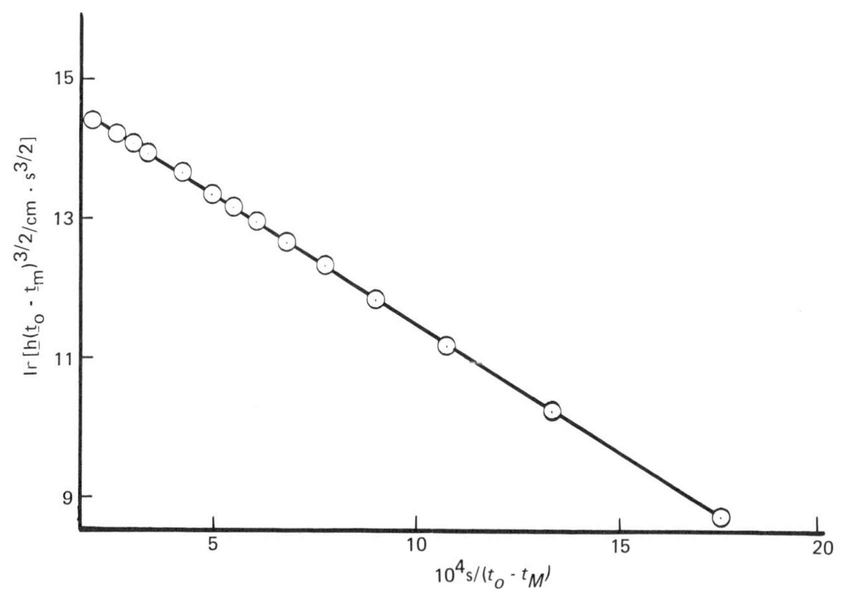

Figure 8 An example of plotting Eq. (41) for the diffusion of CH_4 (0.5 cm^3) into He ($\dot{V}$ = 0.28 cm^3 sec^{-1}) at 296 K and 2.03 atm. A straight diffusion column L (61 cm X 4 mm inner diameter), and a column l' + l = 40 + 40 cm X 4 mm inner diameter were used. (From Ref. 24.)

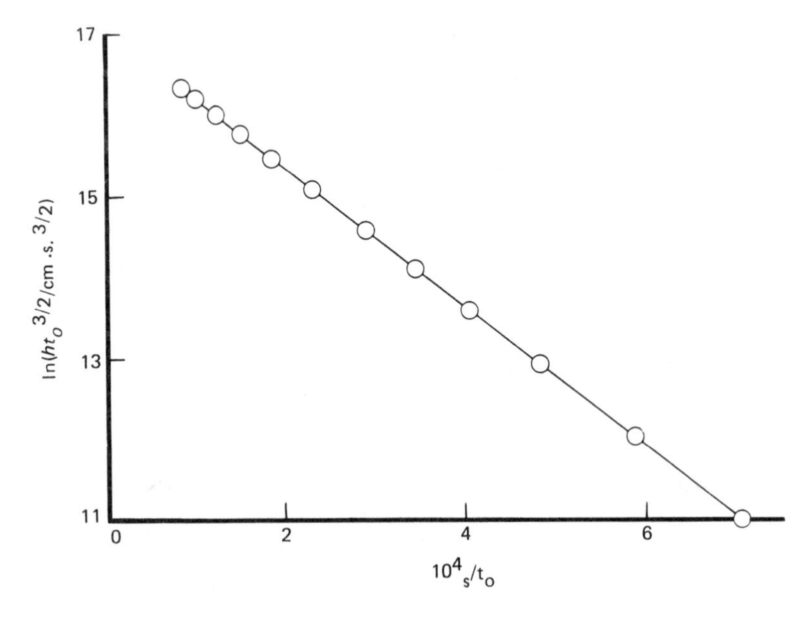

Figure 9 Plot of Eq. (42) for the diffusion of C_2H_6 into N_2 at 388.5 K and 1 atm, with a coiled diffusion column L (111.4 cm X 4 mm inner diameter) placed inside the chromatographic oven. The two lengths l' + l were 99.4 + 99.7 cm X 4 mm inner diameter. (From Ref. 25.)

standard errors in the same table a precision better than 1% is found for all but two values (1.5 and 1.2%). From the standard errors in Table 2 a precision better than 1% is calculated for all but one value (2.2%).

Comparison of the diffusion coefficients found with those calculated theoretically, either using the Hirschfelder–Bird–Spotz equation [30] or the Fuller–Schettler–Giddings equation [31], permits the calculation of the method's accuracy defined as

$$\text{Accuracy (\%)} = \frac{|D_{\text{found}} - D_{\text{calcd}}|}{D_{\text{found}}} 100 \qquad (43)$$

This is given in the last column of Tables 1 and 2. With the exception of two pairs containing methane as solute in Table 1 and three values for the pair C_2H_4–He in Table 2, the accuracy is better than 7.5% in all other 55 cases. The high deviation of the experimental from the calculated values in the methane-containing pairs is probably due to the approximations used in the calculated values. Also, it must be noted that the Fuller–Schettler–Giddings equation gives values closer

Table 1 Diffusion Coefficients of Five Solutes into Three Carrier Gases, at Ambient Temperatures and Reduced to 1 atm Pressure[a].

Solute gas	T (K)	$\dot{V}$ (cm^3 sec^{-1})	P (atm)	10^3D (cm^2 sec^{-1}) Experimental	Calculated[c]	Literature[d]	Accuracy[b] Percentage
Carrier gas N_2							
CH_4	296.0	0.260	1.96	272 ± 4	214	—	21.3
C_2H_6	293.0	0.267	1.99	142 ± 0.03	144	148	1.4(2.7)
n-C_4H_{10}	295.5	0.300	2.15	98 ± 0.2	98.6	96	0.3 (2.7)
C_2H_4	296.0	0.120	1.49	168 ± 2	156	163	7.1 (4.3)
	292.0	0.268	2.00	156 ± 0.4			0
	292.0	0.538	2.71	161 ± 0.4			3.1
C_3H_6	298.0	0.260	1.96	124 ± 0.4	120	—	3.2
Carrier gas H_2							
CH_4	293.0	0.287	1.70	699 ± 3	705	730	0.9 (3.4)
C_2H_6	297.0	0.267	1.56	548 ± 5	556	540	1.5 (3)
n-C_4H_{10}	296.0	0.273	1.60	386 ± 3	373	400	3.4 (6.8)
C_2H_4	293.0	0.300	1.75	525 ± 5	559	602	6.5 (7.1)
C_3H_6	296.0	0.273	1.60	485 ± 3	486	—	0.2

Table 1 (continued)

Carrier gas He							
CH_4	295.7	0.250	1.78	527 ± 3	669	—	26.9
	295.0	0.283	2.03	520 ± 1			28.7
	296.0	0.283	2.03	514 ± 0.2			30.2
	296.7	0.283	2.03	522 ± 3			28.2
C_2H_6	295.6	0.300	2.15	518 ± 3	507	—	2.1
n-C_4H_{10}	290.0	0.283	2.03	333 ± 3	330	364	0.9 (9.3)
C_2H_4	296.0	0.283	2.03	558 ± 4	544	—	2.5
C_3H_6	291.0	0.283	2.03	412 ± 4	440	—	6.8

[a]The coefficients were determined using Eq. (41). The errors in this and following tables are "standard errors" calculated by regression analysis.
[b]Defined by Eq. (43). Numbers in parentheses are the accuracies of the respective literature values.
[c]Calculated with Hirschfelder–Bird–Spotz equation [30].
[d]Obtained from Ref. 32.
Source: Ref. 24.

Table 2 Diffusion Coefficients of Three Solutes in Two Carrier Gases, at Various Temperatures and 1 atm Pressure[a]

		$10^3 D$ (cm^2 sec^{-1})		
Solute gas	T (K)	Experimental	Calculated[c]	Accuracy[b] (%)
Carrier gas He				
C_2H_6	296.7	491 ± 2	456	7.1
	322.6	556 ± 2	528	5.0
	344.0	618 ± 3	590	4.5
	364.4	684 ± 3	653	4.5
	385.3	745 ± 6	720	3.4
	407.3	807 ± 4	793	1.7
	426.3	878 ± 8	859	2.2
	447.3	941 ± 5	935	0.6
C_2H_4	296.8	525 ± 4	478	9.0
	322.9	599 ± 1	554	7.5
	336.0	649 ± 1	594	8.5
	348.1	674 ± 3	632	6.2
	361.3	726 ± 2	674	7.2
	373.9	780 ± 6	716	8.2
	399.9	860 ± 19	806	6.3
	426.9	932 ± 3	903	3.1
	476.5	1112 ± 10	1096	1.4
C_3H_6	345.0	528 ± 0.7	500	5.3
	365.5	584 ± 1	553	5.3
	388.0	642 ± 1	614	4.4
	407.7	690 ± 1	670	2.9
	428.0	750 ± 2	730	2.7
	449.4	819 ± 3	795	2.9

Table 2 (Continued)

Solute gas	T (K)	$10^3 D$ (cm^2 sec^{-1}) Experimental	Calculated[c]	Accuracy[b] (%)
Carrier gas N_2				
C_2H_6	322.8	172 ± 0.2	170	1.2
	345.7	193 ± 0.2	191	1.0
	365.0	214 ± 0.7	210	1.9
	388.5	242 ± 0.3	234	3.3
	407.6	256 ± 0.2	255	0.4
	427.5	282 ± 0.4	277	1.8
	449.3	303 ± 0.5	302	0.3
C_2H_4	322.8	189 ± 0.08	179	5.3
	344.7	213 ± 0.1	200	6.1
	364.2	234 ± 0.3	221	5.6
	387.6	260 ± 0.3	246	5.4
	407.5	286 ± 0.4	269	5.9
	428.9	306 ± 0.3	294	3.9
	449.8	335 ± 0.9	319	4.8
C_3H_6	322.8	143 ± 0.2	138	3.5
	344.6	164 ± 0.1	155	5.5
	387.4	202 ± 0.2	190	5.9
	406.4	220 ± 0.4	206	6.4
	428.9	243 ± 0.3	227	6.6
	459.0	266 ± 0.2	255	4.1

[a]The coefficients were determined using Eq. (42).
[b]Defined by Eq. (43).
[c]Calculated with the Fuller–Schettler–Giddings equation [31].
Source: Ref. 25.

to those found experimentally than other theoretical equations. Finally, the accuracies of the present method can be compared with those of the values determined by broadening techniques [32]. These are given in parentheses in Table 1 and are defined again by Eq. (43), with D_{lit} in place of D_{found}. This comparison leads to the conclusion that, with the exception of C_2H_4–N_2, the values of diffusion coefficients determined by the reversed-flow method are closer to the theoretically calculated values than are the experimental values found by broadening techniques, under similar conditions of temperature and pressure.

By plotting ln D versus ln T, as shown in Fig. 10, we calculated the exponent n in the relationship

$$D = AT^n \tag{44}$$

which all theoretical and semiempirical equations lead to for the dependence of D on T. For carrier gases helium and nitrogen, mean values of 1.61 ± 0.01 and 1.74 ± 0.02, respectively, were found [25]. The mean values of n found from similar plots of calculated diffusion coefficients [30] are 1.679 ± 0.001 and 1.808 ± 0.004 for helium and nitrogen, respectively. The mean values for n, from diffusion coefficients determined by the present method, lie between the 1.5, suggested by

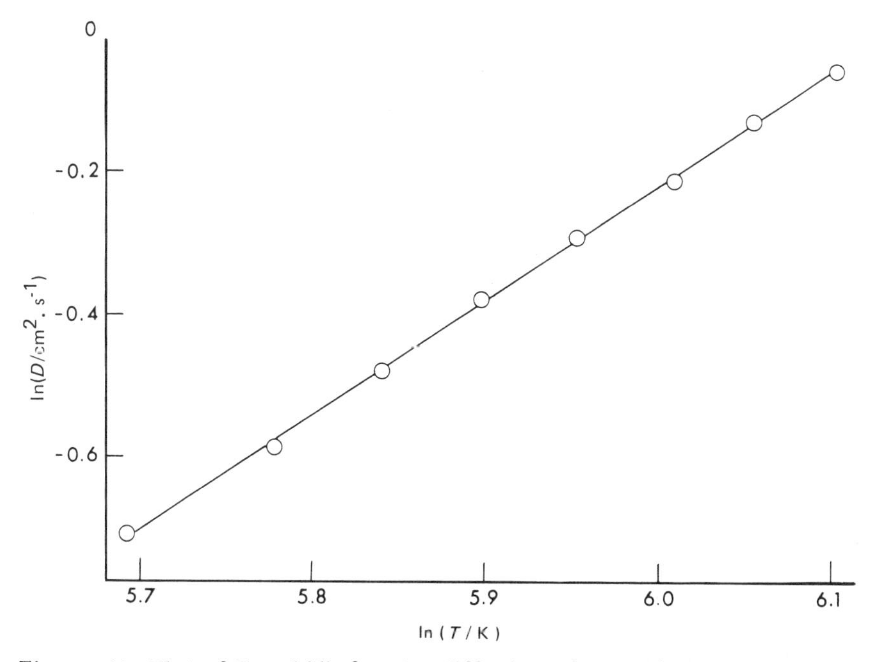

Figure 10 Plot of Eq. (44) for the diffusion of C_2H_6 into He. (From Ref. 25.)

Stefan–Maxwell, Gilliland, and Arnold equations [33], and 1.81, predicted by the Chen–Othmer equation [33]. A value of 1.75 is also predicted by the Huang [32] and Fuller–Schettler–Giddings equations [31].

D. Diffusion Coefficients in Multicomponent Gas Mixtures

The reversed-flow method for measuring gas diffusion coefficients can be extended to simultaneous determinations of diffusion coefficients in multicomponent gas mixtures [26], an experimental problem which has practical as well as theoretical importance. This extension of the method is done by filling the column section l (see Figs. 4 and 5) with a chromatographic material, for example, silica gel, which can effect the separation of some or all components of the gas mixture. When the chromatographic sampling is then performed, two or more sample peaks appear in the chromatogram (Fig. 11). These correspond to two or more different components of the mixture, provided that the components have sufficiently different t_R values in the filled column l. For each of these components Eq. (42) holds true, and therefore the maximum height h of each peak, measured from the ending base line, can be plotted in the form $\ln(ht_0^{3/2})$ versus $1/t_0$ to yield, from the slope $-L^2/4D$, its *effective* binary diffusion coefficient in the mixture. Table 3 lists some results found [26] together with theoretically calculated values [30] for the diffusion of each hydrocarbon in pure carrier gas. A comparison between the experimental and the calculated values shows a difference ranging from 0.3 to 7.9%, with one exception ($n\text{-}C_4H_{10}$ in N_2, 16%). These differences are of about the same magnitude as the accuracies for the diffusion of the same hydrocarbons in pure carrier gases (see Table 1).

These findings are in accord with a limiting case of the Stefan–Maxwell equations [34], which predict that, for small mole fractions of components 1 and 2 in nearly pure carrier gas, the effective diffusion coefficient in the ternary mixture is equal to the diffusion coefficient of each component in pure carrier gas.

The presence of chromatographic material in column l does not seem to influence the results, as shown by a binary mixture (H_2 + C_2H_6) included in Table 3. The D_1 found not only coincides with the theoretically calculated value, but also is not significantly different from the value 0.548 $cm^2\ sec^{-1}$ of Table 1, found with the column l empty.

IV. DETERMINATION OF ADSORPTION EQUILIBRIUM CONSTANTS

If the diffusion column L of Fig. 4 is partly filled with a solid adsorbent, on which the solute gas A is reversibly adsorbed (see Fig. 12), the diffusion current into the carrier gas stream is delayed, and the diffusion coefficient comes out smaller than that found with an empty

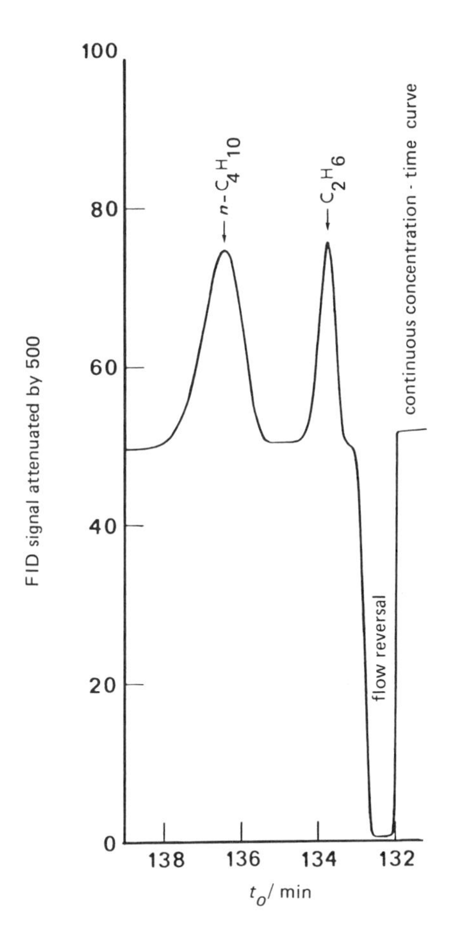

Figure 11 A reversed-flow chromatogram for the simultaneous determination of diffusion coefficients in a ternary mixture of two solutes (C_2H_6 + n-C_4H_{10}) and a carrier gas (H_2). The temperature of the column L (66.5 cm X 4 mm inner diameter) was 292 K and the pressure in it was 1.7 atm. Section l (40 cm X 4 mm inner diameter) was filled with activated silica gel and held at 454 K. The duration of the backward flow (t') was 24 sec. (From Ref. 26.)

Table 3 Effective Diffusion Coefficients Reduced to 1 atm in Some Ternary Mixtures Consisting of a Carrier Gas and Two Hydrocarbons[a]

Carrier gas	Solute gases		T(K)	$10^3 D_1$ ($cm^2 sec^{-1}$)		$10^3 D_2$ ($cm^2 sec^{-1}$)	
	1	2		Experimental	Calculated	Experimental	Calculated
H_2	C_2H_4	C_2H_6	298	554 ± 15	593	600 ± 7	557
	C_2H_4	n-C_4H_{10}	296	586 ± 37	584	381 ± 13	379
	C_2H_6	n-C_4H_{10}	292	534 ± 9	538	379 ± 8	372
	C_2H_6	n-C_4H_{10}[b]	292	503 ± 17	538	360 ± 14	372
	C_2H_6	—	297	556 ± 13	556	—	—
He	C_2H_6	n-C_4H_{10}	294	494 ± 7	506	354 ± 7	338
N_2	C_2H_4	n-C_4H_{10}	296	166 ± 4	156	118 ± 3	99

[a]The experimental values are effective diffusion coefficients in the ternary mixtures, while the calculated ones refer to diffusion in pure carrier gas.
[b]The mixture injected was 1.5 cm^3, whereas in all other cases it was 0.5 cm^3 (1:1 w/w).

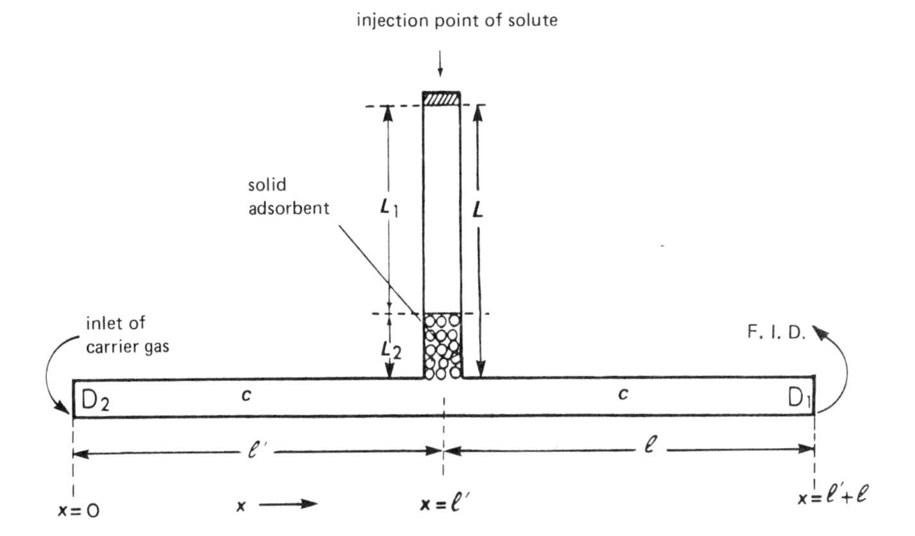

Figure 12 The two sections (L_1 and L_2) of the diffusion column L connected to an empty column l' + l for measuring partition ratios.

L. This phenomenon can be used to determine the adsorption equilibrium constant of the solute between the solid and the gas phase.

The experimental setup and the procedure are the same as in the determination of gas diffusion coefficients (see Sec. III.A), except for the following. The lower part L_2 of the diffusion column, containing the solid adsorbent, is placed inside the chromatographic oven, while the empty upper part L_1 is kept at ambient temperature. Before the experiments, the adsorbent is activated by heating it at a sufficiently high temperature with carrier gas flowing in the column l' + l.

A. Complementary Theory

The concentration of A at x = l' as a function of time, $c(l', t_0)$, is now found, assuming instantaneous equilibration of the solute between the gas and the solid phase and a linear adsorption isotherm. Following a procedure similar to that outlined in Sec. III.B, although a little more complicated [28], we find, using the same approximations as before,

$$c(l', t_0) = \frac{N' \exp(-L_{eff}^2 / 4Dt_0)}{t_0^{3/2}} \tag{45}$$

where

$$N' = 2mL_{eff}/\dot{V}(\pi D)^{1/2}\left[1 + (1 + k_A)^{1/2}\left(\frac{T_1}{T_2}\right)^{n/2}\right] \tag{46}$$

and L_{eff} is an effective length for the diffusion column defined as

$$L_{eff} = L_1 + L_2(1 + k_A)^{1/2}(T_1/T_2)^{n/2} \tag{47}$$

In these equations D is the diffusion coefficient of A into B measured with an empty column L at temperature T_1, as described in Sec. III; k_A is the partition ratio of A between the solid and the gas phase B; T_1 and T_2 are the temperatures of the two column sections L_1 and L_2, respectively; and n is the exponent in Eq. (44). If this is unknown, the value 3/2 of the simple kinetic theory can be used.

Clearly, Eq. (45) has exactly the same form, with Eq. (37), of the simple diffusion with an empty column L, and reduces to that when $k_A = 0$, that is, no adsorption of A on the solid, and $T_1 = T_2$.

Also, an equation like Eq. (41) can be written

$$h_{\tau'=t_M} = \frac{N' \exp[-L_{eff}^2/4D(t_0 - t_M)]}{(t_0 - t_M)^{3/2}} \tag{48}$$

and plotted linearly as $\ln[h(t_0 - t_M)^{3/2}]$ versus $1/(t_0 - t_M)$. As before, the slope b will be equal to $-L_{eff}^2/4D$, and, with the knowledge of D, L_1, L_2, T_1, and T_2, the partition ratio is calculated:

$$k_A = \frac{[2(-Db)^{1/2} - L_1]^2}{L_2^2(T_1/T_2)^n} - 1 \tag{49}$$

Finally, from the value of k_A the adsorption equilibrium constant K_A can be found using the relation $K_A = k_A\varepsilon/(1 - \varepsilon)$, ε being the void fraction in the filled column section L_2. This is determined by a standard method [35].

B. Results and Discussion

Partition ratios and adsorption equilibrium constants at various temperatures, determined by the method outlined above, have been reported elsewhere [28]. They refer to the adsorption of methane, ethane, n-butane, ethene, propene, and but-1-ene on aluminium oxide from helium.

The k_A values for methane are worth mentioning, because they are small, but not zero, as usually assumed by chromatographers, who use this gas to determine gas holdup time in gas solid chromatography

(GSC) with a flame ionization detector. At 323 K, for example, the k_A value of CH_4 is 0.792, which means that a gas holdup time 79% greater than the actual value would be determined.

From the variation of the adsorption equilibrium constants with temperature, the differential enthalpies (heats) and entropies of adsorption can be determined using the relation [28]

$$\ln\left(\frac{K_A}{T}\right) = \ln R + \frac{\Delta S^{\theta}}{R} - \frac{\Delta S^{\theta}}{R}\frac{1}{T} \qquad (50)$$

A plot of $\ln(K_A/T)$ versus $1/T$ will give ΔH^{θ} from the slope and ΔS^{θ} from the intercept, it being understood that the range of T is narrow enough for ΔH^{θ} and ΔS^{θ} to be regarded as independent of temperature. Some results are presented in Table 4. The values of ΔS^{θ} given should be regarded as relative. They are referred to a hypothetical adsorbed standard state of 1 mol/cm^3 of solid, and a partial pressure of 1 atm as the gas standard state.

The merits of the present method for determining adsorption equilibrium constants are reflected in the differential heats of adsorption found, which correspond to isosteric of zero coverage more closely than values determined from conventional gas-chromatographic data by plotting $\ln V_N$ or $\ln(V_N/T)$ versus $1/T$. This is because the total amount of solute injected into the system is of the order of 10^{-5} mol, and, owing to its diffusional distribution along the column $L_1 + L_2$, only a small fraction of this is located over the adsorbent bed, and this is almost uniformly distributed all the time. By constrast, in usual chromatographic elution the whole amount of solute moves along the column as a relatively narrow band, with much higher local concentrations at a certain region of the adsorbent and zero concentration at others.

Table 4 Enthalpies and Entropies of Adsorption of Various Hydrocarbons on Al_2O_3 from Helium

Substance	$-\Delta H^{\theta}$ (kJ mol^{-1})	$-\Delta S^{\theta}$ (J K^{-1} mol^{-1})
Methane	13 ± 2	143 ± 6
Ethane	14.1 ± 0.4	129 ± 1
n-Butane	24 ± 3	144 ± 7
Ethene	23 ± 1	149 ± 3
Propene	38 ± 4	189 ± 12
But-1-ene	43 ± 2	190 ± 6

Another aspect of the data in Table 4 is the absence of a compensation effect, that is, a linear dependence of ΔS^{θ} on ΔH^{θ}, which is almost always observed with ΔS and ΔH values determined chromatographically. An explanation of the compensation effect [36] is that it is due to the chromatographic process itself combined with the heterogeneity of the surface. The absence of a compensation effect in the present case can thus be due to the fact that there is no chromatography inside the diffusion column. The enthalpies and entropies of adsorption in Table 4 have been determined by gas-chromatographic instrumentation, but "without chromatography."

V. KINETICS OF THE DRYING STEP OF CATALYSTS

Another rate process, which can be studied using the reversed-flow sampling technique, is the removal of organic solvents from impregnated catalytic supports, like γ-Al_2O_3 or other solids.

The same experimental setup as in the determination of gas diffusion coefficients (see Sec. III.A) is used, with a slight modification. This consists in filling all the diffusion column L (see Figs. 4 and 5) with the impregnated solid and placing it inside the chromatographic oven. This filled column L need not be so long as in diffusion studies and can easily be accomodated inside an oven of usual size.

The concentration of the solvent in the gaseous phase at $x = l'$, $x(l',t_0)$ has been found for a desorption following a simple first-order law and steady-state conditions [29]:

$$c(l',t_0) = \frac{\rho k q_0 L}{v} \exp(-kt_0) \tag{51}$$

where ρ is the volume ratio of solid to gas phases, k is the rate constant for the removal of solvent from the solid surface, and q_0 is the initial concentration of the solvent on the solid.

The relation giving the height of the sample peaks for $t' > t_M + t'_M$ is obtained from Eq. (26) by substituting the right-hand side of Eq. (51) for $c_2(l',t_0)$:

$$h_{\tau = t'_M} = \frac{\rho k q_0 L}{v} \exp[-k(t_0 + t' - t'_M)] \tag{52}$$

Thus, a plot of ln h versus the total time $t_{tot} = t_0 + t'$ should be linear, if the phenomenon follows a first-order law and the other assumptions made hold true. The value of k is found from the slope. An example is given in Fig. 13.

Rate constants for the removal of various solvents from impregnted γ-Al_2O_3 at various temperatures have been determined by this method [29] and found to obey the Arrhenius equation. From standard

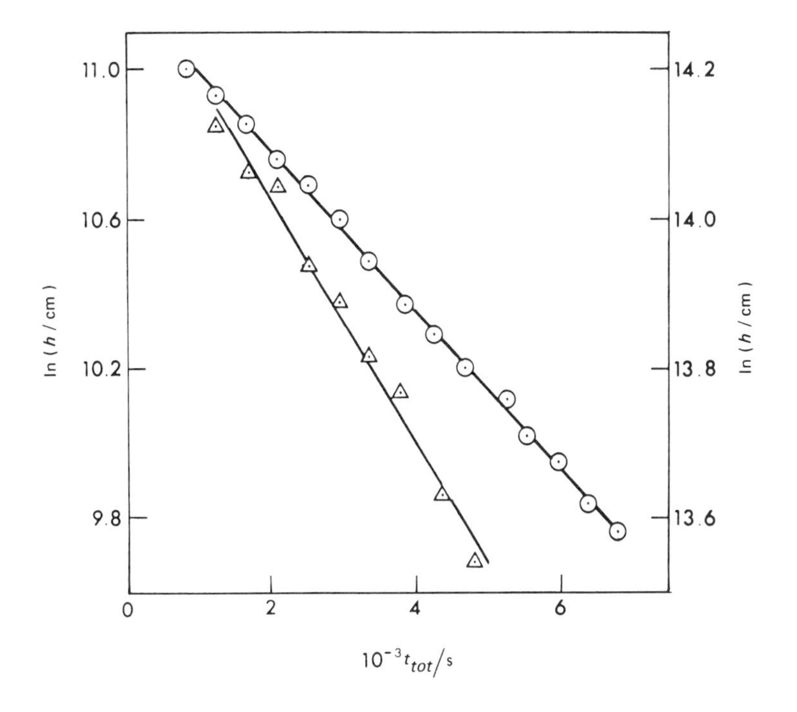

Figure 13 Examples of plotting Eq. (52): removal of n-C_6H_{14} at 333.0 K (○, left-hand ordinate) and of n-C_5H_{12} at 317.5 K (Δ, right-hand ordinate) from γ-Al_2O_3 impregnated with 0.45 $cm^3\ g^{-1}$ of each solvent. (From Ref. 29.)

plots of this equation, energies and entropies of activation for the phenomenon of drying were calculated; they are given in Table 5. It is seen from Table 5 that the removal of the solvents studied from a porous solid is an activated process, in spite of the fact that such a large amount of them is deposited on the solid. The values of activation energy found are small enough to place the process in the domain of physical adsorption. The increase in the Arrhenius parameters with the amount of the solvent used, as indicated by the values for n-C_6H_{14}, is probably due to the partial sintering (taking place during drying) brought about by the excess of solvent.

VI. KINETICS OF SURFACE-CATALYZED REACTIONS

As a last application of the reversed-flow technique, we examine the rate of a slow chemical reaction taking place on the surface of a solid catalyst.

Table 5 Energies E_a and Entropies $\Delta S^{\neq}$ of Activation for the Removal of Various Solvents from γ-Al_2O_3 (Houdry, 60–70 mesh) Impregnated with 0.45 cm^3 of Solvent per Gram

Solvent	E_a ($kJ\ mol^{-1}$)	$-\Delta S^{\neq}$($J\ K^{-1}\ mol^{-1}$)
n-C_5H_{12}	18 ± 1	262 ± 3
n-C_6H_{14}	21.4 ± 0.6	252 ± 2
n-C_6H_{14}[a]	31 ± 3	230 ± 8
n-C_7H_{16}	25 ± 2	245 ± 7
CH_3COCH_3	15.4 ± 0.8	273 ± 2
C_2H_5OH	6.7 ± 0.7	298 ± 2

[a]Nondry impregnation (0.9 cm^3 of solvent per gram).
Source: Ref. 29

A. Experimental

If the reactant A is strongly adsorbed on the active surface, and the catayst can also function as chromatographic material to separate the products from one another and from the reactant, only the chromatographic column l' + l is employed, filled with the catalyst. The reactant is injected as a pulse at a middle position of the column, as shown in Fig. 14, and the experimental setup is depicted in Fig. 15.

After conditioning the catalyst by heating it in situ at an appropriate temperature, under carrier gas flowing in either direction (F or R), some preliminary injections of the reactant in both directions are made to establish constant catalytic activity. Kinetic experiments are conducted after the last reactant introduced has been exhausted to a

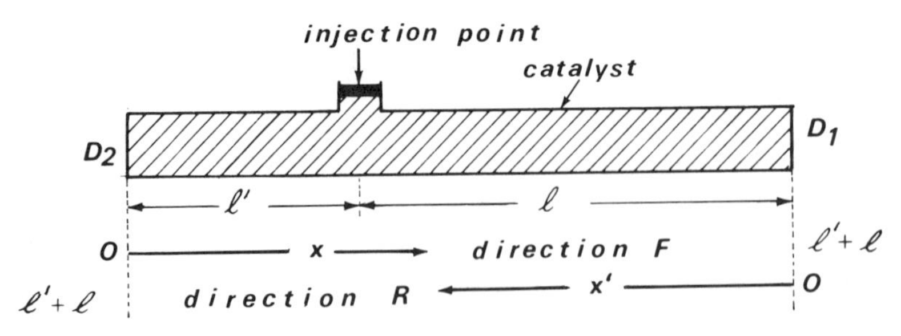

Figure 14 Representation of a catalytic–chromatographic column l' + l to study heterogeneous catalysis of a reactant strongly adsorbed on the active surface. (From Ref. 19.)

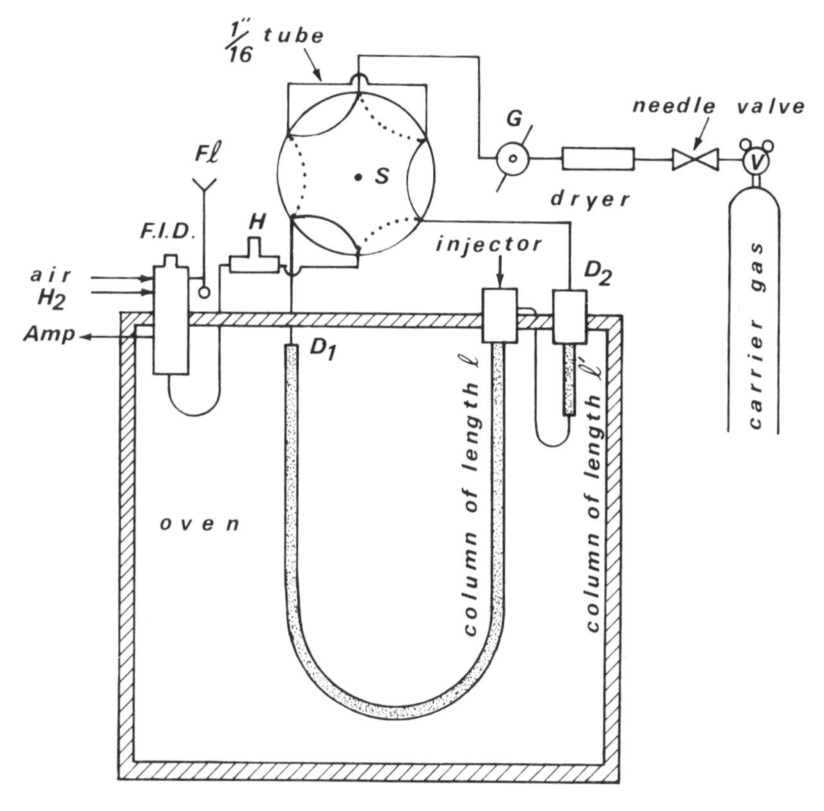

Figure 15 Schematic arrangement of the experimental setup for studying the kinetics of a heterogeneously catalyzed reaction of a substance strongly adsorbed on the catalytic surface: V, two-stage reducing valve and pressure regulator; G, gas flow controller for minimizing variations in the gas flow rate; S, six-port gas-sampling valve; H, restrictor; Fl, bubble flowmeter; and Amp, signal to amplifier. (From Ref. 19.)

negligible amount. A few cubic millimeters (0.5 to 10) of liquid reactant, using a microsyringe, or a few cubic centimeters of gas reactant at atmospheric pressure, using a gas-tight syringe, are introduced onto the column length l with the carrier gas flowing in direction F. The product(s) is recorded by the detector as an asymmetrical elution curve, and the chemical reaction can be sampled in the usual way by reversing the direction of the carrier gas flow. Either case analyzed in Sec. II.B.3, namely, $t' > t_R + t'_R$ or $t' < t_R + t'_R$, can be used, giving one sample peak after *each* flow reversal in the first case, or a higher (almost double) sample peak after *two* successive reversals in the second case.

The foregoing arrangement depicted in Figs. 14 and 15 cannot be applied when the reactant is not retained on the catalytic bed for a sufficiently long time, but is eluted together with the product(s). Also, it cannot be applied with *two* gaseous reactants, unless one of them is used in a great excess of the other as a carrier gas, for example hydrogen in hydrogenation reactions. In situations like these a continuous feed of reactant(s) into the catalytic bed through the injection point (see Fig. 14) is required, and possibly a chromatographic material different from the catalyst itself to separate the various products and reactants from each other.

Instead of using commercially prepared mixtures from gas cylinders or employing mixing chambers, saturators, and similar devices, we can combine the experimental arrangement used to study diffusion coefficients (see Figs. 4 and 5) with that outlined in Figs. 14 and 15. The reactant or a mixture of reactants is then introduced as a pulse at the tope of the column L and allowed to diffuse slowly onto the catalyst retained within a short section near the junction of columns L and l' + l, and heated to a temperature T_1. The catalytically inert chromatographic material, used for separation, fills a part or the whole of column l' + l and is heated to a temperature T_2. An example of this kind, is shown in Fig. 16. The flow reversal at a certain time now samples all substances present at x = l', and the material filling l separates these substances, revealing their relative concentrations. Thus the diffusion feed outlined above ensures the presence of reactants over the catalyst for a long time period, and also that the space within which reaction takes place is confined near the exit of L, that is, near x = l', and is not spread with time along the column l' + l.

If the pressure drop in the filled column l of Fig. 16 is small, the pressure change in tube L at each flow reversal is samll. Moreover, this pressure change lasts only for a short time equal to the duration of the backward flow t'. In cases of large pressure drops along l, the column section l' should also be filled with the same material and be of the same length as l to minimize pressure changes at the exit of L.

Another way of accomplishing this is depicted in Fig. 17. Some noticeable features of this arrangement are as follows: (1) The flow reversals are "confined" in the sampling column l' + l placed inside the catalytic oven, while in the separation column found in the chromatographic oven the carrier gas flows always in the same direction, from injector toward the detector. As a result of this, the pressure changes at the exit of L are negligible, since the catalyst bed is of very short length and the pressure drop along it very small. (2) Reference substances for identification purposes can be chromatographed in the separation column by introducing them through the injector of the chromatographic oven; (3) the catalyst can be heated at a quite different temperature than that of the separation column; (4) a restrictor like H

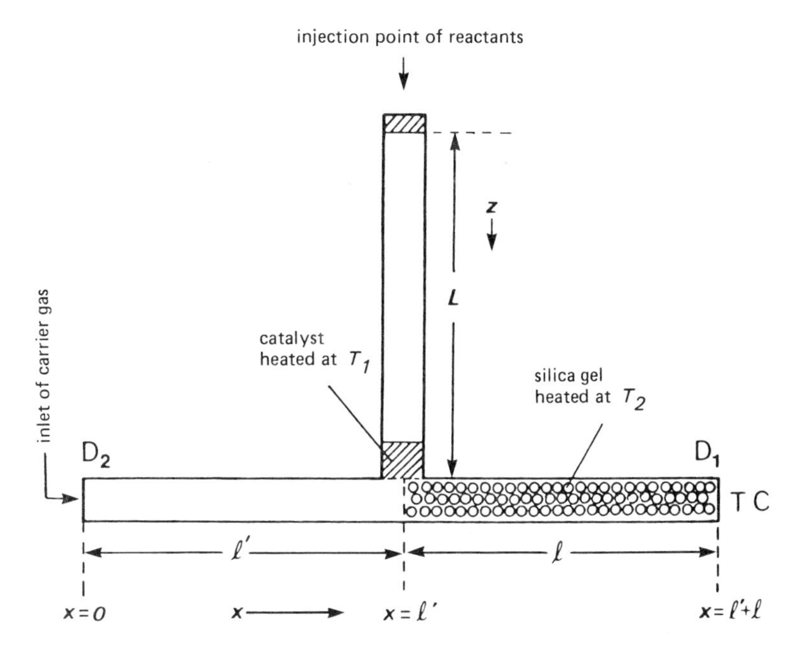

Figure 16 Schematic outline of the cell (with a thermal conductivity detector TC) used to study the oxidation of carbon monoxide with oxygen gas over Co_3O_4 containing catalysts. The feeding of the catalytic bed is done by diffusion of the reactants from a remote point of injection. (Reproduced from Ref. 23 by permission of the National Research Council of Canada.)

in Fig. 15 is no longer necessary; and (5) the diffusion feed described in the previous paragraph can be based not only on a pulse of gaseous reactant(s) at the top of the diffusion column L, but also on a deposit of volatile liquid reactant(s) at the upper part of L, as shown in Fig. 17.

B. Complementary Theory

As before, in order to apply the chromatographic sampling equation, Eq. (19), and other derived relations, like Eqs. (21)–(27), the concentration(s) at $x = l'$ as a function of time, $c(l',t_0)$, is required. In the present problem these concentrations are generally determined by (1) the method used to feed the reactant(s) to the catalytic bed, (2) the rate of the chemical reaction on the catalyst $r(t_0)$, (3) the mass of the catalyst M, and (4) the volume flow rate $\dot{V}$ of the carrier gas in column $l' + l$. Let us examine some representative cases.

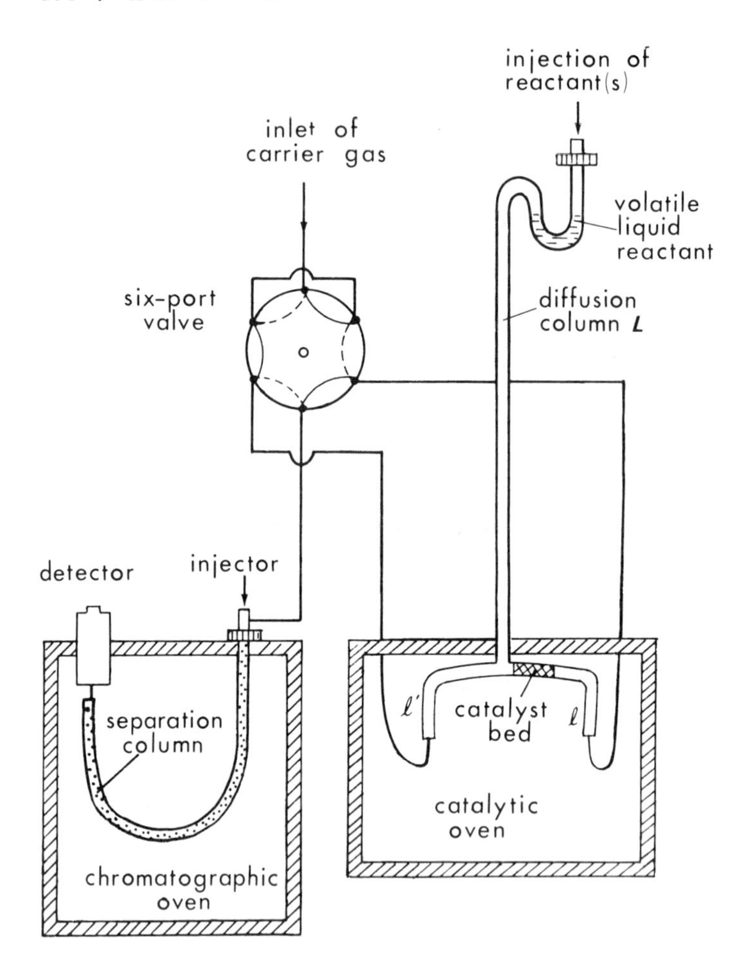

Figure 17 Outline of an experimental setup for catalytic studies using a diffusion feed and a separation column other than l' + l.

If we have the experimental arrangement described by Figs. 14 and 15, where the reactant is injected as a pulse at a middle position of column l' + l and retained there because of its high adsorption equilibrium constant on the catalyst, the rate of the reaction is written as $r(t_0)\,\delta(x - l')$, the delta function representing the x distribution of the reaction rate. Then $r(t_0)$ is only a function of time expressed in mol cm^{-2} sec^{-1}, that is, per unit cross section a of the void space in the column. Identifying $r(t_0)\,\delta(x - l')$ with the last term in Eq. (1), $vc(l',t_0)\,\delta(x - l')$, we obtain

$$c(l', t_0) = \frac{r(t_0)}{v} = \frac{r(t_0)a}{\dot{V}} \qquad (53)$$

This is a product concentration due to the chemical reaction. The gaseous concentration of the reactant is practically zero, since it has a high adsorption equilibrium constant.

If the reactants are not retained for a long time by the catalyst bed and a diffusion feed is used, as described in Sec. VI.A (see Figs. 16 and 17), it is more appropriate to use a mean reaction rate $r(t_0)$ for the whole catalyst bed expressed in mol sec^{-1} g^{-1} of catalyst. Then the concentration of a product will be

$$c(l', t_0) = \frac{r(t_0)M}{\dot{V}} \qquad (54)$$

The concentation of a reactant will depend, not only on the reaction rate, but also on the rate of its diffusion into the carrier gas stream. It is calculated by using the diffusion equation, Eq. (28), the appropriate initial conditions, and the necessary boundary conditions at both ends of the diffusion column L, as exemplified in Sec. III.B. However, the boundary condition expressed by Eq. (33) must now take into account the chemical reaction at the exit of L and be written as

$$-D\left(\frac{\partial c_z}{\partial z}\right)_{z=L} = vc(l', t_0) + \frac{r(t_0)M}{a} \qquad (55)$$

If the initial condition is that of Sec. III.B, that is, the reactant is introduced as a pulse at the closed end of L, the solution of Eq. (28) with the boundary condition (55), carried out as before, is

$$c(l', t_0) = \frac{N \exp(-L^2/4Dt_0)}{t_0^{3/2}} - \frac{r(t_0)M}{\dot{V}} \qquad (56)$$

where N is again given by Eq. (38).

C. Results and Discussion

1. Dehydration of Alcohols

An example of sample peaks due to a chemical reaction, obtained with the experimental setup of Figs. 14 and 15, is shown in Fig. 18. This should be compared with the theoretical elution curves of Fig. 2.

If an assumption concerning the mechanistic model of the reaction can be made, the reaction rate $r(t_0)$ can be expressed as an analytic function of time and this can be substituted directly in Eq. (53). The

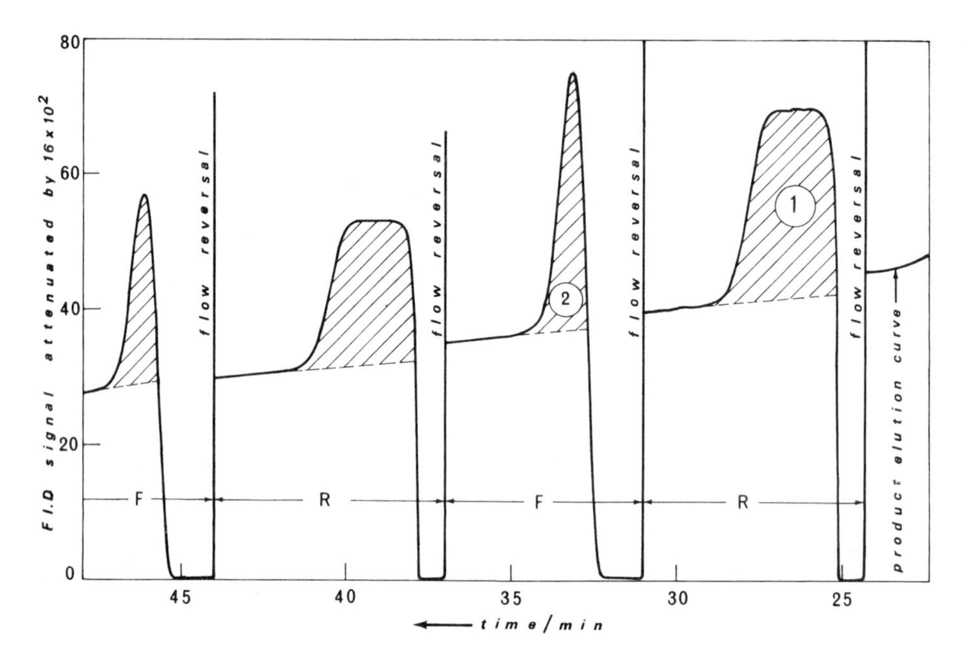

Figure 18 Sample peaks obtained with a 3.8 mm inner diameter glass column of lengths l' = 7 and l = 100 cm, filled with 80–100 mesh 13X molecular sieve activated at 665 K for 21 hr. The carrier gas was nitrogen ($\dot{V}$ = 0.446 $cm^3\ sec^{-1}$), the column temperature 477.3 K, and the injected reactant 0.5 mm^3 propan-2-ol, giving diisopropyl ether as the main product. (From Ref. 19.)

simplest possible case of mechanistic models is

$$A + S \underset{K_A}{\overset{fast}{\rightleftharpoons}} A - S \xrightarrow{k} D - S \underset{K_D}{\overset{fast}{\rightleftharpoons}} D + S \tag{57}$$

that is, the reactant A is rapidly adsorbed on the active centers S, giving the adsorbed reactant A – S. This is then decomposed by a first-order step with a rate constant k to give the adsorbed product D – S, which rapidly equilibrates with the gaseous product D. Here K_A and K_D are adsorption equilibrium constants of A and D, respectively, and if $K_A \gg K_D$, the rate of formation of D in mol $cm^{-2}\ sec^{-1}$ is given by

$$r(t_0) = \frac{kmg}{a} \exp(-kt_0) \tag{58}$$

where m is the mass of A injected (mol) and g is its fraction on *reactive* sites of the surface. The right-hand side of Eq. (58) can now be substituted for $r(t_0)$ in Eq. (53), and the result

$$c(l',t_0) = \frac{kmg}{\dot{V}} \exp(-kt_0) \tag{59}$$

is used in Eq. (21) to express c_1 and c_2 analytically:

$$c = \frac{kmg}{\dot{V}} \{\exp[-k(t_{tot} + \tau)]u(\tau) + \exp[-k(t_{tot} - \tau)][u(\tau) - u(\tau - t'_R)]\} \tag{60}$$

where c is the concentration of the product D at the detector, $t_{tot} = t_0 + t'$, $\tau = t - t_R$ [Eq. (20)], t_R and t'_R being the retention times of D on the column sections l and l', respectively.

The area under the R-sample peaks, shown shaded in Fig. 18, is obtained by applying Eq. (23). In place of $C'(l,p_0)$ we put $C(l',p_0)$, which is the Laplace transform of Eq. (59). The final result is obtained by reversing the transformation with respect to p_0:

$$f = mg[\exp(kt_R) - 1]\exp(-kt_0) \tag{61}$$

For the F peaks t'_R must be substituted for t_R. The physical meaning of this equation becomes obvious if it is written as $f = mg \exp[-k(t_0 - t_R)] - mg \exp(-kt_0)$. It means that when the direction of the gas flow is reversed at time t_0, the product formed between the times $t_0 - t_R$ and t_0 is exhibited as a sample peak. It follows that f is proportional to t_R (or t'_R for the F peaks) and thus a short column length l or l' produces small narrow peaks, such as peak 2 in Fig. 18.

According to Eq. (61), a plot of ln f versus t_0 gives k from the slope of the straight line obtained, if the reaction conforms to the model (57). An example is given in Fig. 19. The rate constant k, calculated from the plots of R and F peaks in Fig. 19, is $(2.5 \pm 0.1) \times 10^{-4}$ and $(2.6 \pm 0.1) \times 10^{-4}$ sec^{-1}, respectively. Rat constants at other temperatures and other alcohols over a 13X molecular sieve and γ-aluminium oxide have been calculated and reported [20]. They all obey Eq. (61). A "t-test" of significance, performed on the coefficients of regression of ln f on t_0, shows that these are significant at a level better than 1%, indicating that the probability for the corresponding t value being exceeded is less than 1%. This is a measure of the goodness of fit of the experimental data by linear plots like those of Fig. 19.

In Fig. 20 Arrhenius plots for the dehydration of one alcohol are shown. From such plots the activation parameters given in Table 6 have been calculated [20]. Note that these are "true" activation parameters, not involving heats of adsorption, since they are derived from rate constants pertaining to the surface step of the reactions. It

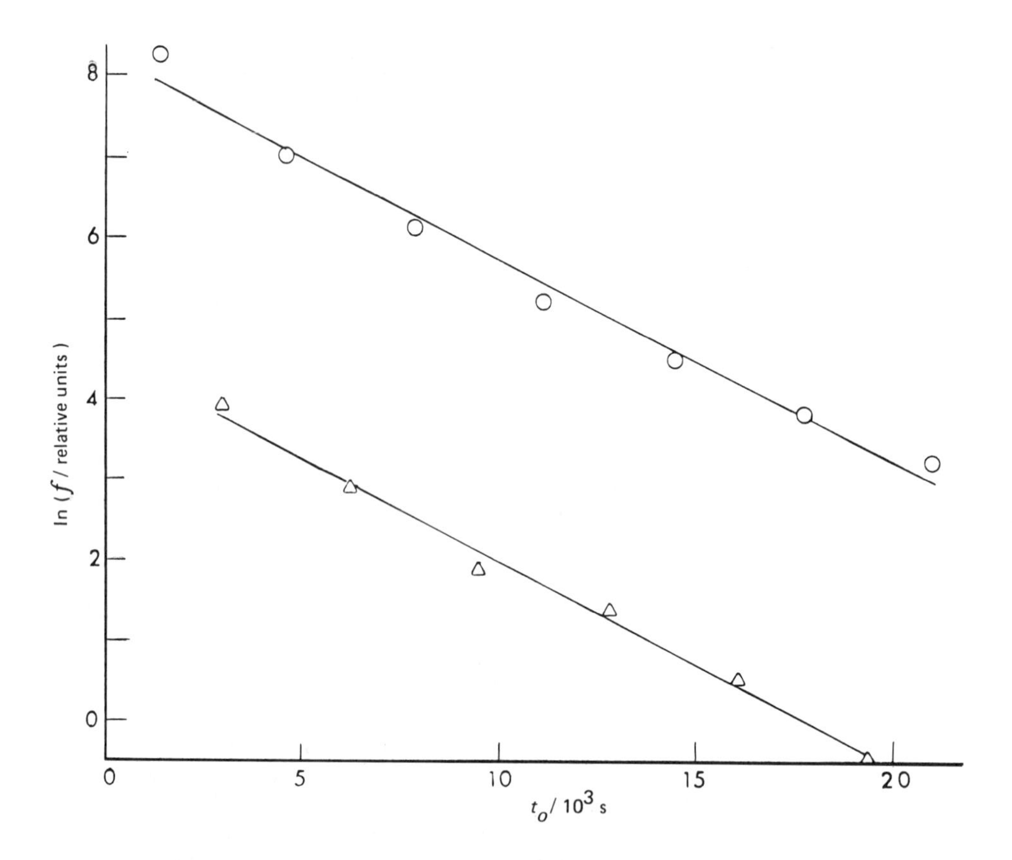

Figure 19 Plots of Eq. (61) for the dehydration of propan-2-ol (0.5 mm^3) to diisopropyl ether, at 452 K, over a 13X molecular sieve (80–100 mesh) activated at 673 K for 12 hr. The lengths of the glass column were l' = 7 and l = 100 cm, and the carrier gas nitrogen ($\dot{V}$ = 0.446 $cm^3\ sec^{-1}$): ○, R peaks, and △, F peaks. (From Ref. 19.)

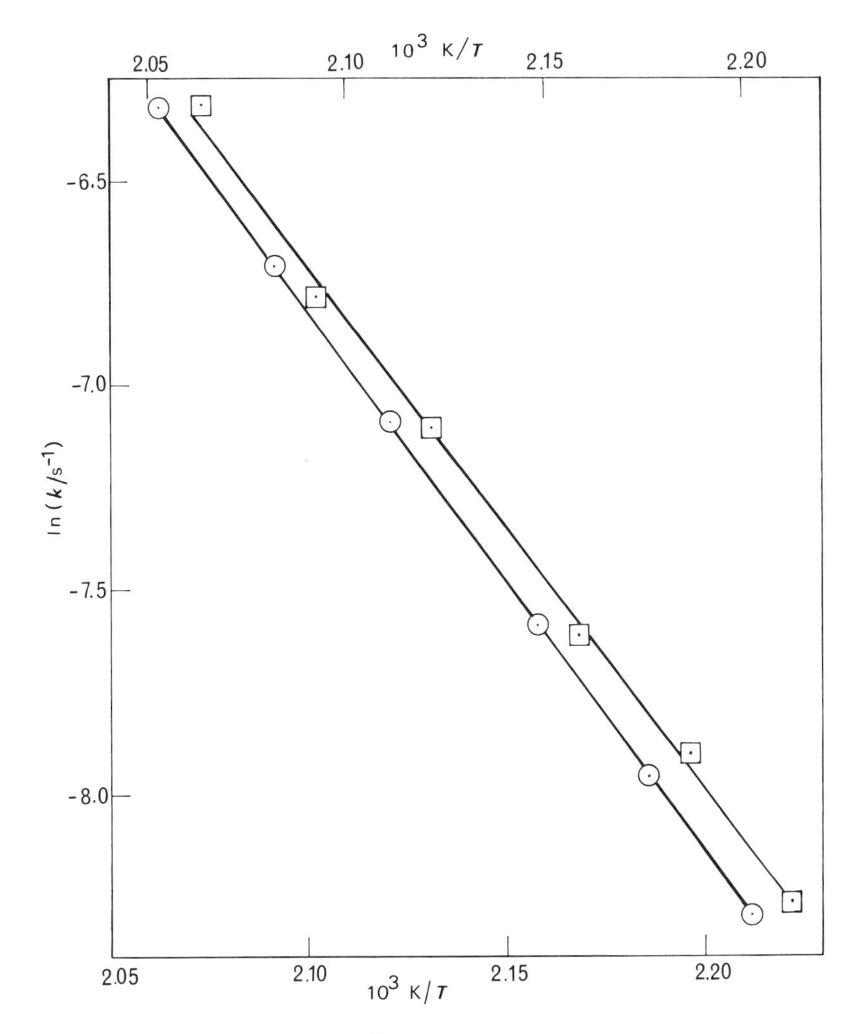

Figure 20 Arrhenius plots for the dehydration of propan-2-ol to diisopropyl ether over a 13X molecular sieve. The experimental conditions are those of Fig. 19: ○, rate constants determined from R peaks (lower abscissa), and □, rate constants from F peaks (upper abscissa). (From Ref. 19.)

Table 6 Activation Energies E_a and Frequency Factors A for the Dehydration of Three Alcohols over a 13X Molecular Sieve and γ-Aluminum Oxide

Alcohol/Catalyst	Main product	l' (cm)	l (cm)	E_a (kJ mol^{-1})		ln (A/sec^{-1})	
				R peaks	F peaks	R peaks	F peaks
Propan-1-ol/13X	Propene	3.0	108	96 ± 4	98 ± 4	16 ± 1	16.1 ± 0.9
Propan-2-ol/13X	Diisopropyl ether	7.0	100	110.6 ± 0.3	107 ± 3	21.1 ± 0.1	20.1 ± 0.7
Butan-1-ol/13X	But-1-ene	1.3	45	118 ± 8	115 ± 8	19 ± 2	19 ± 2
Propan-1-ol/Al_2O_3	Propene	1.5	45	78 ± 5	82 ± 6	10 ± 1	10 ± 1
Propan-2-ol/Al_2O_3	Propene	1.5	45	78 ± 6	81 ± 7	12 ± 1	13 ± 2

Source: Ref. 20.

is evident from Table 6 that activation parameters determined from R peaks coincide, within the limits of experimental error, with those found from F peaks. This is important, because the lengths l and l' of the chromatographic column responsible for the R and F peaks, respectively, differ by a factor of 14 or more. This coincidence indicates that secondary reactions of the detected product or irreversible adsorption of it are negligible.

The activation energies found are consistent with some literature values. Thus Gentry and Rudham [37] determined an activation energy of 110 kJ mol^{-1} for the dehydration of propan-2-ol to diisopropyl ether over a 13X molecular sieve surface similar to ours.

The amount of the reactant alcohol and the volume flow rate of the carrier gas seem to have no significant effect on the rate constants and the activation energies [20].

When plotting Eq. (61) to find k from the slope, a relative value for the expression $mg[\exp(kt_R)-1]$ having the same units as f is obtained. If the response of the detecting system for the product measured can be estimated, for example, by injecting known amounts of the pure product into an empty column under the same experimental conditions, then the absolute value (mol) of the above expression can be found. From the known values of m, t_R, and k, the fraction of active catalytic surface g can be computed. This was done for the system propan-2-ol/13X of Table 6, and the results at five temperatures between 452 and 485 K show that the g value does not seem to change significantly with temperature or from R to F peaks [20]. Its mean value is around 0.25, and this explains why the percentage of conversion to products deviates considerably from 100%, in spite of the fact that the reactant alcohol is never eluted from the catalytic column: It is due to the fact that percentage conversion is simply equal to 100g. Thus, if we divide f by t_R, an average reaction rate is obtained, and then

$$\text{Conversion} = \frac{1}{m}\int_0^\infty \frac{f}{t_R}\,dt_{tot} = \frac{g[\exp(kt_R)-1]}{kt_R} \simeq g \qquad (62)$$

The last approximation is based on the relation $\exp \simeq 1 + x$ for small x.

That the relatively small value of g is actually due to the small fraction of the total surface that is catalytically active, and not to irreversible adsorption of the product on the solid catalyst, is proved by the fact that two column sections (l' and l) containing such different amounts of catalyst give virtually the same value of g.

2. Deamination of Primary Amines

The same experimental setup (Figs. 14 and 15) was used to study the kinetics of the deamination of some primary amines, namely, (1) of

1-aminopropane and 2-aminopropane to propene over a 13X molecular sieve [21], and (2) of aminocyclohexane to cyclohexene over a 13X molecular sieve [21] and over γ-aluminium oxide [19].

The reactions over a 13X molecular sieve conform to the same mechanistic model as that applied to the dehydration of alcohols [Eq. (57)]. Equation (61) was used again to determine rate constants at various temperatures. From these the activation parameters collected in Table 7 were computed.

The differences between activation parameters determined from R and from F peaks lie again within the limits of experimental error, showing that secondary reactions of the detected product or irreversible adsorption of it are negligible. This is because the two lengths l and l' of the column are very different (by a factor of 16 to 36).

No comparison of our results with literature values can be made, since to the best of our knowledge no deaminations over zeolites have been previously studied.

An increasing conversion of aminocyclohexane to cyclohexene with increasing working temperature, as judged form g, is observed here. This, again, is not due to irreversible adsorption of the product on the solid catalyst, as the g values calculated from R and F peaks are not significantly different.

The reaction of aminocyclohexane over γ-aluminium oxide follows a different mechanistic model, since the time dependence of the peak height at $\tau = 0$ is that shown in Fig. 21. This suggests consecutive first-order reactions:

$$A + S \underset{K_A}{\overset{\text{fast}}{\rightleftharpoons}} A - S \xrightarrow{k_1} B - S \xrightarrow{k_2} D - S \underset{K_D}{\overset{\text{fast}}{\rightleftharpoons}} D + S \tag{63}$$

with a surface intermediate B − S.

The rate of production of the final product D is now proportional to the concentration of the adsorbed B − S, and this is given by the classic equation for an intermediate in consecutive reactions. Thus

$$r(t_0) = \frac{k_1 k_2 m g}{a(k_2 - k_1)} [\exp(-k_1 t_0) - \exp(-k_2 t_0)] \tag{64}$$

As before, the right-hand side of this expression is substituted for $r(t_0)$ in Eq. (53), and the result

$$c(l', t_0) = Q[\exp(-k_1 t_0) - \exp(-k_2 t_0)] \tag{65}$$

where

$$Q = \frac{k_1 k_2 m g}{\dot{V}(k_2 - k_1)} \tag{66}$$

Table 7 Activation Energies E_a and Entropies $\Delta S^{\neq}$ for the Deamination of Three Amines over a 13X Molecular Sieve

			E_a (kJ mol^{-1})		$-\Delta S^{\neq}$(J K^{-1} mol^{-1})	
Amine	l' (cm)	l (cm)	R peaks	F peaks	R peaks	F peaks
1-Aminopropane	3	108	71 ± 2	66 ± 9	212 ± 3	218 ± 17
2-Aminopropane	0.5	8	144 ± 3	138 ± 3	84 ± 6	94 ± 6
Aminocyclohexane	1.3	46	162 ± 11	156 ± 5	51 ± 17	59 ± 8

Source: Ref. 21.

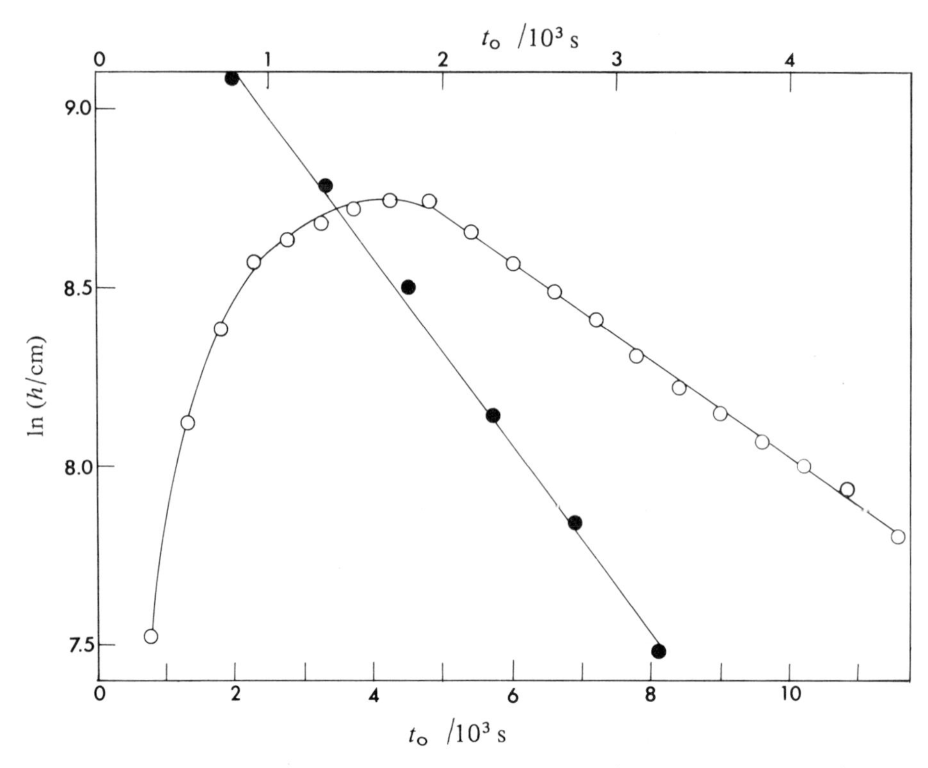

Figure 21 Time dependence of the ln h values for R peaks at $\tau = 0$ for the deamination of aminocyclohexane (1 mm^3) to cyclohexene at 525 K over γ-aluminium oxide (60–70 mesh, Ho-415 Houdry-Hüls) activated at 673 K for 2 hr. Both lengths l' and l of the glass column (inner diameter 4 mm) were 40 cm, and the carrier gas was nitrogen ($\dot{V} = 0.30\ cm^3\ sec^{-1}$): ○, (lower abscissa) experimental points h, and •, (upper abscissa) values of h' obtained by subtracting h from the corresponding extrapolated values $Q \exp(-k_1 t_0)$ of the last linear part back to the times of the ascending experimental curve. (From Ref. 19.)

is used in Eq. (21) to write c_1 and c_2 explicitly:

$$c = Q\{\exp[-k_1(t_{tot} + \tau)] - \exp[-k_2(t_{tot} + \tau)]\}u(\tau)$$
$$+ Q\{\exp[-k_1(t_{tot} - \tau)] - \exp[-k_2(t_{tot} - \tau)]\}[u(\tau) - u(\tau - t'_R)] \tag{67}$$

where $t_{tot} = t_0 + t'$. To determine the rate constants k_1 and k_2, we can employ Eq. (25), which, by using Eq. (65) in place of c_2, gives

$$h = Q[\exp(-k_1 t_{tot}) - \exp(-k_2 t_{tot})] \tag{68}$$

in agreement with the general appearance of the experimental curve of Fig. 21. From the slope of the last linear part (after the induction period) the smaller rate constant, say, k_1, was found equal to $(1.36 \pm 0.02) \times 10^{-4}\ sec^{-1}$. Then the term $Q\exp(-k_2 t_{tot})$ is calculated from the difference $h' = Q\exp(-k_1 t_{tot}) - h$, that is, by extrapolating back the last linear part and subtracting from it the experimental values of h. If now ln h' is plotted versus t_{tot}, k_2 is found from the slope of this new linear plot, as $(6.5 \pm 0.2) \times 10^{-4}\ sec^{-1}$. Analogous results were found from the F peaks [19].

The absolute value of $Q/mol\ cm^{-3}$ in Eq. (66) can be determined by the relation

$$Q_{abs}/mol\ cm^{-3} = (Q_{rel}/cm)/\dot{V}S \tag{69}$$

where $S/cm\ sec\ mol^{-1}$ is the response of the detecting system. From Q_{abs} and Eq. (66), g was found equal to 0.116, showing again that the active catalytic surface is only a small fraction of the total surface. This possibility has been pointed out many times in the literature. A relation analogous to Eq. (62), but based on the peak height h, is

$$\text{Conversion} = \frac{1}{m}\int_0^{\infty} h\, dV = \frac{1}{m}\dot{V}\int_0^{\infty} h\, dt_{tot} = g \tag{70}$$

where Eq. (68) was used for h.

3. Oxidation of Carbon Monoxide

In this reaction the cell of Fig. 16 was used, since there are *two* gaseous reactants, namely, carbon monoxide and oxygen, neither of which can be retained in the catalytic bed for a long time. It is then necessary to use a diffusion feed of the reactants and a separate column of silica gel to separate the product carbon dioxide from the two reactants [23]. The catalysts were Co_3O_4 supported on γ-Al_2O_3 doped with Ca^{2+} in various amounts [38].

After conditioning the whole system and stabilizing the catalyst under carrier gas helium flowing from D_2 to D_1 (see Fig. 16) with $\dot{V} = 0.28\ cm^3\ sec^{-1}$, 2 cm^3 of a $3CO:1O_2$ mixture at atmospheric pressure were rapidly introduced with a gas-tight syringe at the top of the diffusion column L.

The various concentrations at x = l' were sampled by adopting the situation of Fig. 3A, that is, the duration of the backward flow t' (12 sec) was smaller than both the gas holdup time t'_M in the empty section l' and the retention time t_R of all substances in column l. Thus sample peaks were recorded after two flow reversals, that is, after restoration of the helium flow to its original direction. Since t_R is different for the reactans $CO + O_2$ and the product CO_2, two fairly symmetrical peaks appear in each sampling, as shown in Fig. 22.

The pressure change in tube L at each flow reversal was small because of the small pressure drop in the filled column l (0.2 atm), and also because of the short reversal duration (12 sec).

According to Eq. (27), the height of the maximum of the sample peaks, measured from the steady continuous signal, for the reactants h_r and the product h_p, gives

$$h_r = 2c_r(l',t_0) \quad \text{and} \quad h_p = 2c_p(l',t_0) \tag{71}$$

Then, from Eq. (54), the rate of the reaction is

$$r(t_0) = \frac{h_p \dot{V}}{2M} \tag{72}$$

In the dehydration of alcohols and the deamination of primary amines a mechanistic model was assumed and the relevant *integrated*

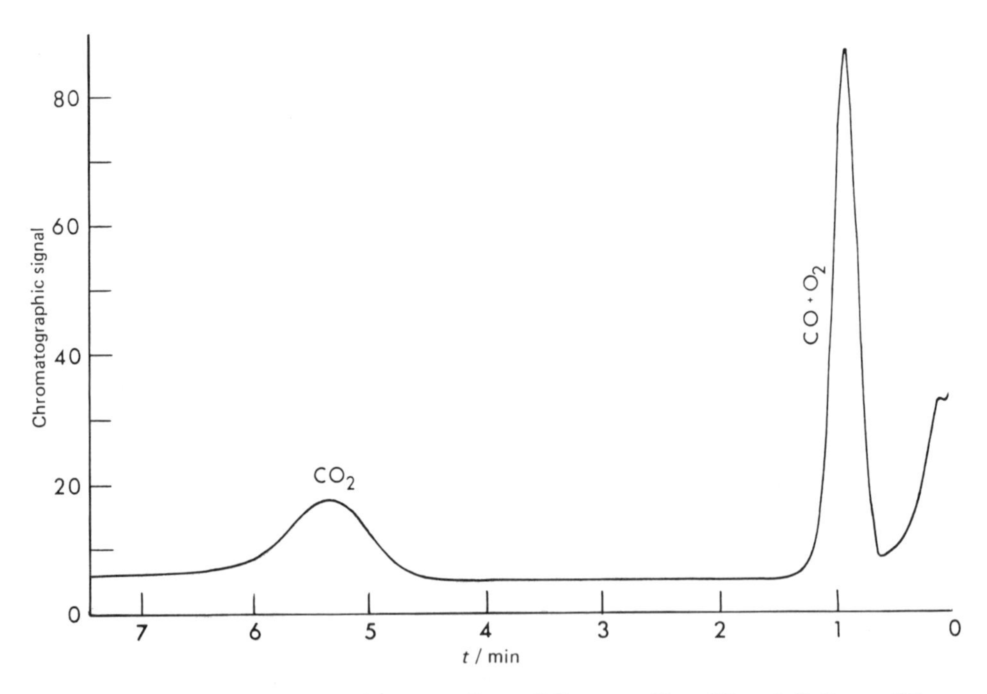

Figure 22 Chromatographic sampling of the reaction $CO + 1/2\ O_2 \rightarrow CO_2$ at 552.2 K, 44 min after injection of the reactants. The three arms of the cell of Fig. 16 had lengths L = 70.5 cm, l = l' = 40 cm, and a common inner diameter of 4 mm. The catalytic bed contained 0.35 g of catalyst. (Reproduced from Ref. 23 by permission of the National Research Council of Canada.)

rate equation [Eqs. (58) and (64)] was substituted in Eq. (23) to plot the experimental data according to this integrated equation. This is equivalent to the method of integration in homogeneous kinetics. Here another approach, equivalent to a *differential* method, is used to compare the experimental data with a given rate low. It is done by calculating the concentrations of the reactants [CO] and $[O_2]$ from h_r and h_p on the basis that (1) steady-state conditions are established after a certain time, (2) the stoichiometry of the reaction is $CO + 1/2\ O_2 \rightarrow CO_2$, (3) the initial composition of the reaction mixture was $3CO:1O_2$, and (4) the effective diffusion coefficients of CO and O_2 in He were found approximately equal. The above lead to the relations [23]

$$[CO] = \frac{3}{8} h_r + \frac{1}{16} h_p \quad \text{and} \quad [O_2] = \frac{1}{8} h_r - \frac{1}{16} h_p \tag{73}$$

Calculating [CO] and $[O_2]$ by Eq. (73) and $r(t_0)$ by Eq. (72), an apparent rate constant k can be computed for any chosen rate law and then checked for constancy with time. Thus, assuming linear Henry-type adsorption isotherms because of the small concentrations involved, a second-order rate law $r(t_0) = k[CO][O_2]$ is expected for a simple Langmuir—Hinshelwood or Rideal mechanism. However, k values calculated using this rate law increase with time continuously without limit. This excludes a mechanism involving a reaction between adsorbed carbon monoxide and adsorbed or gaseous oxygen or vice versa. By contrast, another mechanism involving a reaction between adsorbed carbon monoxide and surface lattice oxygen [39] would follow a phenomenological rate law $r(t_0) = k[CO]$, if the mass of the catalyst is big enough. Rate constants, calculated on the basis of this rate law for all catalysts studied and at various temperatures, initially increase but then reach a limiting value and remain constant after a certain time [23]. Activation energies can be calculated from this limiting k values in the usual way.

4. Other Reactions

The experimental arrangement of Figs. 14 and 15 was also used to study the kinetics of the cracking of cumene over zeolites LaY, HY, and γ-aluminium oxide [40].

An arrangement analogous to that of Fig. 16, with both column sections l and l' filled with silica gel, was employed in the hydrogenation of propene over platinum supported on alumina [41].

Finally, the experimental setup of Fig. 17 was used in the hydrodesulfurization of thiophene over catalysts containing MoO_3 and Co_3O_4 supported on alumina [42].

D. Conclusion

The reversed-flow method for studying heterogeneous catalysis is a differential technique, having all the advandages of the stopped-flow method [3], but not suffering from its limitations mentioned in Sec. I. Compared to the sample vacancy chromatography [43], it is simpler experimentally, since it is not necessary to pass the reactants, mixed with the carrier gas, continuously through the catalyst. Moreover, a whole range of reactant compositions is possible within one single experiment by feeding the catalyst through a diffusion column.

ACKNOWLEDGMENT

The authors thank Mrs. Margaret Barkoula for her assistance.

REFERENCES

1. R. J. Laub and R. L. Pecsok, *Physicochemical Applications of Gas Chromatography*, Wiley, New York, 1978.
2. J. R. Conder and C. L. Young, *Physicochemical Measurement by Gas Chromatography*, Wiley, Chichester, 1979.
3. C. S. G. Phillips, A. J. Hart-Davis, R. G. L. Saul, and J. Wormald, J. Gas Chromatogr. *5*, 424 (1967).
4. I. Hadzistelios, H. J. Sideri-Katsanou, and N. A. Katsanos, J. Catal. *27*, 16 (1972).
5. A. Lycourghiotis, N. A. Katsanos, and I. Hadzistelios, J. Catal. *36*, 385 (1975).
6. A. Lycourghiotis and N. A. Katsanos, React. Kinet. Catal. Lett. *4*, 221 (1976).
7. N. A. Katsanos, J. Chromatogr. *152*, 301 (1978).
8. A. Lycourghiotis, N. A. Katsanos, and D. Vattis, J. Chem. Soc. Faraday Trans. I *75*, 2481 (1979).
9. G. Karaiskakis, A. Lycourghiotis, D. Vattis, and N. A. Katsanos, React. Kinet. Catal. Lett. *15*, 413 (1980).
10. A. Lycourghiotis, D. Vattis, and N. A. Katsanos, Z. Phys. Chem. *126*, 259 (1981).
11. D. Vattis, N. A. Katsanos, G. Karaiskakis, A. Lycourghiotis, and M. Kotinopoulos, J. Chromatogr. *214*, 171 (1981).
12. N. A. Katsanos and A. Tsiatsios, J. Chromatogr. *213*, 15 (1981).
13. A. Tseremegli, N. A. Katsanos, and I. Hadzistelios, Z. Phys. Chem. *129*, 21 (1982).
14. N. A. Katsanos and I. Hadzistelios, J. Chromatogr. *105*, 13 (1975).
15. N. A. Katsanos, G. Karaiskakis, and I. Z. Karabasis, J. Chromatogr. *130*, 3 (1977).

16. G. Karaiskakis and N. A. Katsanos, J. Chromatogr. *151*, 291 (1978).
17. N. A. Katsanos, G. Karaiskakis, D. Vattis and A. Lycourghiotis, Chromatographia *14*, 695 (1981).
18. N. A. Katsanos and I. Georgiadou, J. Chem. Soc. Chem. Commun. 242 (1980).
19. N. A. Katsanos, J. Chem. Soc. Faraday Trans. I *78*, 1051 (1982).
20. G. Karaiskakis, N. A. Katsanos, I. Georgiadou, and A. Lycourghiotis, J. Chem. Soc. Faraday Trans. I *78*, 2017 (1982).
21. M. Kotinopoulos, G. Karaiskakis, and N. A. Katsanos, J. Chem. Soc. Faraday Trans. I *78*, 3379 (1982).
22. N. A. Katsanos, ACHEMA 82 International Meeting of Chemical Engineering, Frankfurt-am-Main, 1982.
23. G. Karaiskakis, N. A. Katsanos, and A. Lycourghiotis, Can. J. Chem. *61*, 1853 (1983).
24. N. A. Katsanos and G. Karaiskakis, J. Chromatogr. *237*, 1 (1982).
25. N. A. Katsanos and G. Karaiskakis, J. Chromatogr. *254*, 15 (1983).
26. G. Karaiskakis, N. A. Katsanos, and A. Niotis, Chromatographia *17*, 310 (1983).
27. N. A. Katsanos, G. Karaiskakis, and A. Niotis, *1983 World Chromatography/Spectroscopy Conference*, London, 1983.
28. G. Karaiskakis, N. A. Katsanos, and A. Niotis, J. Chromatogr. *245*, 21 (1982).
29. G. Karaiskakis, A. Lycourghiotis, and N. A. Katsanos, Chromatographia *15*, 351 (1982).
30. R. B. Bird, W. E. Stewart, and E. N. Lightfoot, *Transport Phenomena*, Wiley, Chichester, 1960, p. 511.
31. E. N. Fuller, P. D. Schettler, and J. C. Giddings, Ind. Eng. Chem. *58*, 19 (1966).
32. V. R. Maynard and E. Grushka, Adv. Chromatogr. *12*, 99 (1975).
33. J. C. Giddings, *Dynamics of Chromatography*, Marcel Dekker, New York, 1965, p. 239.
34. R. B. Bird, W. E. Stewart, and E. N. Lightfoot, *Transport Phenomena*, Wiley, Chichester, 1960, p. 570.
35. A. S. Foust, L. A. Wenzel, C. W. Clump, L. Maus, and L. B. Anderson, *Principles of Unit Operations*, Wiley, New York, 1960, p. 474.
36. N. A. Katsanos, A. Lycourghiotis, and A. Tsiatsios, J. Chem. Soc. Faraday Trans. I *74*, 575 (1978).
37. S. J. Gentry and R. Rudham, J. Chem. Soc. Faraday Trans. I *70*, 1685 (1974).

38. A. Lycourghiotis, A. Tsiatsios, and N. A. Katsanos, Z. Phys. Chem. *126*, 95 (1981).
39. D. Mehanddjiev and E. Nikolova-Zhecheva, J. Catal. *65*, 475 (1980).
40. M. Kotinopoulos, thesis, University of Patras, Patras, Greece, 1983.
41. N. A. Katsanos, G. Karaiskakis, and A. Niotis, *Proc. 8th Intern. Congress on Catalysis*, W. Berlin, 1984, Vol. III, p. 143.
42. N. A. Katsanos, G. Karaiskakis, and A. Niotis, unpublished results.
43. C. S. G. Phillips and C. R. McIlwrick, Anal. Chem. *45*, 782 (1973).

6

Development of High-Speed Countercurrent Chromatography

Yoichiro Ito / National Heart, Lung, and Blood Institute, Bethesda, Maryland

I. INTRODUCTION

Countercurrent chromatography (CCC) is a support-free partition chromatography and offers the advantage over liquid chromatography of eliminating all complications arising from the use of solid supports [1-5]. The method has been termed after two classic partition techniques, that is, the countercurrent distribution method (CCD) and liquid chromatography, since it shares all the merits of these parent methods. By retaining the merits from CCD, CCC can yield pure fractions at a high reproducibility without the risk of adsorptive loss of samples, while it permits continuous elution, monitoring, and fractionation as are performed in liquid chromatography.

A number of problems encountered in developing this novel technique has been solved by using a coil as a separation column. A variety of CCC schemes developed in the past may be reduced to two basic systems. The first uses a hydrostatic equilibrium of two solvent phases in a stationary coil and is named "hydrostatic equilibrium system" (HSES). The second uses a hydrodynamic equilibrium of two solvent phases in a rotating coil and is accordingly named "hydrodynamic equilibrium system" (HDES). (Details of these basic systems are described in the following section.) Because of the simplicity of the system, HSES has been quickly developed to produce several useful CCC schemes such as helix (toroidal coil) CCC [1, 6–9], droplet CCC [6,7,10], and locular CCC [6,7]. These CCC schemes based on HSES generally share common features of stable retention of the stationary phase, limited flow rates, and relatively long separation times.

On the other hand, development of HDES has taken a much longer time course because of the complexity involved in the hydrodynamic motion of the two solvent phases in a rotating coil. In addition, establishment of efficient CCC schemes necessitated development of a variety of flow-through centrifuge devices which effect planetary motion of the coiled column to facilitate both retention of the stationary phase and mixing of the two phases in the coil. In each of these planetary centrifuges, the centrifugal force field and the column orientation must be optimized. Eventually, these efforts have been rewarded in producing various efficient CCC schemes which require much shorter separation times. For example, the horizontal flow-through coil planet centrifuge [11–17] has yielded sizable separations of various biological samples in overnight runs, instead of days which would have been required in earlier CCC schemes. The method, however, has a limited capacity of holding the stationary phase in the column, which restrains further reduction of the separation times by application of higher flow rates of the mobile phase.

A great advancement in CCC technology has been recently made in the discovery that a coiled column, if mounted coaxially around the holder, retains an exceedingly large volume of the stationary phase against an extremely high flow of the mobile phase [18]. In the coaxial orientation of the coil, the column capacity can be conveniently increased by forming multiple layers of the coil around the holder as tubing compactly wound on a reel. This multilayer coil has produced highly efficient separations of a number of biological samples in a few hours of elution (19,20). Current studies suggest that the separation time may be further shortened without sacrificing partition efficiency.

With advent of the multilayer coil, CCC is no longer a slow separation method. It is expected that in the near future the performance of CCC will become comparable with that of high-performance liquid chromatography in both partition efficiency and separation speed in laboratory-scale preparative separations.

II. BASIC PRINCIPLES OF COUNTERCURRENT CHROMATOGRAPHY [3,4]

A. Two Basic CCC Systems

As described earlier, a variety of CCC schemes have stemmed from two basic CCC systems, that is, the hydrostatic equilibrium system (HSES) and the hydrodynamic equilibrium system (HDES). Mechanisms of these basic CCC systems are illustrated in Fig. 1.

The HSES (Fig. 1, left) uses a stationary coiled tube. The coil is first filled with one phase of an equilibrated two-phase solvent system. The other phase introduced at one end of the coil percolates through the first phase on one side of the coil. This leaves a segment of the first phase stationary in the first coil unit. The process repeats in each coil unit until the mobile phase reaches the outlet of the coil. Thereafter the mobile phase displaces only itself, leaving the first phase stationary in each coil unit. Consequently, solutes introduced locally at the inlet of the coil are subjected to a partition process between the two phases and separated according to their relative partition coefficients.

The HDES configuration is the same as that of the HSES, with the exception that the coil is slowly rotated around its own axis. Introduction of this simple motion, however, creates amazingly complex cf fects on the behavior of the two phases in the coil. The rotation introduces functional asymmetry to the coil. When a water-filled coil is rotated, any object, whether it is lighter (air bubble) or heavier (glass bead) than water, tends to move toward one end of the coil. This end is called the "head", and the other end the "tail" of the coil. The two immiscible solvents confined in such a coil undergo a complex motion but finally reach a hydrodynamic equilibrium state wherein each coiled turn is occupied by a nearly equal amount of each solvent and any excess of either solvent remains at the tail. This hydrodynamic behavior of the two solvent phases can be efficiently used for solute partitioning. In Fig. 1 (right) the coil is filled with one phase and the

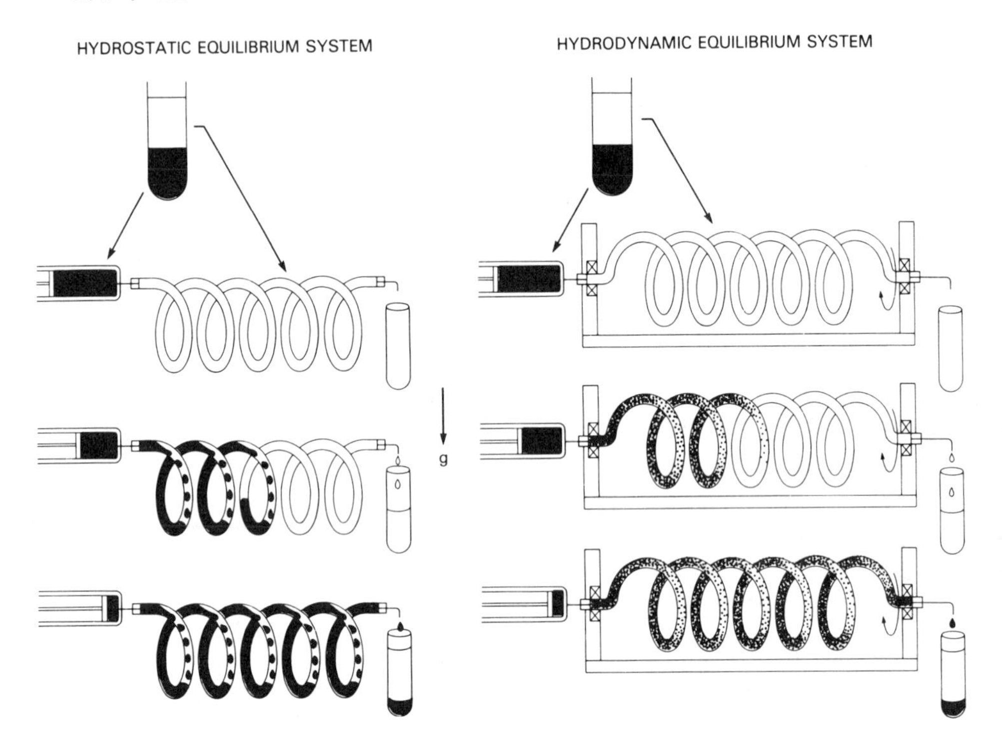

Figure 1 Principle of two basic countercurrent chromatography systems. (From Ref. 2.)

other phase is introduced at the head of the coil while the coil is slowly rotated around its own axis. As the mobile phase reaches the first coil unit, hydrodynamic equilibrium is quickly established between the two phases and a large amount of the stationary phase is retained in the coil. This process repeats itself in each coil unit until the mobile phase reaches the tail. Thereafter the mobile phase displaces only the mobile phase, leaving the stationary phase in each coil unit. Consequently, solutes locally introduced through the head of the coil are subjected to an efficient partition process between the two phases and eluted in order of their partition coefficients as in liquid chromatography, but in the absence of solid supports.

B. Comparison of Two Basic CCC Systems

While both basic systems are capable of performing CCC, each offers its own specific advantage. Retention of the stationary phase is much more stable in the HSES. The HSES usually yields the retention in close to 50% of the total column space under the proper flow rate, whereas the HDES produces variable retention rates according to the operational conditions, such as the rotational speed of the coil and various physical properties of the applied solvent phases. On the other hand, both the interfacial area of the two phases and the degree of mixing are far greater in the HDES. In the HDES the efficient partition process takes place throughout the length of the coil, whereas in the HSES one-half of the column space is entirely occupied by the mobile phase and therefore becomes nonefficient free space. These characteristic features of the two basic CCC systems are summarized in Table 1.

Because of the rotatory motion of the coil, the HDES becomes much more complex, but it promises greater potential and endless versatility in performing CCC. In fact, the basic HDES has recently brought forth a new CCC system which literally permits countercurrent flow of two immiscible solvent phases through a coiled column. This so-called "dual countercurrent system" described in the following section has opened up a rich domain in CCC technology.

III. DUAL COUNTERCURRENT SYSTEM

A. Mechanism of the Dual Countercurrent System [18]

The dual countercurrent system is distinguished from other CCC systems by its unique capability of yielding true countercurrent flow of two solvent phases through the coil. This new system has stemmed from the basic HDES, which utilizes a coaxially rotating coil. In the HDES previously described, slow rotation of the coil establishes a hydrodynamic equilibrium of the two solvent phases so that each solvent phase occupies nearly half the space in every coiled turn, while any excess of either phase is isolated at the tail of the coil. It has been observed, however, that distribution of the two phases is altered by the rotational speed of the coil. As the rotational speed is increased, one of the phases, typically the heavier phase, tends to occupy a greater space on the head side of the coil and, in a certain critical range of the rotational speed, the two phases are occasionally seen completely separated along the length of the coil, one phase entirely occupying the head side, and the other phase the tail side. With further increases in the rotational speed, the magnitude of the radially produced centrifugal force field eventually overcomes gravity, resulting in separation of the two phases across the opening of the tube in such a way that the heavier phase is held in the outer portion, and the lighter phase in the inner portion of the coil. Consequently, the two phases tend to distribute themselves throughout the coil at the volume ratio of the two phases present in the whole coil. Among the above-mentioned experimental conditions, the dual countercurrent process becomes possible at the critical rotational speed of the coil

Table 1 Characteristic Features of Two Basic CCC Systems

	Hydrostatic equilibrium system (HSES)	Hydrodynamic equilibrium system (HDES)
Motion of the coil	Stationary	Slow rotation around its own axis
Symmetry of the coil	Symmetrical	Asymmetrical with the head and the tail
Retention of the stationary phase	Stable	Variable
Mixing of the solvents	Mild	Efficient
Interfacial area	Limited	Broad
Efficient column space	50% (one side of the coil)	100%
Partition efficiency	Low	High

Source: Ref. 2.

which produces complete separation of the two phases along the length of the coil.

Let us assume a pair of immiscible solvent phases A and B. When phase A is distributed on the head side, and phase B on the tail side of the coil, continuous elution is possible in the following three ways:

Scheme 1 (normal elution mode: head → tail). The coil is filled with phase A (stationary phase), and phase B (mobile phase) is eluted through the head toward the tail of the coil.

Scheme 2 (reversed elution mode: tail → head). The coil is filled with phase B (stationary phase), and phase A (mobile phase) is eluted through the tail toward the head of the coil.

Scheme 3 (dual countercurrent mode: head ⇄ tail). The dual countercurrent operation is initiated by filling the coil with either phase A, phase B, or any mixture of these (the initial composition of the column contents is not important). This is followed by simultaneous elution of phases A and B through the tail and head of the coil, respectively. This operation requires an additional flow tube at each end of the coil for the collection of effluents. If desirable, another flow tube may be connected at the middle of the coil for continuous sample feeding.

All the above schemes are used for both continuous extraction and CCC. In extraction, the sample present in a large volume of the mobile phase is extracted into the stationary phase (schemes 1 and 2) or into the other mobile phase (scheme 3). In CCC, the sample is locally introduced at the inlet of the coil (schemes 1 and 2) or continuously fed through the sample feed line at the middle of the coil (scheme 3).

Because of the late discovery of this dual countercurrent system, none of these schemes have been fully studied yet; however, the relatively simple one-way elution schemes 1 and 2 have been successfully developed to perform continuous countercurrent extraction (see Sec. V) and high-speed CCC (see Sec. VI). In the following a simple preparative CCC method is described. The method utilizes a coil assembly slowly rotating in the unit gravitational field and is considered to be the simplest model of the one-way elution scheme based on the dual countercurrent system.

B. Preparative CCC with a Coaxially Rotating Coil [21]

This simple CCC method is an extension of the one-way elution schemes based on the dual countercurrent system described earlier. The method utilizes a large-bore coil coaxially mounted around the rotary holder. Hydrodynamic motion and distribution of the two immiscible solvent phases confined in such a coil is conveniently studied with a closed-end coil mounted on a controllable-speed rotary device. Experiments were performed with a single-layer coil (0.55 cm inner diameter, 4-m-long fluorinated ethylene propylene (FEP) tubing) having helical diameters of 3, 10, and 20 cm, as illustrated in Fig. 2 (left). Preliminary results were obtained with a two-phase solvent system composed of chloroform, acetic acid, and 0.1 N hydrochloric acid (2:2:1), which was later used for separation of dinitrophenyl (DNP) amino acid samples.

Figure 3 illustrates the distribution of the two solvent phases in the coaxially rotated coils measured at their hydrodynamic equilibrium. In each phase distribution diagram, the volume percentage of each phase distributed at the head side of the coil is plotted against the applied rotational speed. Three diagrams obtained from different helical diameters show common features characteristic of the coaxially rotated coil.

In the slow rotational speed between 0 and 30 rpm, the two phases distribute fairly evenly in the coil (stage I). When the rotational speed is increased, the lower nonaqueous phase tends to occupy more space on the head side of the coil and, at a critical speed between 60 and 100 rpm, the two phases are almost completely separated along the length of the coil, the lower phase occupying the head side, and the upper phase the tail side of the coil (stage II). After this critical range, the amount of lower phase on the head side tends to decrease rather sharply, crossing below the 50% line (stage III). A further increase in the rotational speed again yields an even distribution of the two

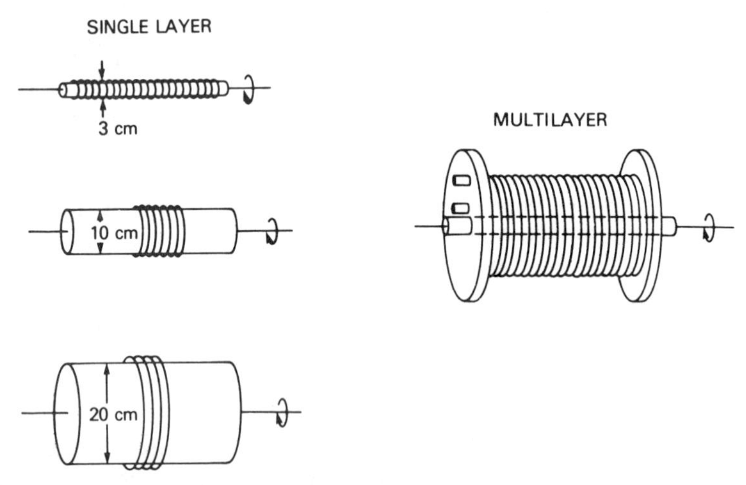

Figure 2 Coaxially rotated coiled columns with various dimensions: single-layer coils (left) and multilayer coil (right). (Adapted from Ref. 21.)

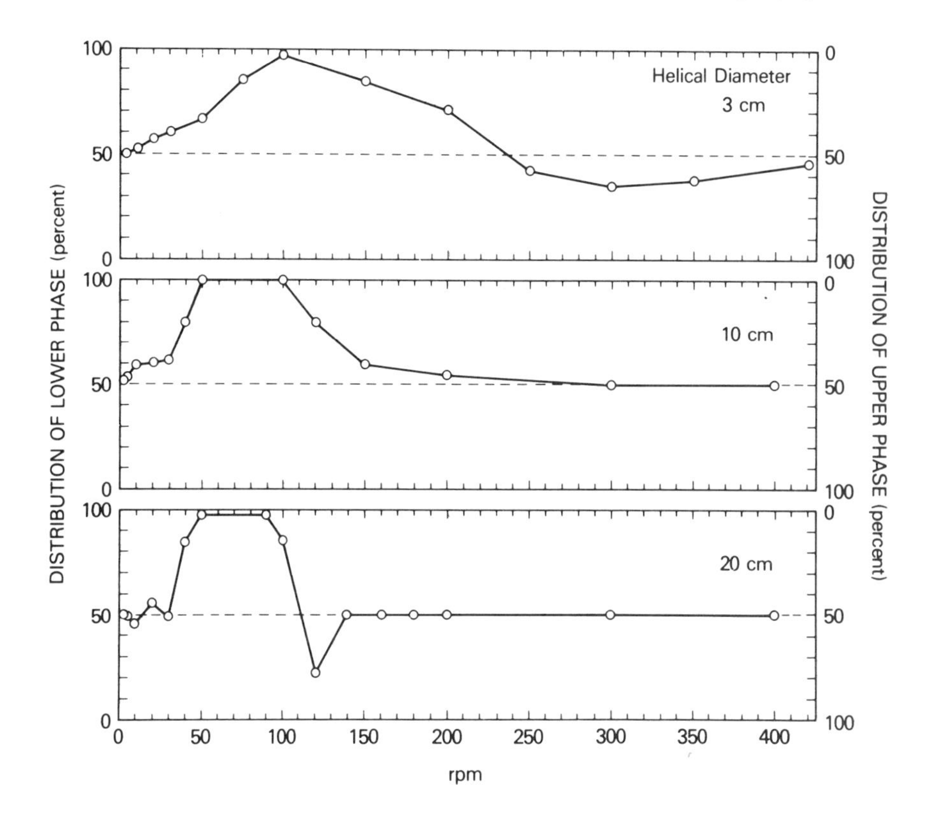

Figure 3 Phase distribution diagrams for coaxially rotated coils with three different helical diameters. (From Ref. 21.)

phases in the coil (stage IV). As the helical diameter increases, all these stages tend to shift toward the lower rpm range, apparently owing to the enhanced centrifugal force field. For performing preparative CCC, stage II is considered to be of the greatest interest, because the system permits retention of a large amount of the stationary phase for either phase under the proper mode of elution. The uneven distribution of the two phases in the coil at the critical rpm permits an extremely high level of stationary phase retention when either the aqueous phase is eluted in the normal mode (head → tail) or the nonaqueous phase is eluted in the reversed mode (tail → head).

The partition efficiency of the coaxially rotated coils was measured by a set of DNP amino acid samples and the two-phase solvent system used for phase distribution studies. The experimental results with the three different helical diameters are summarized in Fig. 4, in which in-

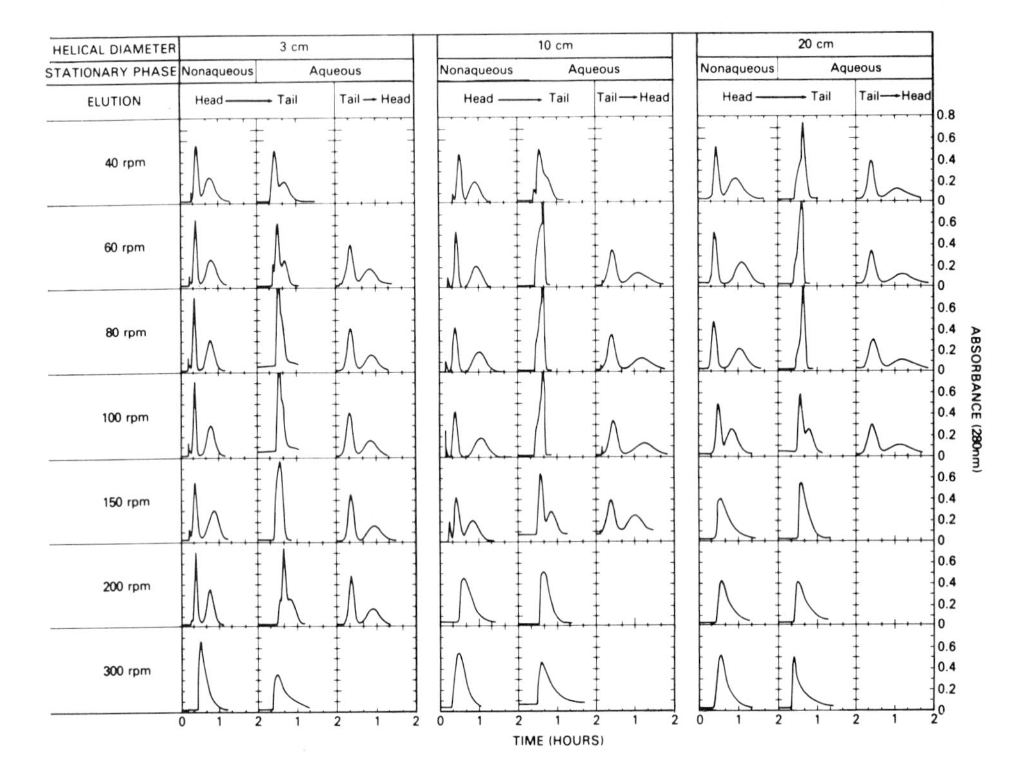

Figure 4 Effects of helical diameters of the coaxially rotated coil on the separation of DNP amino acids at various rotational speeds. (From Ref. 21.)

dividual charts are arranged according to various operational conditions. In addition to the normal elution mode (head → tail), the reversed elution mode (tail → head) was also applied for the nonaqueous mobile phase at the critical rpm range (stage II) to achieve a high level of aqueous phase retention as expected from the phase distribution diagrams shown in Fig. 3.

The overall results of these experiments with the coaxially rotated coils clearly indicate that the best peak resolutions are obtained at the critical rotational speed, ranging between 60 and 100 rpm, when either the aqueous phase was eluted in the normal mode (head → tail) or the nonaqueous phase was eluted in the reversed mode (tail → head).

Comparison of the results obtained from the three different helical diameters illustrated in Fig. 4 reveals the considerable shift of the optimal conditions toward the lower rpm range in the larger helical diameter coil, this being consistent with the finding from the phase distribution diagrams in Fig. 3. However, the satisfactory peak resolution observed in a wide range of rotational speeds in each helical diameter suggests that the choice of a proper rotational speed, such as 80 rpm, would produce good separations in the coil with any helical diameter between 3 and 20 cm. In light of these experimental data with the coaxially rotated coils, a new column configuration called the multilayer coil (Fig. 2, right) has been devised for performing large-scale preparative CCC.

The preparative capability of the multilayer coil has been demonstrated in the separation of 1 g of the DNP amino acid mixture under the optimal operational conditions determined by the preliminary studies on short coils as described above. Figure 5 shows the chromatograms

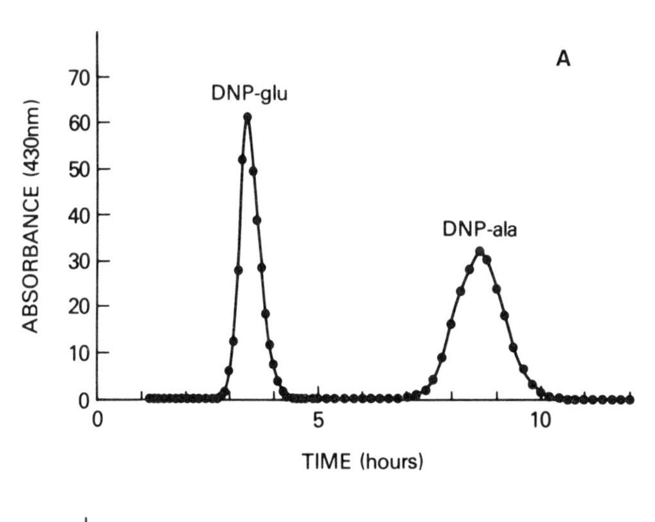

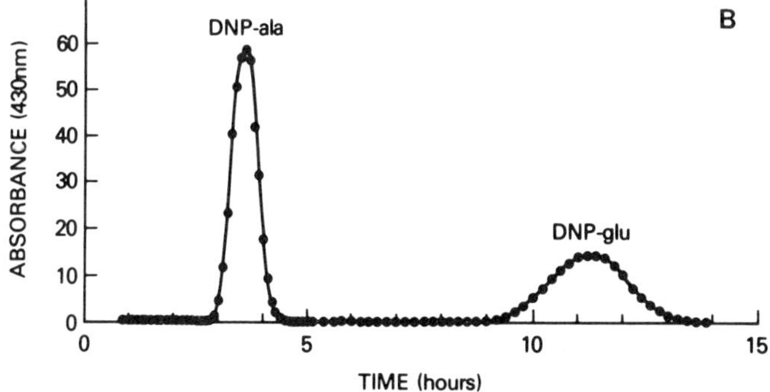

Figure 5 Chromatograms of DNP amino acids (1 g) obtained with the multilayer coil: (A) elution with the upper aqueous phase in the normal mode (head → tail) and (B) elution with the lower nonaqueous phase in the reversed mode (tail → head). (From Ref. 21.)

obtained with the multilayer coil in which both the upper aqueous phase (Fig. 5A) and the lower nonaqueous phase (Fig. 5B) are used as the mobile phase in the suitable elution mode. The two DNP amino acids are well resolved and eluted out as symmetrical peaks. The retention of the stationary phase measured after the separation was 68% in Fig. 5A and 84% in Fig. 5B. The separations obtained with this multilayer coil with a 750-ml capacity are much better than those from the eccentrically mounted glass coils with a similar internal diameter having a total capacity of 900 ml [22].

This simple preparative CCC method has a number of advantages over many other CCC methods, primarily because of its large capacity of holding the stationary phase under efficient mixing of the two phases. On the other hand, the method has inherent shortcomings, in that the stationary phase is sustained in the coil solely by gravity against the flow of the mobile phase. During elution the stationary phase is retained in the coil under a subtle balance between the two forces, that is, the rotational acceleration, which sends the stationary phase toward the inlet of the coil, and the flowing mobile phase, which carries the stationary phase toward the outlet of the coil. Consequently, application of a higher flow rate of the mobile phase in the present method would sharply decrease the retention capacity of the coil, thus resulting in loss of peak resolution and decreased sample loading capacity. In order to establish a faster and more efficient partition method, it is natural to consider enhancement of the acceleration field by centrifugation.

IV. FLOW-THROUGH CENTRIFUGE SCHEMES FOR PERFORMING CCC [2,3]

A. Rotary-Seal-Free Flow-Through Centrifuge Schemes

In the past a series of flow-through centrifuge schemes have been developed for performing CCC. These centrifuge schemes are equipped with a novel flow-through mechanism which entirely eliminates the need for the conventional rotary-seal device, thus providing the following advantages:

1. The system is leak free under high pressure up to several hundred psi.
2. They readily facilitate the use of multiple flow channels.
3. There is no mechanical damage to samples.
4. Corrosive solvents do not affect the systems.
5. There are no heat dissipation problems at the seals.
6. They eliminate contamination problems.

Because of these advantages, these schemes can become a reliable elution system, comparable to other chromatographic schemes. Consequently, the method provides broad application for the separation and purification of various biological materials.

Figure 6 shows the principle of various types of seal-free flow-through centrifuge schemes and their mutual relationship. Each diagram indicates the orientation and motion of the cylindrical coil holder with a bundle of flow tubes tightly supported at a point marked by a black circle on the central axis. These schemes are divided into three classes according to the mode of their planetary motion, that is, synchronous, nonsynchronous, and nonplanetary, as indicated at the top of Fig. 6. In the synchronous series, rotation and revolution of the holder are synchronized in such a way that one revolution (around the

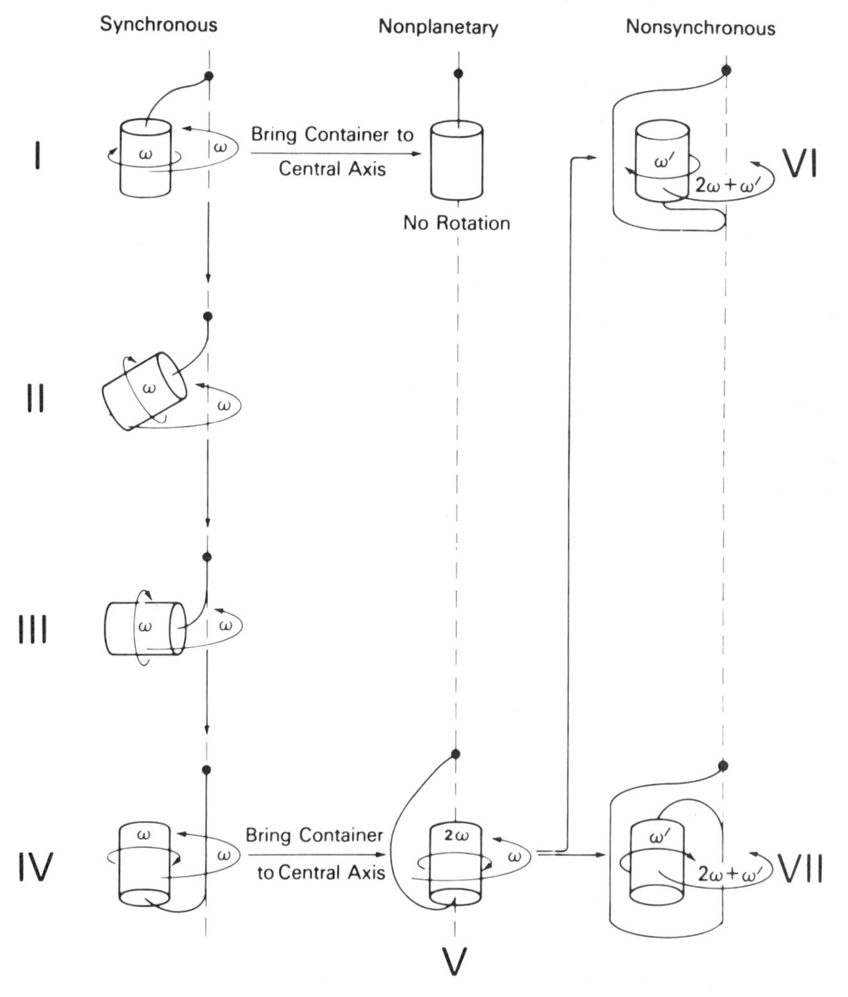

Figure 6 Various types of seal-free flow-through centrifuge schemes and their mutual relationship. (From Ref. 2.)

central axis of the apparatus) produces one rotation (around its own axis) of the holder. In the nonsynchronous series, the rates of rotation and revolution are independently adjusted. In the nonplanetary series, rotation and revolution of the holder share a common axis, resulting in either nonrotation or a single rotation around the central axis of the apparatus.

Scheme I in the synchronous series has a vertical holder which counter-rotates about its own axis while revolving around the central axis of the apparatus. This motion of the holder is similar to that of a beaker when a chemist tries to mix its contents with his hand. This simple synchronous planetary motion of the holder eliminates the need for the rotating seal [23]. This same principle can be applied to other synchronous schemes with tilted (scheme II), horizontal (scheme III), and even inverted orientation (scheme IV) of the holder.

Scheme IV is a transitional form to a nonplanetary scheme, scheme V. When the holder of scheme I is brought to the centrifuge axis, the counter-rotation of the holder cancels out the revolutional effect, resulting in zero rotation of the holder. When the holder of this transitional form is brought to the central axis, it gains an angular velocity ω because the rotation of the holder is added to the revolutional effect. This shift also causes revolution of the flow tubes around the central axis of the apparatus. Scheme V, in turn, provides a base for the two nonsynchronous schemes, schemes VI and VII. The holder of scheme V is again moved away from the central axis to undergo a synchronous planetary motion as in schemes I and IV. Although this planetary motion is synchronous with respect to the base, the net revolutional rate of the holder is the sum of the revolutional rates of the top and the base. Since the revolutional rate of the base is independent of the top planetary motion, the rotation/revolution ratio of the holder becomes freely variable. In this scheme the centrifugal force field rotates around the axis of the holder as in scheme I, but at any desired rate.

Each centrifuge scheme illustrated in Fig. 6 displays its own characteristic acceleration field. The complexity of the pattern of the acceleration produced by these schemes also varies to a great extent. Scheme V yields a stable radial acceleration field, as in the conventional centrifuge systems, whereas synchronous schemes II [24] and III [25] produce extremely complex patterns of acceleration which involve three-dimensional fluctuation of the vector during each revolutional cycle. In the rest of the schemes, the acceleration vectors are always confined in a plane perpendicular to the holder axis so that the acceleration can be analyzed mathematically on a simple coordinate system.

All these centrifuge schemes, except for scheme VII, have been constructed and successfully used to perform CCC. Recently it has been discovered that scheme IV can produce an ideal hydrodynamic phase distribution for the dual countercurrent system in an enhanced acceleration field.

B. Analysis of Acceleration Produced by Typical Synchronous Planetary Motions [2,3]

A simple mathematical analysis has been applied to two typical synchronus planetary motions, that is, schemes I and IV, illustrated in Fig. 6. Figure 7 shows the schematic diagram of each scheme (Fig. 7A), the coordinate system for analysis of acceleration (Fig. 7B), the orbits of arbitrary points on the holder (Fig. 7C), and centrifugal force vectors at various locations on the holder (Fig. 7D).

In scheme I, the column holder revolves around the central axis (center of revolution) and simultaneously counter-rotates around its own axis (center of rotation), both at the same angular velocity ω (Fig. 7A, left). For analysis of acceleration, the coordinate system is chosen in such a way that the center of revolution is located at the center of the coordinate system and the center of rotation is on the x axis at time 0 (Fig. 7B, left). An arbitrary point initially located at P_0 forms angle θ_0 with the x axis; then after time t, location of the point, P(x,y), is expressed by

$$x = R\cos\theta + r\cos\theta_0 \tag{1}$$

$$y = R\sin\theta + r\sin\theta_0 \tag{2}$$

From these equations, the orbit of the arbitrary point is easily obtained by eliminating the variable, θ, and

$$(x - r\cos\theta_0)^2 + (y - r\sin\theta_0)^2 = R^2 \tag{3}$$

This represents a circle with radius R centered at point ($r\cos\theta_0$, $r\sin\theta_0$). The fact that the radius of the circle, R, is independent of both θ_0 and r indicates that every point located on the holder has a circular orbit with radius R (Fig. 7C, left).

The net magnitude of acceleration, α, and its acting angle formed from the x axis, γ, at the arbitrary point on the holder are further computed from the second derivatives of Eqs. (1) and (2), and

$$\alpha = \left[\left(\frac{d^2x}{dt^2}\right)^2 + \left(\frac{d^2y}{dt^2}\right)^2\right]^{1/2} = R\omega^2 \tag{4}$$

$$\gamma - \pi = \tan^{-1}\left[\left(\frac{d^2y}{dt^2}\right)\right] \Big/ \left[\left(\frac{d^2x}{dt^2}\right)\right] = \omega t = \theta \tag{5}$$

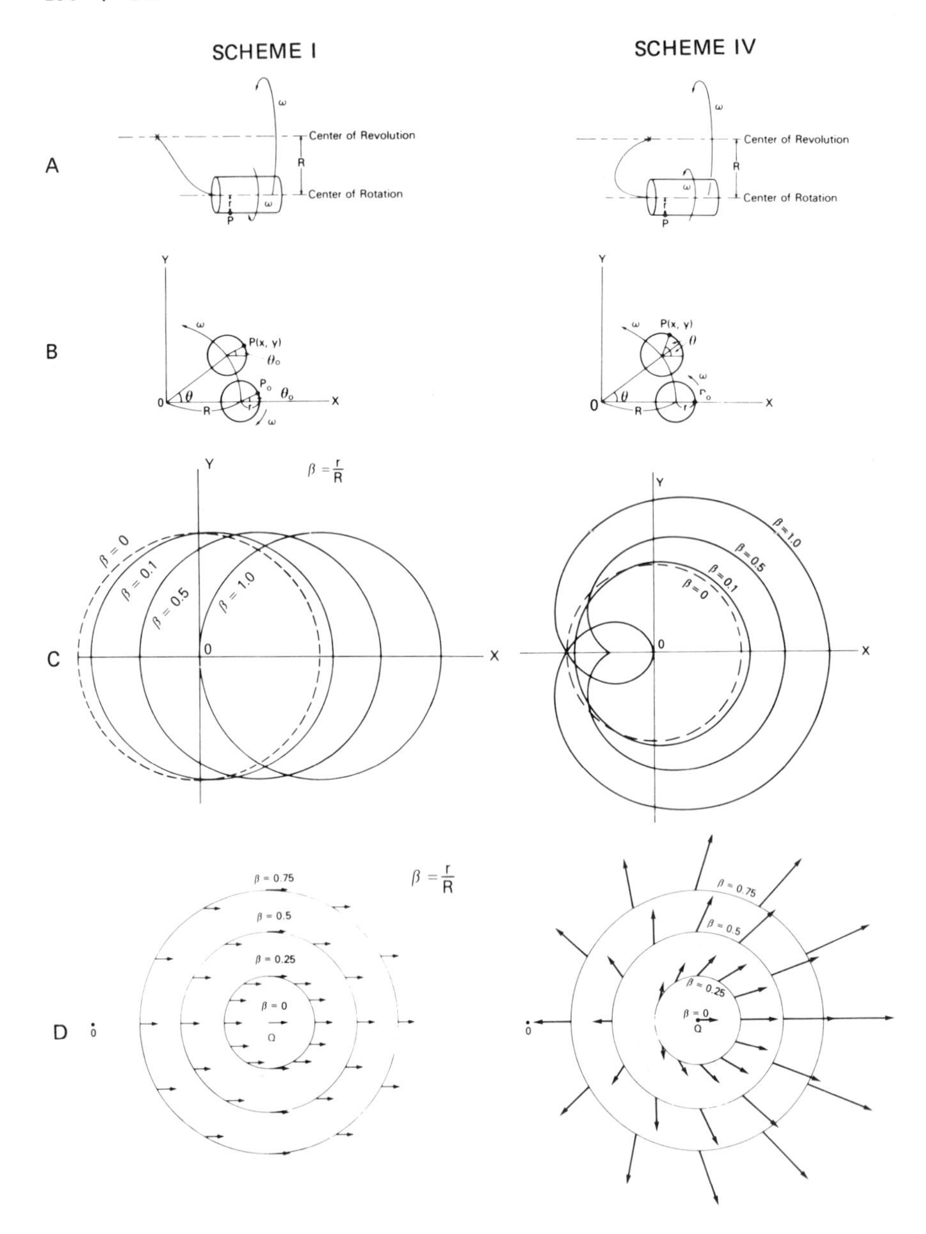

Figure 7 Analysis of acceleration for schemes I (left) and IV (right): (A) diagram of the synchronous planetary motion, (B) coordinate system for analysis, (C) orbits of arbitrary points, and (D) distribution of centrifugal force vectors. (From Ref. 2.)

This shows that the points are subjected to a constant magnitude of acceleration, $R\omega^2$, which rotates around each point at a uniform rate of ω. The acceleration is found to be free of r and θ_0, both of which determine the location of the point on the holder. This indicates that at any given moment every point on the holder is subjected to the identical centrifugal force field acting in a plane perpendicular to the axis of the holder (Fig. 7D, left).

The acceleration analysis is similarly carried out on the planetary motion produced by scheme IV (Fig. 7, right). In Fig. 7A (right), the holder revolves around the central axis of the apparatus (center of revolution) and simultaneously rotates about its own axis (center of rotation) at the same angular velocity in the same direction. In order to simplify the analysis, the coordinate system is selected so that the center of revolution is located at the central point O, whereas both the center of rotation and the arbitrary point are initially located on the x axis (Fig. 7B, right). After the lapse of t, the center of rotation circles around the point O by $\theta = \omega t$, and the location of the arbitrary point, P(x,y), is given by

$$x = R\cos\theta + r\cos 2\theta \tag{6}$$

$$y = R\sin\theta + r\sin 2\theta \tag{7}$$

The orbit of the point computed from Eqs. (6) and (7) displays great variety in shape, depending on the locations of the point on the holder, which are conveniently expressed by the ratio of the radius of rotation r to the radius of revolution R, or $\beta = r/R$ (Fig. 7C, right). When $\beta < 0.25$, the orbit is a single circular loop. As β increases, it becomes heart shaped ($\beta = 0.5$) and then forms a double loop ($\beta = 1.0$) which gradually approaches a double circle with greater β values. These results suggest that the acceleration field also fluctuates periodically during each revolutional cycle of the holder.

The acceleration acting on the arbitrary point is further calculated from the second derivatives of Eqs. (6) and (7) as

$$\alpha = \left[\left(\frac{d^2x}{dt^2}\right)^2 + \left(\frac{d^2y}{dt^2}\right)^2\right]^{1/2} = R\omega^2(1 + 16\beta^2 + 8\beta\cos\theta)^{1/2} \tag{8}$$

acting at the angle γ_x, relative to the x axis, that is,

$$\gamma_x = \pi + \tan^{-1}\left[\left(\frac{d^2y}{dt^2}\right)\Big/\left(\frac{d^2x}{dt^2}\right)\right]$$

$$= \pi + \tan^{-1}\left(\frac{\sin\theta + 4\beta\sin 2\theta}{\cos\theta + 4\beta\cos 2\theta}\right) \tag{9}$$

where $\beta = r/R$, provided that $R \neq 0$.

From Eqs. (8) and (9), the relative centrifugal force vectors acting at various locations on the holder are obtained. In Fig. 7D (right) three circles centered at Q correspond to β values of 0.25, 0.5, and 0.75, as labeled, while point Q, the center of rotation, is at $\beta = 0$. A set of arrows around each circle indicates the distribution pattern of the centrifugal force field at a given moment. It is important to note that the arbitrary point on the holder experiences these vectors in sequence during each revolutional cycle. As clearly shown in Fig. 7D (right), when $\beta > 0.25$, the vector is always directed outward from the circle. The vector also undulates in its relative magnitude and direction according to the location on the holder at given β values. As the β value increases, the magnitude of the relative centrifugal force vector becomes greater while the amplitude of angular oscillation around the axis of rotation decreases.

The results of the above analyses clearly illustrate a great contrast between the two types of synchronous planetary motions. In scheme I a uniform acceleration field rotates around the point, regardless of the location on the holder. Scheme IV, on the other hand, gives a heterogeneous acceleration field which varies in both magnitude and direction according to the location of the point on the holder.

In short, scheme IV produces a more complex pattern of acceleration, but it promises greater versatility in application to CCC by allowing one the choice of the orientation and location of the coil on the holder to obtain the desirable hydrodynamic distribution of the two solvent phases in the coil.

C. Phase Distribution Produced by Synchronous Planetary Motion

Studies on the motion of the two solvent phases in a rotating coil provide the base for development of the CCC technology. While no useful mathematical formula is presently available to predict such hydrodynamic phenomena, one can easily observe the motion and distribution of the two solvent phases in the coil by means of actual experiment. In practice, such hydrodynamic studies can be carried out in two different ways, that is, with or without externally introducing the flow of the mobile phase through the coil; however, most of the previous works have been performed under the elution of the mobile phase, since it gives additional data about the net retention of the stationary phase in the coil, which is extremely useful for performing CCC.

In the past the hydrodynamic distributions of various two-phase solvent systems in the rotating coil have been rather intensively studied with several types of seal-free flow-through coil planet centrifuges (CPCs) such as the (vertical) flow-through CPC (scheme I), the horizontal flow-through CPC (scheme IV), the new (combined) horizontal flow-through CPC (schemes I and IV), and the toroidal CPC (scheme IV). These investigations revealed the interesting finding that schemes I and IV give quite different phase distribution patterns, even though this may be expected from the results of the acceleration analyses described earlier. In the following, typical phase distribution diagrams produced by these schemes will be reviewed together with a new set of data currently available in the author's laboratory.

1. Scheme I Planetary Motion

As shown in the acceleration analysis (Fig. 7D, left), scheme I provides a uniformly rotating centrifugal force field at every point on the holder. This indicates that the vertically mounted coil at any location on the holder is subjected to exactly the same field at every moment during revolution. Consequently, one can eliminate one variable, the coil location on the holder, from the experimental conditions for obtaining phase distribution diagrams of various solvent pairs in the coil. Then the major experimental variables to be considered are the radius of revolution, the helical and internal diameters of the coil, and the operational conditions, such as the speed of revolution of the apparatus, the flow rate of the mobile phase, and physical properties of the applied solvent system.

Figure 8 illustrates a set of phase distribution diagrams obtained with the vertical flow-through CPC at a 30.7-cm radius of revolution [26]. In these experiments the coils were wound on 5-mm-diameter

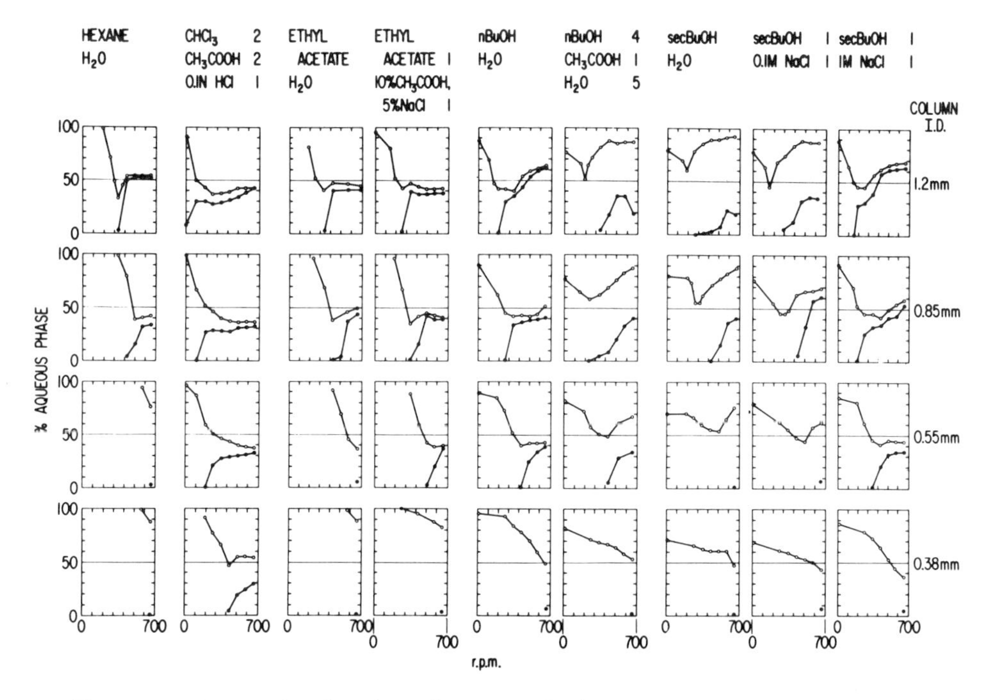

Figure 8 Phase distribution diagrams of nine solvent systems in relation to the internal diameter of the column obtained by the flow-through coil planet centrifuge (scheme I). (From Ref. 26.)

cores, except for the 1.2 mm inner diameter coil, which had a 7-mm core. The mobile phase was always introduced through the head of the coil at a constant linear flow rate of approximately 1 cm/sec. Two-phase solvent systems of a wide hydrophobic spectrum were examined by using both aqueous and nonaqueous phases as the mobile phase.

In each phase distribution diagram (Fig. 8), the ordinate indicates the percentage of the aqueous phase volume occupying the column space, and the abscissa the applied revolutional speed in rotations per minute. Two points were measured for each speed: upper circles for the aqueous phase as the mobile phase and lower circles for the aqueous phase as the stationary phase. As the flow rate is decreased, these two points move closer and will finally reduce to one point which represents the true equilibrium under conditions of no flow. A pair of curves drawn through each point shows some common features between various phase systems examined. The upper curve starts to move down earlier and often displays a sharp spike before it becomes nearly horizontal, whereas the lower curve starts to rise later and smoothly approaches the upper curve.

Effects of the column inner diameter on the phase distribution are visualized by comparing the diagrams from the top (1.2 mm inner diameter) to the bottom (0.38 mm inner diameter) for each phase system. It exhibits a clear trend that a decrease in inner diameter shifts the curves toward faster rotations, especially in phase systems with high interfacial tension apparently due to stronger solvent–wall interaction in the small-bore coil. Therefore the phase distribution diagrams in the top row of Fig. 8 (1.2 mm inner diameter) provide the most informative data for the hydrodynamic behavior of the two solvent phases in the coil. The overall data indicate that the two phases establish a hydrodynamic equilibrium in the coil with a volume ratio (aqueous to nonaqueous) generally ranging between 0.5 and 2. From the trend of the curves at the vicinity of the maximum revolutional speed of 700 rpm, it is quite clear that the above volume ratio would not be radically altered by increasing the coil inner diameter and/or rate of revolution. Thus the complete separation of the two phases (volume ratio of 0 or ∞) required for the dual countercurrent system is not produced with the present set of experimental conditions. Although the two phases may be completely separated around 100 to 200 rpm, the applicable flow rate is limited and efficient mixing of the phases is not attained under such low speeds of revolution.

Similar results have been obtained with a 2.6 mm inner diameter coil on a 1.25 cm diameter core mounted on the pulley-side holder of the combined horizontal flow-through CPC [13,15]. Recently, experiments have been performed in the author's laboratory with a coil of much wider helical diameters. Results obtained from a 1.6 mm inner diameter tube wound around a 10 cm diameter holder with a 20-cm revolutional radius show no significant difference from the data previously obtained with coils of smaller helical diameter. It may be concluded that scheme I synchronous planetary motion usually does not produce complete separation of the two phases along the length of the coil and therefore is not suitable for the dual countercurrent system.

2. Scheme IV Planetary Motion

Analysis of the synchronous planetary motion produced by scheme IV has shown that the centrifugal force vector fluctuates in both magnitude and direction during each cycle of revolution. As illustrated in Fig. 7D (right), the distribution pattern of the force vector depends strongly upon the location of the point on the holder and, with $\beta > 0.25$, the force vector is always directed outward from the circle. Consequently, the phase distribution of the two solvent phases in the coil becomes quite sensitive to the orientation and location of the coil on the holder.

When a coiled column is eccentrically mounted at the periphery of the holder at a location $\beta > 0.25$, two immiscible solvents in the coil are separated by the outwardly directed centrifugal force in such a way that the heavier phase occupies the outer portion, and the lighter phase the inner portion of each turn of the coil, as observed in the basic HSES. However, fluctuation of the centrifugal force field produces vigorous mixing of the two phases at their interface to promote the partition process. With this column geometry, the present scheme has been successfully applied to perform both preparative [11,12,16, 27,28] and analytical [29-31] CCC. In these CCC schemes with eccentric coil orientation, the retention level of the stationary phase is usually below 50% of the total column capacity, as in the basic HSES described in Sec. II.

Performance of the scheme, however, undergoes a radical change when the coil is mounted on the holder in a coaxial orientation. The coaxial coil is simply made by winding a piece of tubing directly onto the holder, forming multiple turns. With this column configuration, the centrifugal force separates the two phases across the diameter of the tube while the fluctuating force field enables one of the phases to pass toward the head of the coil. Consequently, this scheme becomes a modified version of the HDES. Preliminary studies have shown that this particular scheme produces an ideal phase distribution for the dual countercurrent system in a variety of two-phase solvent systems at high speeds of revolution [18].

Recently, the phase distribution produced by the above scheme has been more systematically studied in the author's laboratory on a horizontal flow-through CPC with a 10-cm radius of revolution. Phase distribution diagrams of various solvent systems obtained from a 1.6 mm inner diameter coil with three different β values of 0.25, 0.5, and 0.75 are illustrated in Fig. 9. In each phase distribution diagram, the percentage of retention of the stationary phase is plotted against the

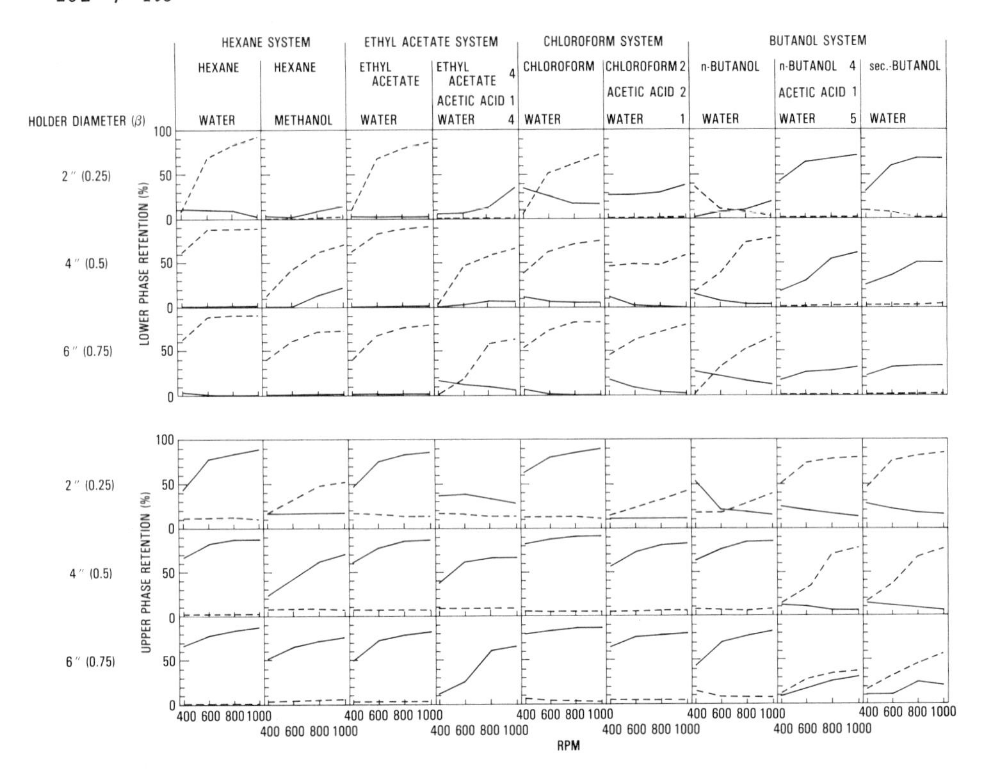

Figure 9 Phase distribution diagrams of nine volatile solvent systems obtained with a horizontal flow-through coil planet centrifuge (scheme IV) with a 10-cm radius of revolution at various β values. The solid line indicates the retention obtained by head-to-tail elution, and the broken line that by tail-to-head elution. A 1.6 mm inner diameter single-layer coil was used.

applied revolutional speeds. The two curves drawn in each chart indicate the results obtained by different elution modes; the solid line is for head-to-tail elution (normal mode) and the broken line for tail-to-head elution (reversed mode). A flow rate of 120 ml/hr was applied in all solvent systems except for the butanol phase systems (right), which were eluted at 60 ml/hr. Both upper and lower phases were used for the mobile phase for each solvent system; the top three rows illustrate the results with the upper phase mobile, and the bottom three rows with the lower phase mobile. All these solvent systems shown in Fig. 9 may be classified into three categories according to their hydrodynamic behavior expressed in the phase distribution diagrams.

The first group characterized by high interfacial tension, low viscosity, and a large density difference between the two phases includes hexane–water, ethyl acetate–water, and chloroform–water, which yield a high percentage of retention of the stationary phase at all β values, provided that the upper phase is eluted in the reversed mode, or the lower phase in the normal mode. The phase distribution patterns of the first group clearly indicate that the upper phase is always distributed on the head side and the lower phase on the tail side, regardless of the β values applied. The second group with intermediate physical properties is typically formed by adding the third solvent to the first group and produces a good retention at β values of 0.5 or greater, with evidence that the upper phase dominates on the head side, and the lower phase on the tail side, as observed in the first group. However, at $\beta = 0.25$, the above relation of the two phases is reversed; the upper phase dominates on the tail side, and the lower phase on the head side, with much lower retention levels. The third group of solvent systems possesses physical properties in contrast to those of the first group, that is, low interfacial tension, high viscosity, and a small density difference between the phases. This group is represented by sec.-butanol–water and n-butanol– acetic acid–water (4:1:5). These phase systems show the highest retention levels of the stationary phase at $\beta = 0.25$, where the upper phase is distributed on the tail side and the lower phase on the head side. It is interesting to note that n-butanol–water, which belongs to the second group, moves into the third group by addition of a small volume of acetic acid. Further experiments have shown that addition of a salt to the third-group solvent systems alters their phase distribution patterns significantly. For example, sec.-butanol–1 M NaCl (1;1) and n-butanol–acetic acid–1 M NaCl (4:1:5) give phase distribution diagrams similar to those in the second-group solvent systems.

Overall results indicate that scheme IV planetary motion provides ideal phase distribution patterns for dual countercurrent system. The data of these phase distribution diagrams would be extremely useful in developing efficient instruments for both countercurrent extraction and high-speed CCC, described in Secs. V and VI, respectively.

V. EFFICIENT COUNTERCURRENT EXTRACTION SYSTEM WITH SINGLE-LAYER COIL

In the separation and purification of biological materials, preliminary extraction may often become the essential initial step. When a small quantity of the sample to be isolated is present in a large volume of fluids, enrichment of the sample is also necessary. A countercurrent extraction scheme with a coil planet centrifuge, based on the dual countercurrent system described earlier, has the capability of efficiently performing both clean-up and sample enrichment in a one-step operation.

A. Apparatus

The design principle of the horizontal flow-through CPC based on scheme IV (Fig. 6) is schematically illustrated in Fig. 10A. A cylindrical coil holder coaxially holds a planetary gear which is coupled to an identical stationary sun gear (shaded) mounted on the central axis of the centrifuge. This gear arrangement produces a synchronous planetary motion (scheme IV in Fig. 6) of the coil holder. The holder revolves around the central axis of the apparatus and simultaneously rotates about its own axis at the same angular velocity and in the same direction. In doing so, the holder always maintains its axis parallel to and at a distance R from the central axis of the apparatus. The coiled column is made by winding a piece of flexible tubing around the holder having a radius r, as shown in Fig. 10A.

The cross-sectional view through the central axis of the prototype fabricated according to the above basic design is illustrated in Fig. 10B. The motor drives the rotary frame around the horizontal stationary pipe (shaded) mounted on the axis of the centrifuge. The rotary frame consists of a pair of aluminum disks rigidly bridged together with multiple links (not shown in Fig. 10) and holds a pair of rotary column holders in symmetrical positions 10 cm away from the central axis of the centrifuge. The bottom holder has a diameter of 15 cm ($\beta = 0.75$), and the top holder a diameter of 10 cm ($\beta = 0.5$). The shaft of each holder is equipped with a plastic planetary gear which is coupled to an identical sun gear (shaded) mounted around the central stationary pipe. In order to provide mechanical stability, a short coupling pipe is mounted coaxially to the free end (right side) of the rotary frame, while the other end of the coupling pipe is supported by a stationary wall member of the centrifuge through a ball bearing. The coiled column was made by winding the desired length of a polytetrafluoroethylene (PTFE) tube around one of the holders while a counterweight was applied to the other holder to balance the centrifuge. A pair of flow tubes from the coiled column is first passed through the center hole of the holder shaft and led through the side hole of the short coupling pipe to reach the opening of the central stationary pipe. These flow tubes are thoroughly lubricated with silicone grease and protected with a piece of plastic tubing at each supported portion to prevent direct contact with metal parts. The speed of revolution can be regulated up to 1000 rpm. The apparatus is a compact tabletop model whose dimensions are approximately 40 X 40 X 42 cm.

B. Preliminary Experiments on DNP Amino Acid Extraction [18]

A series of experiments has been performed to demonstrate the capability of the present scheme to extract a solute present in a large

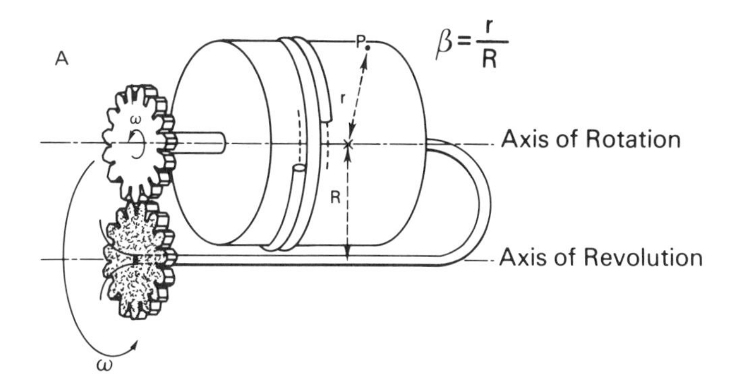

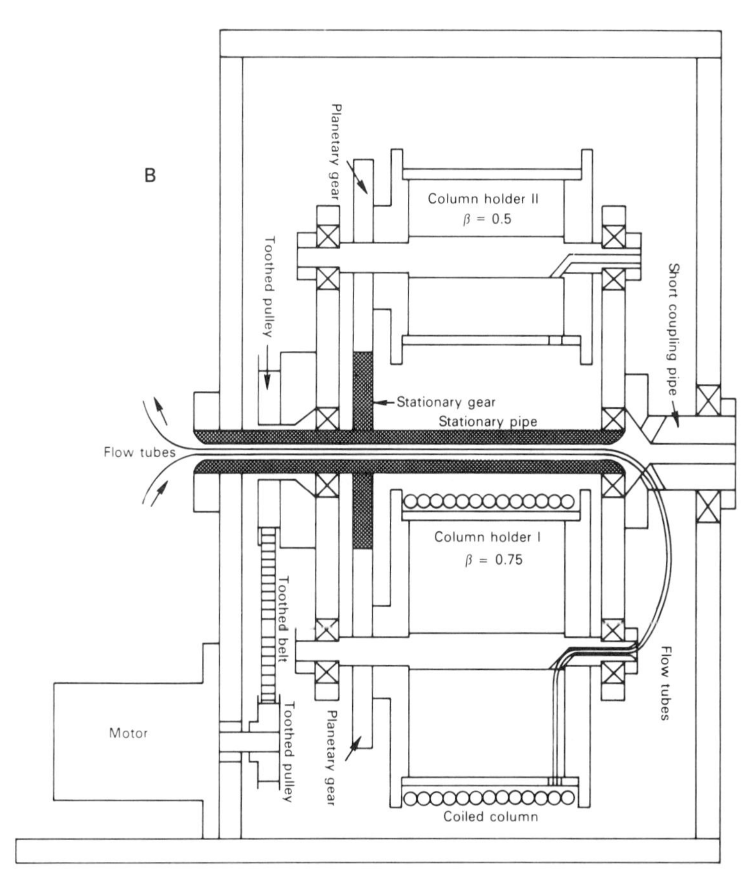

Figure 10 Design of the flow-through coil planet centrifuge for continuous countercurrent extraction: (A) design principle of the apparatus and (B) cross-sectional view of the apparatus. (From Refs. 20 and 18, respectively.)

volume of the mobile phase into a small volume of the stationary phase retained in the coiled column. Successful operation required the following set of conditions:

1. The column should have a capacity to retain a large volume of the stationary phase under a high flow rate of the mobile phase.
2. The sample to be extracted should be partitioned almost exclusively to the stationary phase.

For the present studies a pair of DNP amino acids, N-DNP-L-leucine (DNP-Leu) and δ-N-DNP-L-ornithine (DNP-Orn) were selected, because these colored samples are readily observed through the column wall during the extraction process under stroboscopic illumination. The two-phase solvent systems composed of ethyl acetate–aqueous solution were found to be the most ideal extraction media for these samples.

The phase distribution diagrams of ethyl acetate–aqueous solvent systems are illustrated in Fig. 11. Three curves in each diagram indicate the effects of the different flow rates on the retention of the stationary phase in the 2.6 mm inner diameter coiled column ($\beta = 0.75$). In these diagrams the ideal retention level for extraction may be considered to be over 70% at or near the plateau of the curve, although much lower retention levels can be applied for extraction unless carryover of the stationary phase takes place. Figure 11A and B shows the retention curves of the solvent system composed of ethyl acetate and water where both (A) nonaqueous and (B) aqueous phases were used as the stationary phase. In both cases the ideal retention levels are provided by the revolutional speed of over 600 rpm at all applied flow rates. Figure 11C and D shows similar retention curves for the phase system composed of ethyl acetate–0.5 M NaH_2PO_4 (1:1). In Fig. 11C (nonaqueous phase, stationary) and D (aqueous phase, stationary), retention levels show much improvement over the previous phase system (Fig. 11A and B) owing to a greater difference in density between the two phases produced by addition of the salt. The overall results indicate that these solvent systems provide excellent retention over a broad range of operational conditions for both aqueous and nonaqueous stationary phases.

Extraction of DNP amino acid samples has been performed with the above solvent systems using both aqueous and nonaqueous phases as the extraction medium (stationary phase). A typical extraction procedure may be divided into three steps: extraction, cleaning, and collection. In each operation the column was filled with the stationary phase and the mobile phase containing the sample was eluted through the column in the proper direction while the apparatus was run at 600 rpm. The extraction process was continued until 400 ml of the mobile phase were eluted. Then the mobile phase was replaced by the same phase, but free of solute to wash the column contents. This cleaning process was continued until the additional 100 ml of the mobile phase

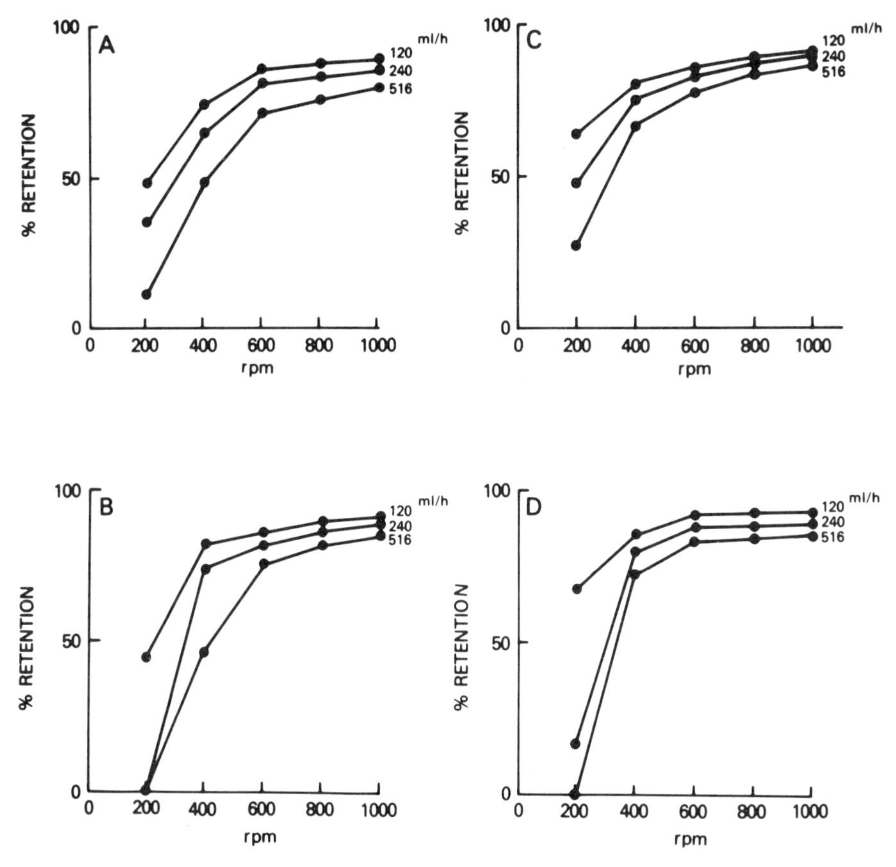

Figure 11 Phase distribution diagrams of ethyl acetate–aqueous systems obtained from a coil planet contrifuge (scheme IV): (A) ethyl acetate–water, retention of the stationary nonaqueous phase by head-to-tail elution of the aqueous phase; (B) ethyl acetate–water, retention of the stationary aqueous phase by tail-to-head elution of the nonaqueous phase; (C) ethyl acetate–0.5 M NaH_2PO_4 (1:1), retention of the stationary nonaqueous phase by head-to-tail elution of the aqueous phase; and (D) ethyl acetate–0.5 M NaH_2PO_4 (1:1), retention of the stationary aqueous phase by tail-to-head elution of the nonaqueous phase. (From Ref. 18.)

were eluted. This would elute out all impurities having partition coefficients of 0.1 or greater. (The partition coefficient is defined here as the solute concentration in the mobile phase divided by that in the stationary phase.) The sample extracted into the stationary phase in the coiled column was collected by eluting with the mobile phase in the opposite direction. This was done by switching the feed and return flow lines either by simply disconnecting the flow lines or the use of a four-way valve. The sample still remaining in the column was then washed out by eluting the column with the other phase originally used as the stationary phase.

The results of the DNP amino acid extraction experiments are summarized in Table 2. In experiments 1 to 3 in Table 2, DNP-Leu was dissolved in 400 ml of the aqueous phase at various concentrations and extracted into the nonaqueous phase retained in the column. The harvested stationary phase volume measured approximately 10 ml, containing over 90% of the original sample. A small amount of the sample still remaining in the column, usually a few percent of the total, was conveniently recovered by eluting the column with several milliliters of the nonaqueous phase. The total sample recovery is always well over 90%, as shown in Table 2. The reduction of the sample concentration from 4 to 0.04 mg % somewhat improved the recovery rate, indicating that no significant sample loss occurs due to the adsorption effects and further reduction of the sample concentration is feasible with high levels of recovery. The mode of elution that uses the nonaqueous phase at the stationary phase has a great advantage in practical extraction, in that the collected solvent is highly volatile and free of salts, facilitating further concentration. It also permits stepwise or gradient elution of the sample by eluting the column with a modified aqueous phase to achieve further purification.

In experiments 4 and 5 in Table 2, the DNP-Orn sample was dissolved in 400 ml of the mobile nonaqueous phase and extracted into the stationary aqueous phase by eluting the column from the tail toward the head. The collected stationary aqueous phase measured approximately 12 ml in volume. This exceeds the volumes in experiments 1 to 3, as expected from the results of the retention studies. The sample still remaining in the column was eluted out with several milliliters of the aqueous phase. The sample recovery ranged over 95%, with an improved figure at the reduced sample concentration, as observed in the previous experiments.

In practice, application of the method to aqueous crude extracts or physiological fluids requires preliminary adjustment of the solvent composition to provide a suitable partition coefficient of the desired material for the applied pair of solvents. In this case pre-equilibration of the two phases may not be essential. Experiment 6 in Table 2 shows an example of operation with such nonequilibrated solvents. The sample DNP-Leu was first dissolved in 400 ml of 0.5 M NaH_2PO_4 aqueous solution containing ethyl acetate at 5%, which is slightly below the saturation level of about 7%. The column was filled with ethyl acetate, followed by elution with the above sample solution. Both extraction and cleaning processes were performed as in other experiments. The sample solution collected from the column measured slightly over 6 ml. This depletion of the stationary phase apparently resulted from use of the nonequilibrated solvent pair, but without any effect on the sample recovery.

The overall results indicate a potential usefulness of the present method in processing a large amount of crude extracts or biological fluids in research laboratories. A small amount of the sample present in several hundred milliliters of the original solution can be enriched in 10 ml of the nonaqueous phase free of salt in 1 hr at a high recovery rate.

C. Extraction of Urinary Drug Metabolites [32]

The present countercurrent extraction method has been applied to the analysis of urinary anthracycline antitumor antibiotics by Nakazawa et al. at the Baltimore Cancer Program, Laboratory of Clinical Biochemistry, National Cancer Institute, Baltimore, Maryland. The solvent system for extraction was selected by a series of sample partition studies. Two original sample drugs, daunorubicin and adriamycin, were equilibrated with three different organic–aqueous solvent pairs containing various kinds of salts at 1 M, and their partition coefficients (sample concentration in the nonaqueous phase divided by that in the aqueous phase) were fluorometrically determined. As shown in Table 3, n-butanol–aqueous systems containing sodium dibasic phosphate (Na_2HPO_4) gave the highest partition coefficient values for both daunorubicin and adriamycin. In consideration of possible crystal formation in a 1 M solution, the solvent system of n-butanol–0.3 M Na_2HPO_4 was used for the extraction studies. The partition coefficients of daunorubicin and adriamycin in this solvent system measured 78 and 31, respectively.

Urine samples collected from patients treated with each drug were separately pooled, and Na_2HPO_4 was added to yield a concentration of 0.3 M followed by saturation with n-butanol for use as the mobile phase. Extraction was performed with a coiled column of 2.6 mm inner diameter, 15 cm helical diameter, and 60-ml capacity. The column was first filled with n-butanol and eluted with 0.3 M Na_2HPO_4 at 500 to 700 ml/hr while the apparatus was run at 650 rpm. After the column contents were equilibrated, the extraction was started by eluting the column with the conditioned urine sample of 1 to 2 liters in volume under the same operational conditions. Harvested extracts from the column were dried, redissolved in methanol, and subjected to reversed-phase high-performance liquid chromatography analysis.

Figure 12 shows typical results obtained from urine samples of patients undergoing daunorubicin chemotherapy. The original urine sample before extraction (Fig. 12A) gave only two anthracycline species, daunorubicin (D_1) and 13-OH-daunorubicin (D_2), with considerable amounts of interfering materials near the solvent front. The urine sample after extraction (Fig. 12B) revealed sharp, enriched peaks for

Table 2 Experimental Conditions and Results of Countercurrent Extraction of DNP Amino Acids

Experiment number	Solvent system	Mobile phase	Stationary phase	Sample (PC)[a]	Sample concentration in mobile phase (mg %)	Extracted mobile phase volume (ml)	Flow rate (direction)	Rotations per minute	Collected stationary phase volume (ml)	Sample recovery (%)
1	Ethyl acetate 1, 0.5 M NaH_2PO_4 2	Aqueous	Non-aqueous	DNP-Leu (<0.01)	4	400	516 ml/hr (head–tail)	600	10.5	94
2	Ethyl acetate 1, 0.5 M NaH_2PO_4 2	Aqueous	Non-aqueous	DNP-Leu (<0.01)	0.4	400	516 ml/hr (head–tail)	600	10.0	97
3	Ethyl acetate 1, 0.5 M NaH_2PO_4 2	Aqueous	Non-aqueous	DNP-Leu (<0.01)	0.04	400	516 ml/hr (head–tail)	600	11.8	100

4	Ethyl acetate 2, 0.5 M NaH_2PO_4 1	Non-aqueous	Aqueous	DNP-Orn (<0.01)	0.4	400	516 ml/hr (tail–head)	600	11.8	97
5	Ethyl acetate 2, 0.5 M NaH_2PO_4 1	Non-aqueous	Aqueous	DNP-Orn (<0.01)	0.04	400	516 ml/hr (tail–head)	600	11.8	100
6	Non-equilibrium system	5% Ethyl acetate in 0.5 M NaH_2PO_4	Ethyl acetate	DNP-Leu (<0.01)	0.4	400	516 ml/hr (head–tail)	600	6.1	99

[a]The partition coefficient (PC) is defined as solute concentration in the mobile phase divided by that in the stationary phase.

Source: Ref. 18.

Table 3 Partition Coefficients of Daunorubicin and Adriamycin in Various Two-Phase Solvent Systems[a]

Salt	n-Butanol—H_2O (1:1)	Chloroform—isopropanol—H_2O (3:1:2)	Ethyl acetate—H_2O (1:1)
	Daunorubicin		
None	2.7	3.2	0.1
$(NH_4)_2SO_4$	5.9	1.6	0.1
NaH_2PO_4	12.4	0.6	0.2
NH_4Cl	16.8	2.3	0.3
KCl	27.3	4.7	0.4
LiCl	20.9	4.4	0.3
CH_3COONH_4	21.6	12.1	1.3
NaCl	24.4	4.6	0.3
Na_2SO_4	36.3	12.5	0.6
Na_2HPO_4	>85.5	—	>6.9
	Adriamycin		
None	2.1	2.3	0.1
$(NH_4)_2SO_4$	2.2	0.4	0.1
NaH_2PO_4	4.3	0.3	0.4
NH_4Cl	7.6	1.1	0.1
KCl	12.3	2.4	0.8
LiCl	10.0	1.8	0.7
CH_3COONH_4	10.1	4.1	0.4
NaCl	11.7	1.8	0.9
Na_2SO_4	10.8	7.7	0.1
Na_2HOP_4	>40.0	—	>1.2

[a]A total of 100 nmol of the sample was added to each solvent system (4 ml) and the partition coefficients (ratio of solute concentration in the nonaqueous phase to that in the aqueous phase) were fluorometrically determined. The concentrations of the salts in the aqueous solution were 1 M in all cases. Various amounts of salt crystals were formed in all solvent systems prepared from 1 M Na_2HPO_4 and no coefficient value could be determined from the chloroform—isopropanol—1 M Na_2HPO_4 system.

Source: Ref. 32.

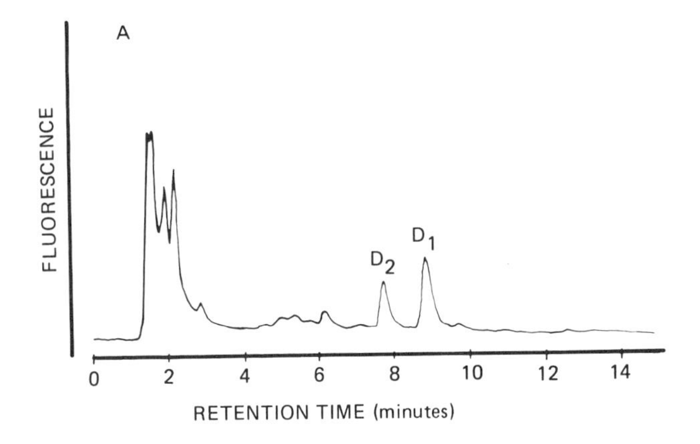

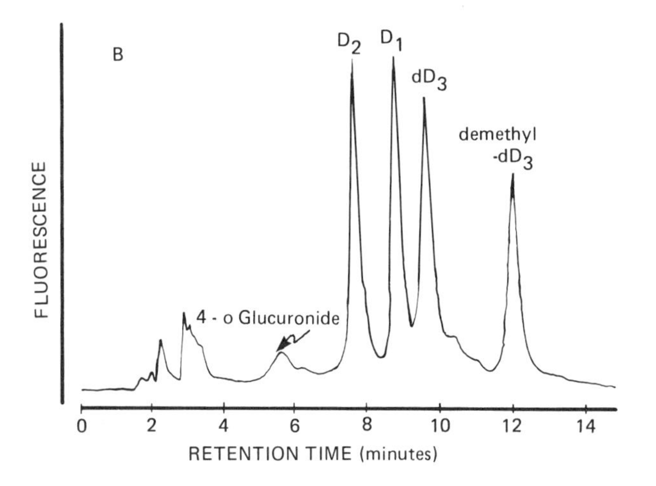

Figure 12 High-performance liquid-chromatographic (HPLC) analysis of daunorubicin and its metabolites from human urine before and after countercurrent extraction. (A) An aliquot of a 24-hr urine sample from a patient treated with daunorubicin was analyzed by reversed-phase HPLC. The early (2 min) column eluate contained considerable interfering materials. Only two anthracycline species, daunorubicin (D_1) and 13-OH-daunorubicin (D_2), were identifiable. (B) The same 24-hr urine sample has been subjected to countercurrent extraction. The early interfering materials were largely excluded from the extract. Sharp, enriched peaks for D_1 and D_2 were evident, and three originally undetectable metabolites of daunorubicin (two aglycones, dD_3 and demethyl-dD_3, and one conjugated species) were made apparent. Anthracycline detection was carried out through fluorescence monitoring. (From Ref. 32.)

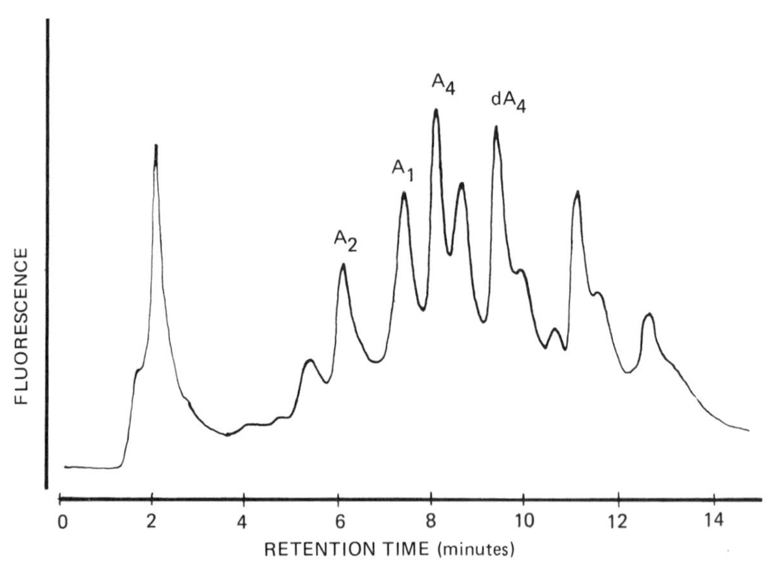

Figure 13 High-performance liquid-chromatographic analysis of adriamycin and its metabolites in a urine sample after countercurrent extraction with the coil planet centrifuge. The patient had been undergoing adriamycin therapy. Adriamycin (A_1) and three metabolites (A_2, A_4, and dA_4) were enriched by extraction. (From Ref. 32.)

D_1 and D_2 together with three originally undetectable metabolites of daunorubicin (two aglycones, dD_3 and demethyl-dD_3, and one conjugated species). Similar results were obtained from the urine sample containing adriamycin and its metabolites (Fig. 13).

VI. HIGH-SPEED PREPARATIVE CCC WITH A MULTILAYER COIL

Countercurrent extraction based on the dual countercurrent system has demonstrated that a short, coiled column coaxially mounted on the holder of a coil planet centrifuge (scheme IV) is capable of yielding an efficient solute partitioning under a high flow rate of the mobile phase [18]. This finding strongly suggstes that the use of a longer coil would produce a chromatographic separation of solutes in a short period of time. Recently, high-speed CCC has been successfully developed with this idea in mind by utilizing a new coil configuration, called a multilayer coil, which compactly accommodates a number of coiled layers around a spool-shaped holder [19,20].

A. Apparatus and Operative Procedure

Figure 14 shows the photograph of our prototype of a high-speed CCC apparatus. The design of the apparatus is essentially the same as that

Figure 14 Tabletop model of a high-speed CCC apparatus with a multilayer coil.

of the coil planet centrifuge used for countercurrent extraction (Fig. 10B), except that the holder is modified into a spool shape to accommodate the multilayer coil, and can be removed from the rotor by loosening a pair of screws at each bearing block.

The column is prepared from a single piece of PTFE tubing of various internal diameters (2.6, 1.6, and 1.0 mm) by winding it directly onto the holder (10 cm in diameter) equipped with a pair of large flanges to support the coiled layers in place. The β value for each column ranges from 0.5 (internal terminal) to 0.8 (external terminal). This range of β values is suitable for most solvent systems, except for some butanol solvent systems, as indicated by the phase distribution diagrams shown in Fig. 9.

The multilayer configuration of the coiled column creates a gradual change of the centrifugal force field along the layers of the coil from the internal terminal toward the external terminal. This centrifugal force field gradient can be effectively used to improve the retention of the stationary phase in the column by applying the proper mode of elution. As clearly shown in the analysis of acceleration (Fig. 7D, right), the centrifugal force vector is always directed outward from the circle for β values greater than 0.25, and therefore the force gradient tends to move the heavier phase from the internal terminal toward the external terminal of the multilayer coil and vice versa for the lighter phase. Since the heavier phase usually moves toward the tail and the lighter phase toward the head in each layer of the coil, the maximum retention of the stationary phase and hence the highest peak resolution can be attained by pumping either the heavier phase through the internal head terminal or the lighter phase through the external tail terminal of the multilayer coil.

Operation of the apparatus is quite simple and requires no particular skill. In each separation the column is first filled with the stationary phase, followed by sample injection through the sample port. Then the column is eluted with the mobile phase in the proper mode described above while the apparatus is run at a given speed of revolution, usually between 800 and 1000 rpm. The eluate is continuously monitored with an ultraviolet monitor and fractionated with a fraction collector as in liquid chromatography. It may be noted here that the recording of the elution curves may be disturbed by a steady carryover of a small amount of the stationary phase, although this does not affect the separation. This problem can be solved by adapting the stream-splitting detection method which has been successfully applied to droplet CCC.

B. Preliminary Experiments on DNP Amino Acid Separation

The potential capability of high-speed CCC with the multilayer coil has been demonstrated on separation of a set of DNP amino acid samples with a two-phase solvent system composed of chloroform—acetic acid—0.1 N hydrochloric acid (2:2:1). Typical chromatograms obtained with a medium-sized column (1.6 mm inner diameter) are shown in Fig. 15, where both aqueous (Fig. 15A) and nonaqueous (Fig. 15B) phases are used as the mobile phase [20]. In both chromatograms all components are well resolved as symmetrical peaks and eluted out in a few hours. More efficient but otherwise similar separations were observed with a small column (1.0 mm inner diameter) under a reduced flow rate of 150 ml/hr.

The sample loading capacity of the preparative column (2.6 mm inner diameter and 400-ml capacity) has been investigated with five DNP amino acid samples and the above solvent system [33]. In the first set of experiments the applied sample volume was varied from 1 to 40 ml, while the sample concentration remained constant at 5 g % in the stationary phase to study the effects of sample dose on the retention of the stationary phase and the partition efficiency. Both aqueous

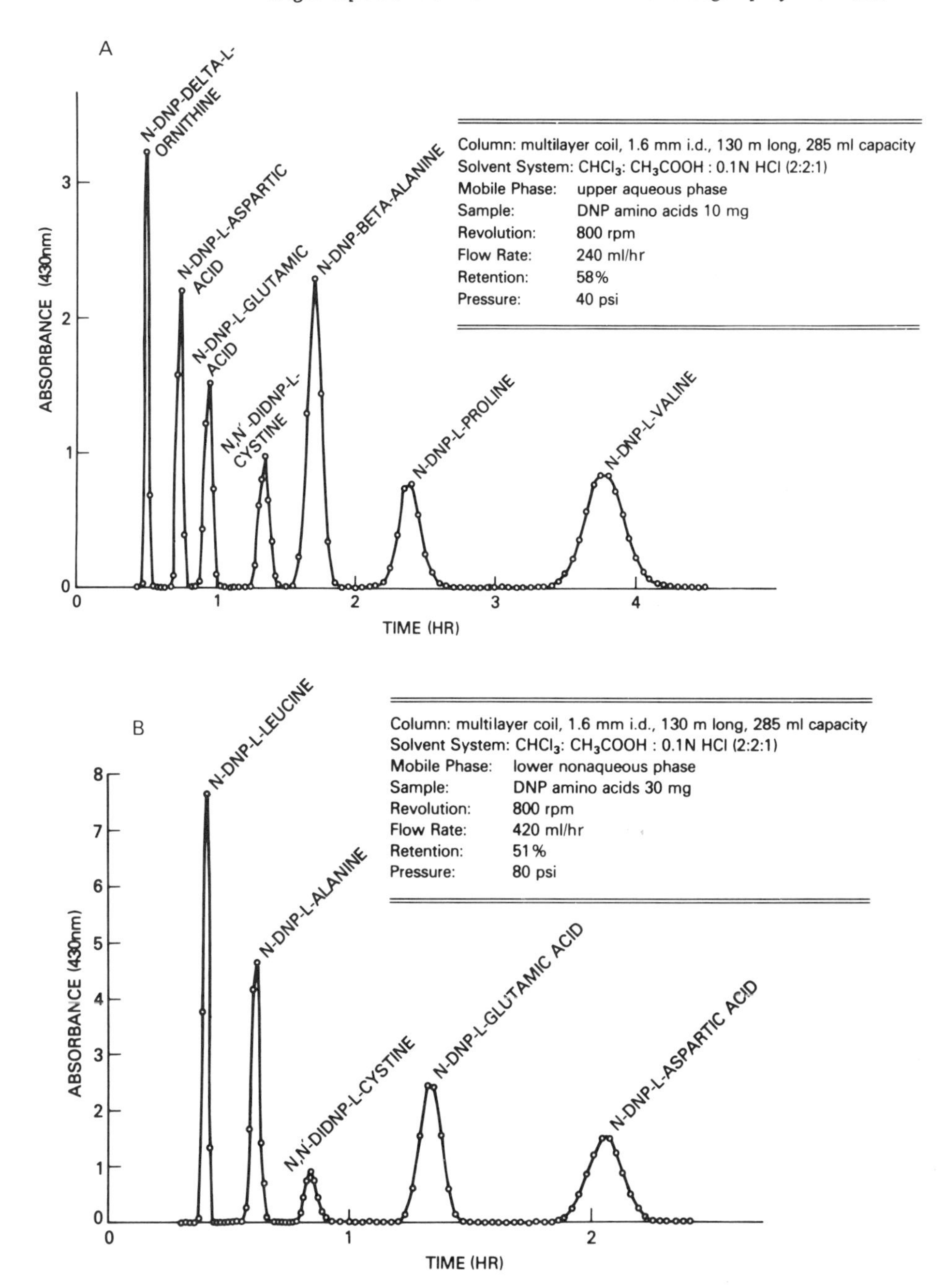

Figure 15 Chromatograms of DNP amino acid samples obtained by high-speed CCC with a medium-sized multilayer coil (1.6 mm inner diameter): (A) aqueous phase mobile and (B) nonaqueous phase mobile. (From Ref. 20.)

and nonaqueous phases were used as the mobile phase at a flow rate of 500 ml/hr.

Figure 16A shows effects of the sample dose on the retention of the sationary phase, where the retention volume, expressed in percentages relative to the total column capacity, is plotted against the applied sample volume. The results clearly indicate a general trend that the retention of the stationary phase decreases with the increased sample size. The results also show that the retention level of the nonaqueous phase (Fig. 16A bottom) is substantially lower than that of the aqueous phase (Fig. 16A top). However, this low retention level of the nonaqueous phase is improved by applying a higher revolutional speed at 1000 rpm, as indicated by the broken line in Fig. 16A bottom. Chromatograms obtained from these experiments are illustrated in Fig. 16B, where individual charts are arranged according to the sample dose and the mobile phase. Although the peak resolution gradually decreases with the increased sample dose, the integrity of the individual peaks is well preserved in all charts, except for the 2-g run with the mobile upper phase. This lowest peak resolution coincides with the lowest retention level of the stationary phase at 28% (Fig. 16A bottom), strongly suggesting that the retention level of the stationary phase may play a critical role in partition efficiency in the present method. As indicated in the phase distribution diagrams (Fig. 9), the retention of the stationary phase is largely altered by physical factors of the solvent systems, such as the interfacial tension, density difference, and viscosity of the two phases. Introduction of a large sample dose will locally alter these factors to result in a lowered retention level of the stationary phase. Under these circumstances, application of a higher revolutional speed and/or lower flow rate may often improve both the retention and peak resolution.

In the second set of experiments, the effects of the sample volume (10, 20, and 40 ml) and the choice of the sample diluents (mobile phase, stationary phase, and the mixture of these) were studied on separation of 1 g of the DNP amino acid mixture. The results showed that the peak resolution was not significantly affected by the sample diluent until the sample volume was increased to 40 ml. With a 40-ml sample volume, the best results were obtained by dissolving the sample in the stationary phase. Dissolving the sample in the equal-volume mixture of the two phases always yielded fair results. The overall experimental results indicated that the present method can yield efficient chromatographic separations of 1-g samples in a few hours of elution.

C. Chromatograms of Various Samples with Typical Two-Phase Solvent Systems

Versatility of high-speed CCC has been demonstrated by separation of a variety of biological samples with two-phase solvent systems, which covers a broad spectrum of hydrophobicity. These separations were most extensively performed with a medium-sized column (1.6 mm inner diameter and 285-ml capacity) at a revolutional speed of 800 rpm [20].

Figure 17 illustrates a typical chromatogram of indole plant hormones obtained with a hydrophobic solvent system composed of chloroform—acetic acid—water (2:2:1). A chromatogram of gramicidins, relatively hydrophobic peptides, on chloroform—benzene—methanol—water (15:15:23:7) is shown in Fig. 18, where three major components, A, B, and C, were further partially resolved into their valine and isoleucine analogs. The relatively poor resolution of the gramicidin C analogs in this chromatogram would be much improved if the lower nonaqueous phase were used as the mobile phase. Purines and pyrimidines were separated with a hydrophilic solvent system composed of n-butanol—1 M potassium phosphate (pH 6.5) by using both nonaqueous (Fig. 19A) and aqueous (Fig. 19B) phases as the mobile phase. In addition to plain elution, a pH gradient elution was also successfully applied to separate a set of dipeptides (Fig. 20). The column was eluted with the aqueous phase to introduce a linear gradient of di-

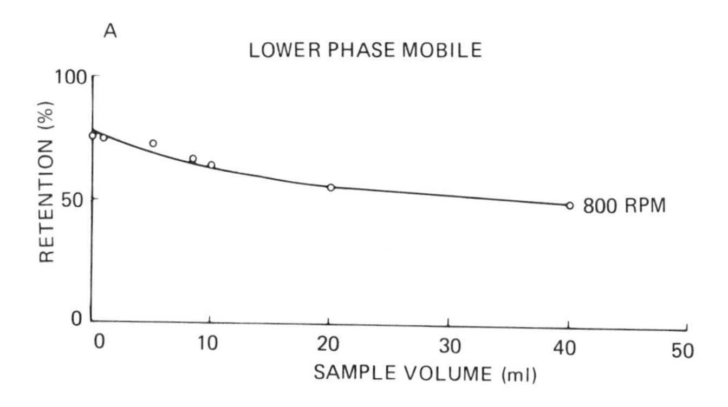

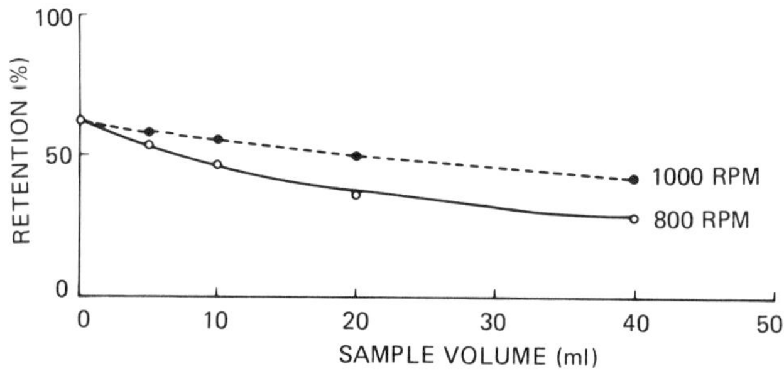

Figure 16A Effects of sample volume on the retention level of the stationary phase. (From Ref. 33.)

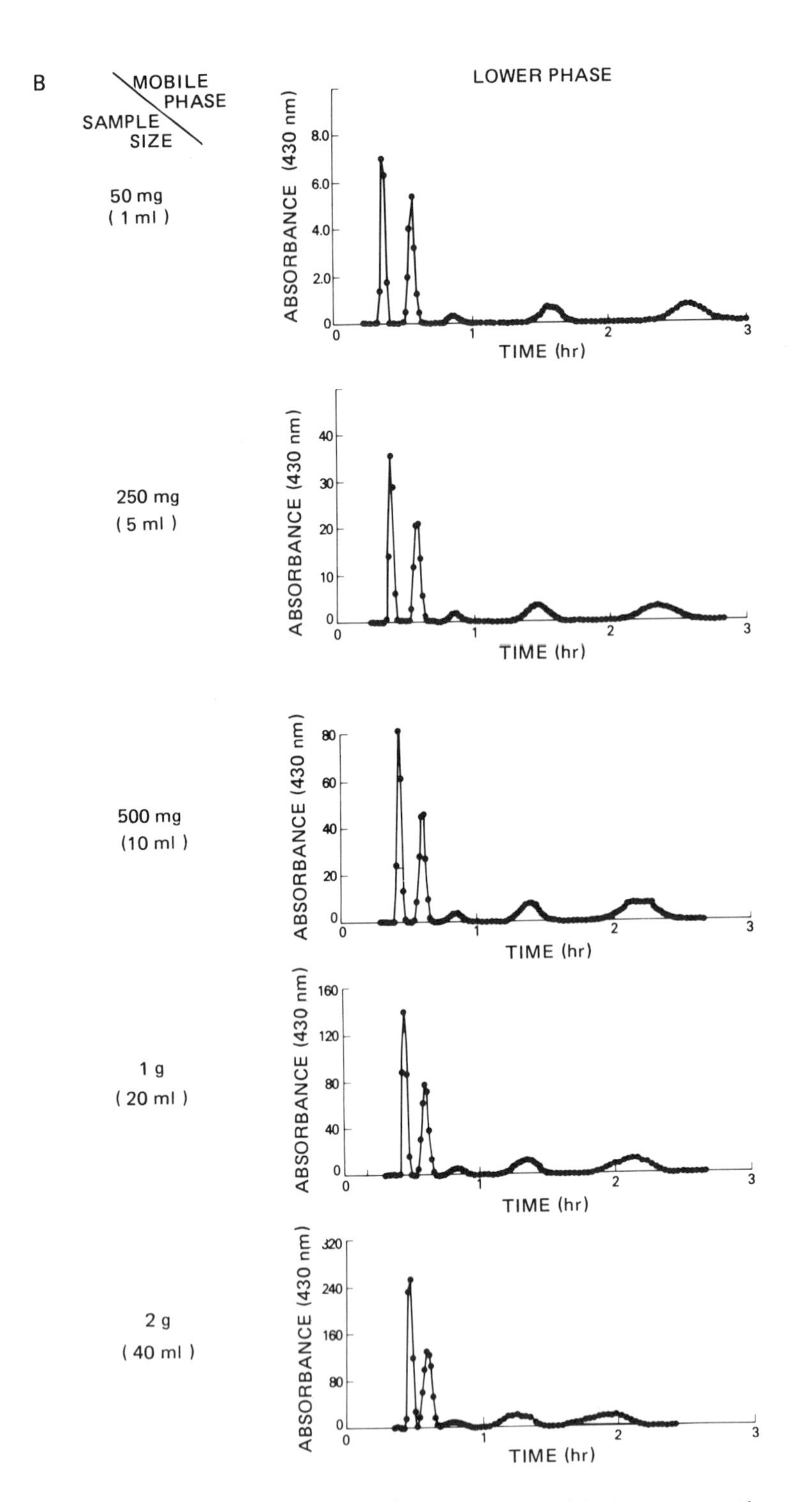

Figure 16B Peak resolution of DNP amino acids in preparative-scale separations with a large multilayer coil (2.6 mm inner diameter). (From Ref. 33.)

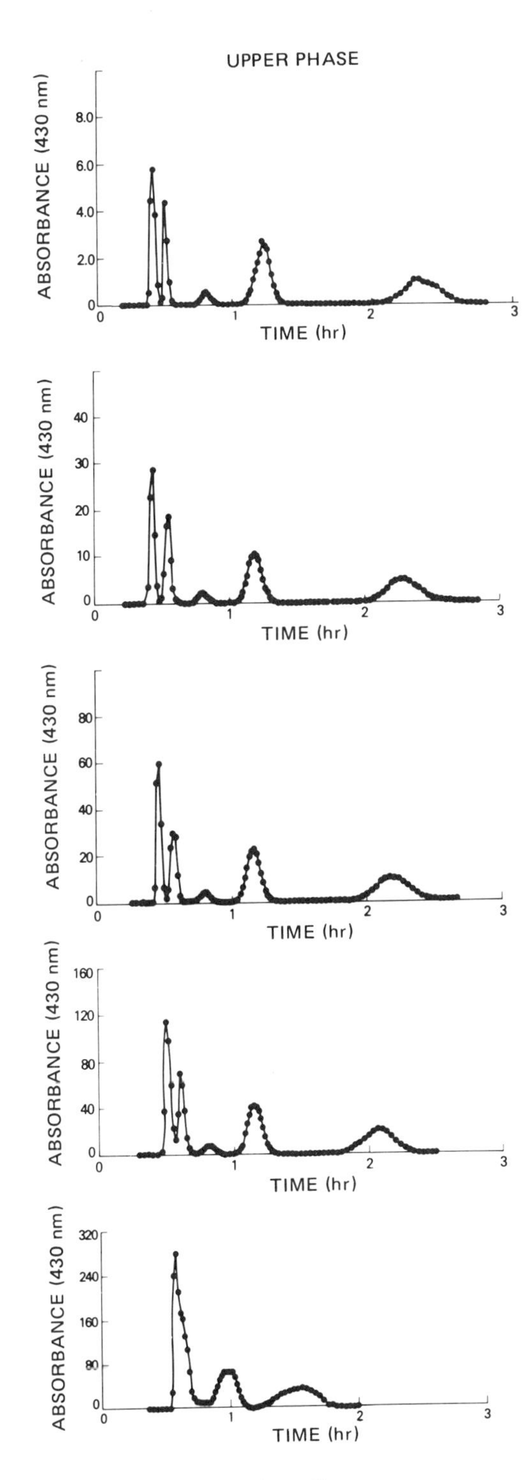

Figure 16B (continued)

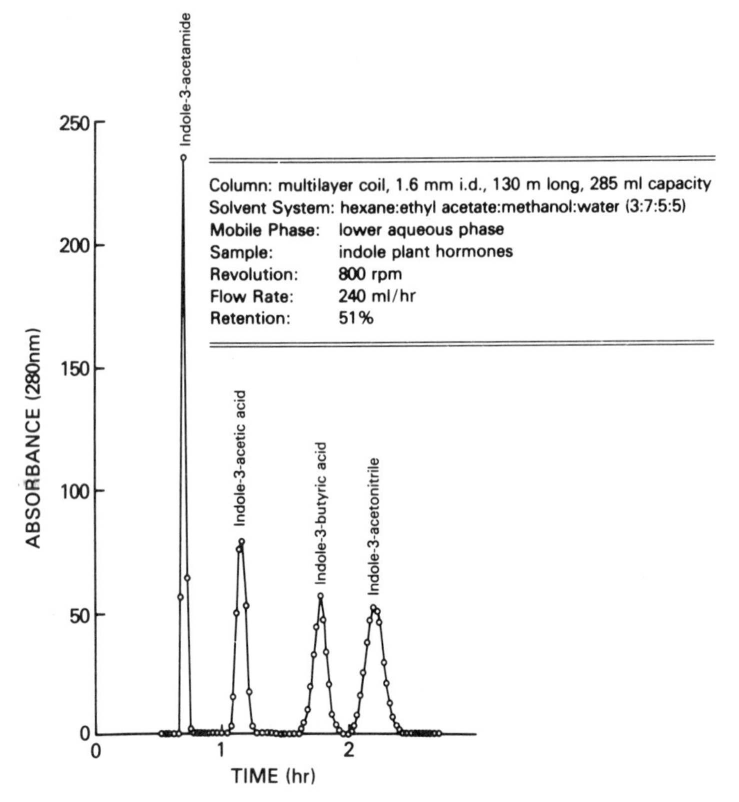

Figure 17 Chromatogram of indole plant hormones with a medium-sized multilayer coil (1.6 mm inner diameter). (From Ref. 20.)

chloroacetic acid concentration between 1 and 0% on the base of n-butanol–0.1 M ammonium formate. The similar gradient elution may be applied to the separation of polypeptides and other compounds.

All chromatograms shown above were obtained from solvent systems which permit satisfactory retention of the stationary phase under the present operational conditions (at β values between 0.5 and 0.8 for a 10-cm radius of revolution). However, the most hydrophilic solvent systems in the third group, such as sec.-butanol–water and n-butanol–acetic acid–water (4:1:5), fail to produce enough retention under these operational conditions. As clearly illustrated in the phase distribution diagrams (Fig. 9), these particular solvent systems can yield the best retention levels at β values around 0.25 and, therefore, the coil holder and the multilayer coil should be modified to meet the above requirement.

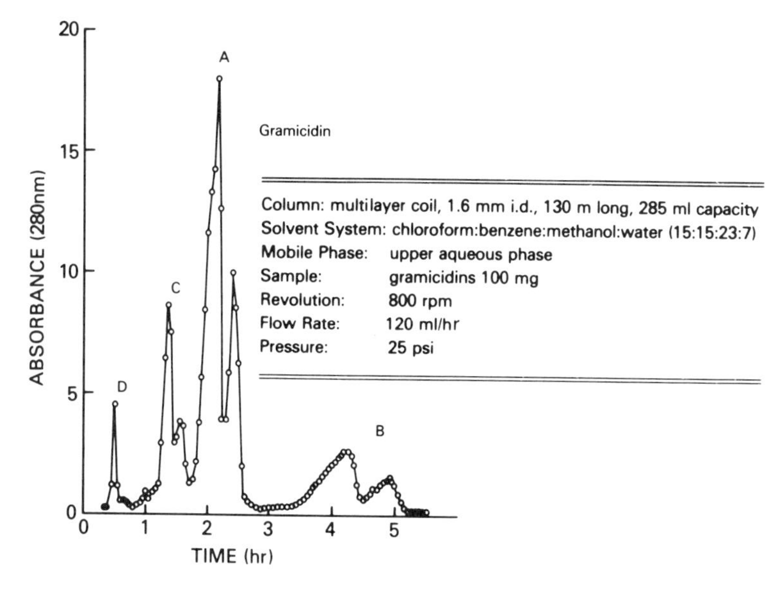

Figure 18 Chromatogram of gramicidins with a medium-sized multilayer coil (1.6 mm inner diameter). (From Ref. 20.)

VII. CONCLUSION

In 1960s the CCC method was developed to take over the role of classic CCD, which was extensively used in 1950s for the separation and purification of various important materials. In virtue of being the genuine liquid–liquid partition technique, CCD and CCC share inherent advantages—such as reliable sample recovery, high purity of yielded fractions, and excellent reproducibility and predictability—over liquid chromatography, which employs the solid supports. On the other hand, both CCD and CCC require longer separation times and have lower partition efficiencies compared with the liquid chromatography. A sizable separation with the Craig CCD apparatus often takes many days of operation. Droplet CCC, developed in 1970, substantially shortened the separation times of CCD without sacrificing partition efficiency. During the succeeding decade the development of CCC has continued to improve both the separation times and partition efficiency to meet the demand of users in the research laboratories. Development of the horizontal flow-through CPC in 1977 has further shortened the separation times in such a way that most separations can be completed with overnight runs at efficiencies exceeding 1000 theoretical plates. The advent of high-speed CCC in 1980 has finally revolutionalized CCC technology to cut the separation times down to a few hours without losing the peak resolution.

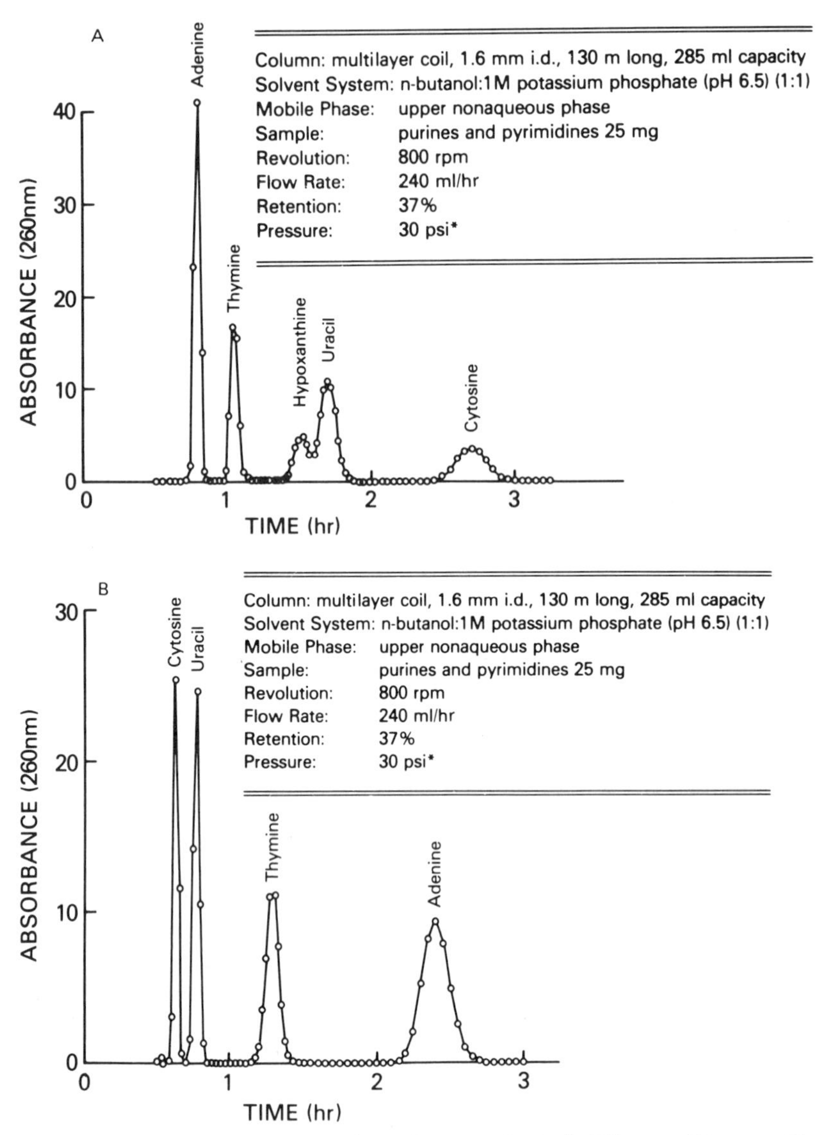

Figure 19 Chromatograms of purines and pyrimidines with a medium-sized multilayer coil (1.6 mm inner diameter) using both (A) upper and (B) lower phases as the mobile phase. (From Ref. 20.)

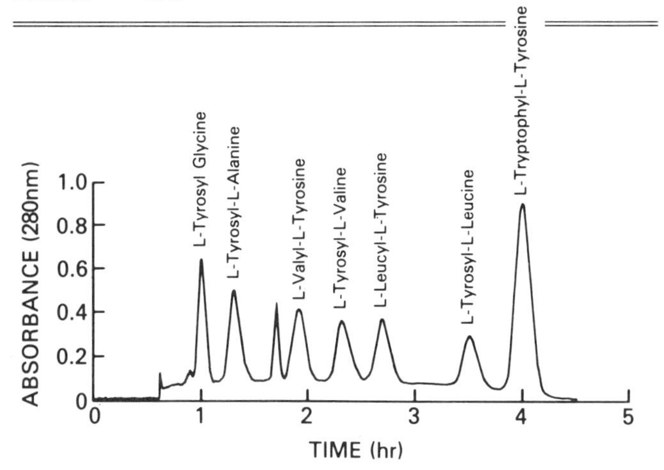

Figure 20 Chromatogram of dipeptides by a gradient elution obtained with a medium-sized multilayer coil (1.6 mm inner diameter). (From Ref. 20.)

While refinement, improvement, and innovation are still actively in progress toward establishment of the dual countercurrent extraction and CCC, we feel that CCC technology has presently reached the stage where many important applications can be successfully planned and performed. As a complementary method (rather than a competitive method) to HPLC, CCC will soon occupy many essential parts in flow diagrams for the separation and purification of various materials in research laboratories.

REFERENCES

1. Y. Ito and R. L. Bowman, Science *167*, 281 (1970).
2. Y. Ito, J. Biochem. Biophys. Met. *5*, 105 (1981).
3. Y. Ito, Protein, Nucleic Acid and Enzyme, *26*, 1020 (1981).
4. N. B. Mandava, Y. Ito, and W. D. Conway, Am. Lab. *14*(October), 62 (1982).

5. N. B. Mandava, Y. Ito, and W. D. Conway, Am. Lab. *14*(November), 45 (1982).
6. Y. Ito and R. L. Bowman, J. Chromatogr. Sci. *8*, 315 (1970).
7. Y. Ito and R. L. Bowman, Anal. Chem. *43*, 69A (1971).
8. Y. Ito and R. L. Bowman, Anal. Biochem. *85*, 614 (1978).
9. I. A. Sutherland and Y. Ito, J. High Resolution Chrom. and Chrom. *1*, 171 (1978).
10. T. Tanimura, J. J. Pisano, Y. Ito, and R. L. Bowman, Science *169*, 54 (1970).
11. Y. Ito and R. L. Bowman, Anal. Biochem. *82*, 63 (1977).
12. Y. Ito and R. L. Bowman, J. Chromatogr. *147*, 221 (1978).
13. Y. Ito, Anal. Biochem. *100*, 271 (1979).
14. Y. Ito, J. Chromatogr. *188*, 33 (1980).
15. Y. Ito, J. Chromatogr. *188*, 43 (1980).
16. Y. Ito and G. J. Putterman, J. Chromatogr. *193*, 53 (1980).
17. E. A. B. Brown and Y. Ito, J. Biochem. Biophys. Met. *3*, 77 (1980).
18. Y. Ito, J. Chromatogr. *207*, 161 (1981).
19. Y. Ito, J. Chromatogr. *214*, 122 (1981).
20. Y. Ito, J. Sandlin, and W. G. Bowers, J. Chromatogr. *244*, 247 (1982).
21. Y. Ito and R. Bhatnagar, J. Liquid Chromatogr. *7*, 257 (1984).
22. Y. Ito and R. Bhatnagar, J. Chromatogr. *207*, 171 (1981).
23. Y. Ito and R. L. Bowman, Science *173*, 420 (1971).
24. Y. Ito and R. L. Bowman, Anal. Biochem. *65*, 310 (1975).
25. Y. Ito, R. L. Bowman and F. W. Noble, Anal. Biochem. *49*, 1 (1972).
26. Y. Ito and R. L. Bowman, J. Chromatogr. Sci. *11*, 284 (1973).
27. M. Knight, Y. Ito, and T. N. Chase, J. Chromatogr. *212*, 356 (1981).
28. H. Nakazawa, P. A. Andrews, N. R. Bachur, and Y. Ito, J. Chromatogr. *205*, 482 (1981).
29. Y. Ito, Anal. Biochem. *102*, 150 (1980).
30. Y. Ito, J. Chromatogr. *192*, 75 (1980).
31. N. B. Mandava and Y. Ito, J. Chromatogr. *247*, 315 (1982).
32. H. Nakazawa, C. E. Riggs, Jr., M. J. Redwood, N. R. Bachur, R. Bhatnagar, and Y. Ito, J. Chromatogr. *307*, 323 (1984).
33. J. L. Sandlin and Y. Ito, J. Liquid Chromatogr. *7*, 323 (1984).

7

Determination of the Solubility of Gases in Liquids by Gas-Liquid Chromatography

Jon F. Parcher, Monica L. Bell, and Ping J. Lin* / University of Mississippi, University, Mississippi

*Current affiliation: Feng Chia University, Taichung, Taiwan

I. INTRODUCTION

Accurate data for the solubility of gases in liquids are required for numerous practical and theoretical applications, especially for design calculations in the petroleum and petrochemical industries. Some gases such as hydrogen sulfide, carbon dioxide, and sulfur dioxide are of particular interest because of the demand for removal of these "acid" gases from refinery and natural gases. Carbon dioxide and light hydrocarbon gases are used for enhanced oil recovery, and accurate data concerning the solubility of these gases in high molecular weight solvents are critical for the optimization of miscible displacement processes. Other light gases are used for extractive distillation and supercritical fluid extraction, and solubility data are needed for initial screening and subsequent design of these processes.

The thermodynamic properties of gases in liquids are also of interest, especially for high molecular weight liquids where a significant size and weight disparity between the solute and solvent can produce unique solution properties. Good models have been developed for some hydrocarbon systems; however, much more data are needed for other gases, such as those already mentioned, as well as water, nitrogen, oxygen, the nitrogen and sulfur oxides, and the inert gases.

Even though the experimental determination of gas solubilities is often difficult and time-consuming, the demand is such that a large number of these measurements have been carried out at room temperature. Collections of such data have been presented by Wilhelm and Battino [1,2] in recent reviews. Unfortunately, gas solubility data are scarce for temperatures other than the standard 25°C and for solvents other than water and hydrocarbons.

Gas–liquid chromatography (GLC) would, at first glance, appear to be the ideal procedure for this type of measurement. The method is fast, accurate, well established, and relatively simple. Mixtures of gases can be studied, and most commercial instruments allow measurements over a wide range of temperatures. The solvent, or even a mixture of solvents, can be used as the stationary liquid phase in a common packed bed column. This simple configuration is only applicable to nonvolatile solvents. Volatile solvents can, however, be used if the carrier gas is presaturated with the solvent [3]. The retention time of the gas injected as a discrete sample can be related directly and simply to the Henry's law constant or the Raoult's law activity coefficient. These measurements have been carried out successfully for hundreds of vapor–liquid systems, but seldom with the gases mentioned above. These gases are very insoluble in most solvents, and the errors in the measurement of the retention time of the solute and the void volume of the column have restricted the *direct* application of GLC in the determination of gas solubilities.

II. ANCILLARY GLC METHODS

Gas chromatography has, however, proven to be a popular technique for the indirect determination of gas solubilities. In these ancillary methods the primary gas-chromatographic data are not retention times but, rather, the peak areas or detector response as a function of sweeping time. Gas–liquid chromatography has been used as an ancillary method for a great many solubility studies, and the investigations discussed below are simply recent examples of each technique.

A. Direct Sampling of One or Both Phases

Postequilibration analysis of one or both phases is the most common use of GLC for gas solubility measurements. Horvath et al. [4] recently described a method in which the liquid was saturated with flowing sample gas. A sample of the liquid was subsequently analyzed with a gas chromatograph. Several other investigators have applied the same general methods [5–7], and Legret et al. [8] used special sampling techniques to measure gas solubilities at high pressures (50 MPa). These methods are simple and accurate, but require calibration of the GLC system and are often time-consuming because of the interval required for the system to attain equilibrium.

B. Inert Gas Stripping Techniques

1. Analytical Methods

In many cases, it is necessary or simply more convenient to analyze the saturated liquid only for the dissolved gases in a known volume or weight of solvent. This is generally the case with high molecular weight, nonvolatile solvents. The earliest applications of this methodology to gas solubility measurements [9–11] involved the use of a stripping chamber in which a known amount of saturated liquid was introduced. The volatile components were then stripped form the solvent directly into a gas–liquid chromatograph by an inert carrier gas. More recently, Cosgrove and Walkley [12] used the same scheme to measure the solubility of several gases in both H_2O and ^2H_2O. These authors presented accurate solubility data at mole fractions as low as 3×10^{-6} at 1 atm. This corresponds to Henry's law constants in the range of 10^5 atm.

2. Elution Methods

Leroi et al. [13] proposed a unique method of inert gas stripping. In this technique the liquid was saturated with the gas of interest and then stripped with an inert gas as in the previously described methods. However, the *rate* of removal of the gases was monitored, rather than the total amount of gas dissolved. The rate of elution of the solute gas could be related to the activity coefficient. Extrapolation to the starting time gave an accurate value of the activity coefficient or the Henry's law constant. The initial work was carried out with benzene and heptane as the solutes. However, the same group [14,15] later measured the solubility of several light hydrocarbons in heavy liquids with this method. The accuracy was limited by the uncertainty in the determination of the volume of the gas phase in the experimental apparatus; however, accurate data were obtained up to values of 200 atm for the Henry's law constants. The method is relatively simple; it does not require calibration of the chromatograph or any analytical methods, but is limited by the same factors encountered in the direct GLC methods, that is, accurate determination of the gas phase volume.

C. Diffusion Methods

Another type of ancillary technique is based upon diffusion of the gases from a saturated solution through a polymer membrane into a gas chromatograph [16]. This method allowed continuous sampling of gases dissolved in liquids. However, the accuracy of the method was affected by several exogenous variables, such as stirring rate, equilibration time, temperature, and membrane properties.

III. DIRECT GAS-CHROMATOGRAPHIC METHODS

A. High Pressure

Engineers recognized the potential applications of GLC for gas–liquid phase equilibrium measurements in the early 1960's. Kobayashi [17–21], in particular, carried out an elegant series of investigations of the solubility of permanent gases and light hydrocarbons in liquids and even liquid mixtures at high pressures. At high pressures few systems obey Henry's law, so engineering data are usually presented in the form of vaporization equilibrium ratios, K, where K is the ratio of the mole fraction of solute in the vapor phase to the mole fraction of solute in the liquid phase. This data format is not commonly used for low-pressure data, but K data can be related to the more common Henry's law constant or mole fraction solubility (at 1 atm) if the total pressure is known.

These investigations are some of the very few GLC studies of this type that have been carried out with *binary* liquid systems. The composition of the liquid in equilibrium with the vapor was determined either from independent isotherm data or by means of tracer pulse chromatography [22]. These workers recognized the significance of the accurate measurement of the void volume of the GLC column and developed several unique methods for this measurement.

B. Low Pressure (Henry's Law)

Very few reliable direct determinations of gas solubility at infinite dilution ($P \to 0$) have been carried out to date because of the problem of the accurate determination of the retention time t_m of an unretained solute needed to determine the void volume of the column. In spite of the early engineering work cited above, few chromatographers have used any of the proposed schemes. Primarily because of the limitations of the common GLC detectors.

In 1969 Ng et al. [23] attempted to use GLC directly to determine the solubility of C_1 to C_3 hydrocarbons in C_{18} to C_{22} hydrocarbon solvents, using the retention time of an air peak as a measure of t_m. This group later studied some of the same systems in a static still [24] and showed that their GLC data were inaccurate for gases such as methane and ethane. The error was presumably due to the finite solubility of air in the solvents. In another investigation, Lenoir et al. [25] recognized the need to allow for the finite solubility of air and used an independent measurement to obtain the required data. This correction procedure was used to obtain the solubility of 12 gases in 19 solvents. Unfortunately, the solubility data obtained by GLC still did not agree with later measurements by static methods [26–28] for some of the least-soluble gases. More recently, Riedo et al. [29] have used the same type of correction, but for neon rather than air, to measure the specific retention volume of the inert gases, hydrogen, nitrogen, and several alkanes in a high molecular weight hydrocarbon solvent. These data could not be compared with static data because of the synthetic solvent; however, this is one of the few direct chromatographic investigations of gas solubility where the significance of the accurate determination of t_m was fully appreciated.

Normally, the factor limiting the accuracy of physicochemical measurements by GLC is the uncertainty in the amount of stationary liquid phase. For gases, however, it is more often the uncertainty in the determination of the column void volume and t_m that is the limiting factor in the accuracy of the method.

C. Determination of the Column Void Volume

The use of flame ionization detectors (FIDs), which do not respond to air or the inert gases, has produced a great deal of controversy and discussion in the literature regarding the exact significance of and the best way to determine the dead time for any system. Many different schemes have been proposed, and several of these have been reviewed recently [30–32].

1. Real Gases

The retention times of real gases such as air, neon, methane, and even tritium [20] have been used at various times as direct measurements of t_m. The use of the retention time of methane for t_m with FIDs has been particularly questioned [33]. However, it has been shown [34,35] that in certain cases methane will give a satisfactory measure of t_m. The critical parameter, of course, is the retention time (t_{ri}) of the solute of interest. If this is large compared to t_m, then the exact evaluation of the void volume is not critical. Unfortunately, it is with gaseous solutes that this evaluation is the most critical, and the use of the retention time of any real gas, even helium or neon, for t_m will induce generally intolerable errors in the measurements.

2. Hypothetical Unretained Gases

a. n-Alkanes. Extrapolation procedures to determine the retention time of a hypothetical unretained solute from the retention data of real n-alkanes have proven to be a popular method for the determination of t_m with FIDs. Wainwright and Haken [30] recently reviewed and evaluated the different procedures that involve the use of n-alkanes. The details differ, but the basic idea is to find the value of t_m that produces the "best" straight line for a plot of $\ln(t_{Ri}-t_m)$ versus the carbon number of a series of n-alkanes. All of the schemes require retention data for at least three n-alkane solutes, and the solutes must have retention times not much greater than those of these gas solutes to avoid a long and uncertain extrapolation. This means that the three alkanes best suited for the determination of t_m in gas solubility measurements are methane, ethane, and propane. The main difficulty with this approach is the unique properties often observed for the first members of any homologous series. The linearity of plots of $\ln(t_{Ri}-t_m)$ versus alkane carbon number is uncertain for these three alkanes. For example, Ettre [32] reported data that indicated that methane would have to be assigned an "effective" carbon number of 1.5 to obtain a linear plot of the alkanes on SF-96. Wainwright et al. [36] showed that the plots were perfectly linear for several liquid phases with methane assigned the rational carbon number of 1.0. Another recent report was given [35] in which methane was assigned an effective carbon number of 0.5 in order to obtain a linear plot. To make matters worse, Haken et al. recently [37,38] reported that the effective carbon number for methane is variable, but usually close to unity. This uncertainty in the validity of the extrapolation equation with methane as a solute makes the use of this method for the evaluation of t_m very questionable at best.

b. Inert Gases. The same types of linearization schemes have been used with the retention times of a series of inert gases. The experimental heats of adsorption and polarizabilities of the gases were first used as the linearization parameters [39–43]. Intermolecular potential parameters such as Lennard–Jones, Buckingham, Kirkwood–Müller, Slater–Kirkwood, and London have also been used for linearization parameters [44–46]. The modified Buckingham [45] parameters have proven to be especially valuable parameters for this type of linearization scheme [47,48]. The use of inert gases for the evaluation of t_m requires the use of a detector which will respond to these gases. This requirement eliminates the use of one of the most popular detectors, the FID. Presently, this linearization scheme with the inert gases is the most accurate method for the determination of the column dead time.

3. Other Methods

In some of the earliest work [17], the void volume was calculated from the physical dimensions of the column, that is, the void volume was determined as the difference between the volume of the empty column and the volume of the firebrick packing.

Another method for the evaluation of t_m involves the use of some gas or other solute with accurately known solubility in the solvent at the experimental temperature. The dead time can then be determined from experimental measurements of the uncorrected retention time t_{Ri} and known solubility of the reference solute. This method has been used with neon [29], air [25], and methane [21]. It can be very accurate, but the accuracy is limited by the quality and availability of reference solubility data. The need for this reference data also severely limits the range of applicability of this method for the evaluation of t_m.

Other workers have used less common methods, such as the leading edge of a methane peak [49] or the "vacancy" peak obtained from injection of helium in a detectable carrier gas, such as methane [50].

D. Discussion

Even if an accurate method is available for the determination of t_m, there are numerous other precautions that must be observed in order to obtain accurate solubility data from chromatographic systems. These requirements have been extensively discussed in the literature [51–53]. Because of the low solubility of most gases in the liquids used as stationary phases, it is often necessary to sacrifice efficiency for retention; that is, very high liquid loadings and long columns must be used to attain satisfactory accuracy in the measurement of t_{Ri}-t_m. This same requirement necessitates the use of gas sampling valves for sample introduction and a fast, accurate data system. A single injection of a mixture of gases is better than individual injections; however, this is only feasible when the gases are all resolved or the detection system is specific. One significant advantage of the chromatographic method is the ability to operate over an extended temperature range to obtain thermochemical data such as the enthalpies and entropies of solution or mixing. Another advantage of this method is the very fast and efficient degassing of the solvent. This is a very laborious and time-consuming procedure for static systems and often involves boiling or freezing of the solvent. In a chromatographic column the solvent is spread in a very thin film and is continuously purged by an inert gas, such as helium. Under these conditions the degassing time is of the order of minutes, rather than several hours, as is often the case for static experiments.

IV. DETECTION AND DATA SYSTEMS

The popular flame ionization and electron capture detectors (EDCs) are not suitable for these types of measurements because of their poor response to many of the gases of interest and the difficulty in the determination of t_m.

A. Thermal Conductivity Detectors

Most of the early studies were carried out with thermal conductivity detectors (TCDs). These detectors are less sensitive than the FID of ECD; however, small samples sizes are not usually required because most gas—liquid systems obey Henry's law at pressures easily detected by TCDs. These detectors can be obtained with very low internal volumes and fast response times. Thus accurate retention time data can be obtained even for the rapidly eluting inert gases needed to evaluate t_m. The detector is common, simple to operate, and sensitive to any gas with a thermal conductivity different from that of the carrier gas, which is usually helium. The main problem with this type of detector is the universal sensitivity; that is, the detector is a bulk property detector. If two or more of the inert gases elute very close together, it is not possible to determine all of the t_{Ri} values at once. Multiple injections of the pure components must be utilized and this introduces unnecessary uncertainty in the t_{Ri} values used in the linearization schemes for t_m. Several authors [54,55] have shown that even small errors in the retention data used in the linearization scheme often lead to very significant errors in the calculated value of the dead time.

B. Mass-Spectrometric Detection

Mass-specific detection for the light gases (other than the carrier gas) may be attained with a simple quadrupole mass spectrometer, such as those available with many gas chromatograph/mass spectrometer (GC/MS) systems or with stand-alone detectors for any gas chromatograph. These detectors are far more complex, expensive, and difficult to operate than the other detectors. However, a simple mass spectrometer is almost the ideal GC detector. Very few nonanalytical applications of mass spectrometers as GC detectors have been published to date, mostly because of the terrible reputation that MS systems have acquired. Future instrumentation will be reduced in size, complexity, and (hopefully) cost. Many more prosaic applications of GC/MS systems for purely chromatographic measurements should appear shortly.

The major advantage of the MS detectors is the mass specificity, which means that overlapping peaks may be identified and the retention times measured. However, the detector is also very sensitive, fast, and usually supported by a data system of some type. No separator is necessary, because only retention times are measured and very high sensitivity is not required.

V. SOLUBILITY DATA

One of the difficulties encountered in the tabulation of solubility data is the multitude of formats used by different workers. This problem has been discussed in the reviews [1,2] cited previously. Engineering data are usually presented as Henry's law constants or the ubiquitous K value. On the other hand, chromatographers seem to prefer specific retention volumes or Raoult's law activity coefficients. With proper caution it is possible to interconvert each of these results. The relation between the specific retention volum V_g^o and the Raoult's law activity coefficient γ^∞ is given by

$$\gamma^\infty = 273.2R/(V_g^o P_i^o M) \tag{1}$$

where P_i^o is the vapor pressure of the pure solute i, and M is the molecular wight of the solvent. The realation between the Henry's law constant H_i and the Raoult's law activity coefficient is simply $H_i = \gamma^\infty P_i^o$, and the engineering K value is the Henry's law constant times the total pressure in the vapor phase. All of these expressions can be corrected for gas-phase imperfections by means of a compressibility factor or an assumed equation of state.

A. Data at High Pressures

Stalkup and Kobayashi [17] first used gas chromatography to determine the solubility (as vaporization equilibrium ratios, K) of ethane, propane, and butane (at infinite dilution) in n-decane and mixtures of methane + n-decane over a range of temperatures. The binary liquid was produced by the use of methane as the carrier gas. The solubility of the solute gases in the binary solvent was determined as a function of the pressure of methane from 10 to 2000 psia. The composition of the liquid at any pressure was determined from independent isotherm data. The chromatographic solubility data agreed relatively well with published static data. Further investigations of this type were carried out by Koonce and Kobayashi [18,19] for mixtures of methane + propane in n-decane and n-heptane. In this case, binary gases and ternary liquid systems were investigated. The composition of the liquid phase was determined by tracer pulse chromatography with ^{14}C-

labeled methane + propane. The same type of data was obtained for methane + ethane in n-heptane and methane + propane in toluene [20]. Probably the best-characterized systems are carbon dioxide and hydrogen sulfide in mixtures of methane + n-octane. These systems have been studied by three different groups [21,43,56], and several comparisons with static data have shown that the chromatographic data are sufficiently accurate. These studies are unique because, while the gaseous solutes were present at very low pressures, the total pressure of the systems varied up to 2000 psia, and the composition of the solvents also varied over wide ranges. This type of data is difficult, if not impossible, to obtain by any nonchromatographic method.

B. Data at Low Pressures

More recent studies of gas solubilities by direct GC have centered on several alkanes that have been used as stationary liquid phases. Riedo et al. [29] measured the solubility of nitrogen, hydrogen, the inert gases, the C_1 to C_4 n-alkanes, neopentane, and tetramethylsilane in a high molecular weight alkane ($C_{87}H_{176}$) from 30 to 230°C. The study was designed to characterize a new liquid phase; however, great care was taken in the determination of the dead time, and there is no reason to doubt the accuracy of the measurements, even though the retention volumes were sometimes very small.

Other workers [36,48] have measured the solubilities of various gases in other alkanes, in particular, n-hexadecane, n-octacosane, and n-hexatriacontane, at three or more temperatures. One of the major advantages of direct GC methods is the range of temperatures available with most common types of instrumentation. In the latter study [48] the heats of solution of the gases in the three alkane solvents were determined, as well as the solubilities. The primary goal of the study was to establish the accuracy of the method, and it was demonstrated that the solubility data were in excellent agreement with static data for nitrogen, carbon dioxide, carbon monoxide, methane, ethane, and propane in n-hexadecane. The heats of solution of the alkane gases were also shown to agree well with literature data. Table 1 gives the data from this study for this solvent.

Propylene carbonate (4-methyl-1,3-dioxolan-2-one) is a common industrial solvent often used for the selective extraction of acid gases. We have recently measured the solubility of several gases in this solvent using the method described in Ref. 48, and the results are given in Table 2, along with literature data for ethane and carbon dioxide. Again there is excellent agreement between the static and chromatographic data. The uncertainty figures quoted are the standard deviations observed for 8 to 10 points and emphasize the increased uncertainty in the data for the least-soluble gases caused by the experimental determination of t_m. Figure 1 is a plot of these data that show the linear dependence of ln H with 1/T. The heats of solution, calculated

Table 1 Solubility of Gases in n-Hexadecane (Henry's Law Constants, atm)

	Temperature		
Solute	25°C	40°C	55°C
Nitrogen	780	730	770
Carbon monoxide	570	540	540
Oxygen	400	390	360
Argon	340	340	340
Methane	180	190	192
Krypton	120	130	140
Carbon dioxide	73	83	92
Xenon	30	35	41
Ethane	28	34	39
Propane	8	10	13

from the slopes of the lines, are also given in Table 2. The inverse temperature dependence of the least-soluble gases is clearly demonstrated, and the selectivity of this solvent for carbon dioxide can be seen by comparison with the data for n-hexadecane.

Another possible application of direct GC determination of gas solubilities is for the study of solvent mixtures. This type of data is given in Tables 3 and 4 for mixtures of propylene carbonate and n-hexadecane. The data are given as Henry's law constants based on an average molecular weight for the binary solvents. If the solvents were immiscible or formed ideal solutions, the Henry's law constant should have a composition dependence given by

$$1/H_{mix} = X_1/H_1 + X_2/H_2 \tag{2}$$

where H_{mix}, H_1, and H_2 are the Henry's law constants for a given solute in the solvent mixture, pure solvent 1, and pure solvent 2, respectively; X_1 and X_2 represent the mole fractions of solvents 1 and 2 in the mixture. This idealized solution behavior is shown in Fig. 2, along with the measured data for two solvent mixtures. It can be seen that these two particular hydrocarbon solvents form nearly ideal solutions, and that GC is an excellent method for the study of these and

Table 2 Solubility of Gases in Propylene Carbonate (Henry's Law Constants, atm)

Solute	Temperature				Uncertainty (atm)	Heats of solution, (kcal/mol)
	10°C	20°C	30°C	40°C		
Argon	2600	2100	1900	1900	±300	+1.9
Methane	1500	1400	1300	1400	±200	+0.5
Krypton	980	950	950	960	±70	+0.1
Ethane	290 (287[a])	320 (315[a])	340 (340[a])	360 (360[a])	±10	−1.3
Xenon	290	320	330	350	±10	−1.1
Propane	106	128	141	161	±3	−2.4
Carbon dioxide		80 (71[a])	89 (85[a])	112 (103[a], 116[b], 113[c])	±2	−3.1

[a]Interpolated data from Ref. 28.
[b]Measured data from Ref. 56.
[c]Literature data cited in Ref. 56.

Table 3 Solubility of Gases in Propylene Carbonate–n-Hexadecane[a] (Henry's Law Constants, atm)

Solute	Temperature		
	20°C	30°C	40°C
Oxygen	870	880	800
Nitrogen	1600	1600	1400
Argon	760	780	720
Krypton	290	304	304
Xenon	73	81	90
Methane	460	480	470
Carbon dioxide	72	83	94
Ethane	68	74	84

[a]Mixture Mole Fractions: Propylene carbonate = 0.70 and hexadecane = 0.30.

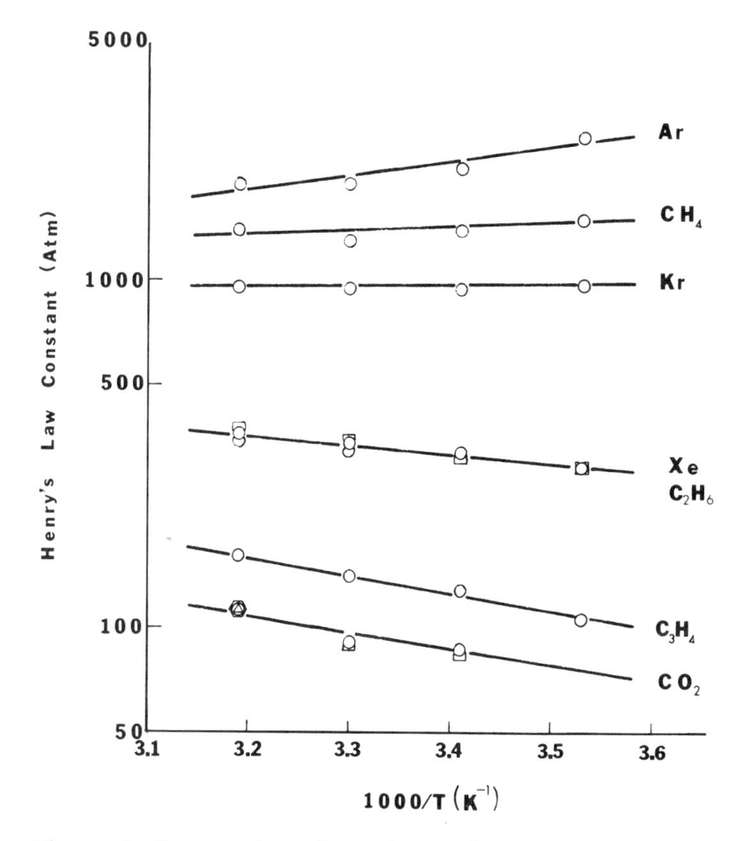

Figure 1 Temperature dependence of the Henry's law constant of several gases in propylene carbonate (○, this work; □, △, ⬡, literature data from Refs. 57 and 58).

other mixed solvents. Figure 2 also indicates the unique selectivity of propylene carbonate for carbon dioxide.

Gas chromatography can also be used to advantage by utilizing the inert atmosphere provided by the gases commonly used as carrier gases. For example, Table 5 gives the data for several common gases in a chloroaluminate molten salt. This particular melt was a mixture of aluminum chloride and 1-methyl-3-ethylimdazolium chloride (44 mol % $AlCl_3$). This mixture is a very polar, ionic solvent which is liquid at room temperature [57]. The major problem with such solvents is that they are unstable in air and decompose to produce HCl and normally must be handled in a dry box with an inert atmosphere. However, the material is also stable when coated on a deactivated support and flushed with helium in a glass column. An additional advantage of handling the molten salt as a GC liquid phase is the rapid, efficient degassing of the solvent by the carrier gas. Normally, degassing of a solvent such as a molten salt is particuarly difficult and time-consuming.

Table 4 Solubility of Gases in proplylene Carbonate–n-Hexadecane[a] (Henry's Law Constants, atm)

	Temperature		
Solute	20°C	30°C	40°C
Oxygen	560	580	600
Nitrogen	1050	1060	1070
Argon	490	500	510
Krypton	172	180	190
Xenon	42	45	51
Methane	250	257	270
Carbon dioxide	70	77	88
Ethane	38	42	49

[a]Mixture Mole Fractions: Propylene carbonate = 0.38 and hexadecane = 0.62.

This particular data set also illustrates one of the abnormal aspects of gas-chromatographic experiments. Both methane and ethane display an unusual temperature dependence in this solvent, as shown in Fig. 3; that is, they do not follow the typical convergent pattern

Table 5 Solubility of Gases in a Molten Salt (Henry's Law Constants, atm)

	Temperature			
Solute	20°C	30°C	40°C	50°C
Methane	1300	1700	1900	2200
Krypton	980	1140	1390	1350
Ethane	620	730	890	800
Xenon	460	500	580	600
Carbon dioxide	120	142	168	195

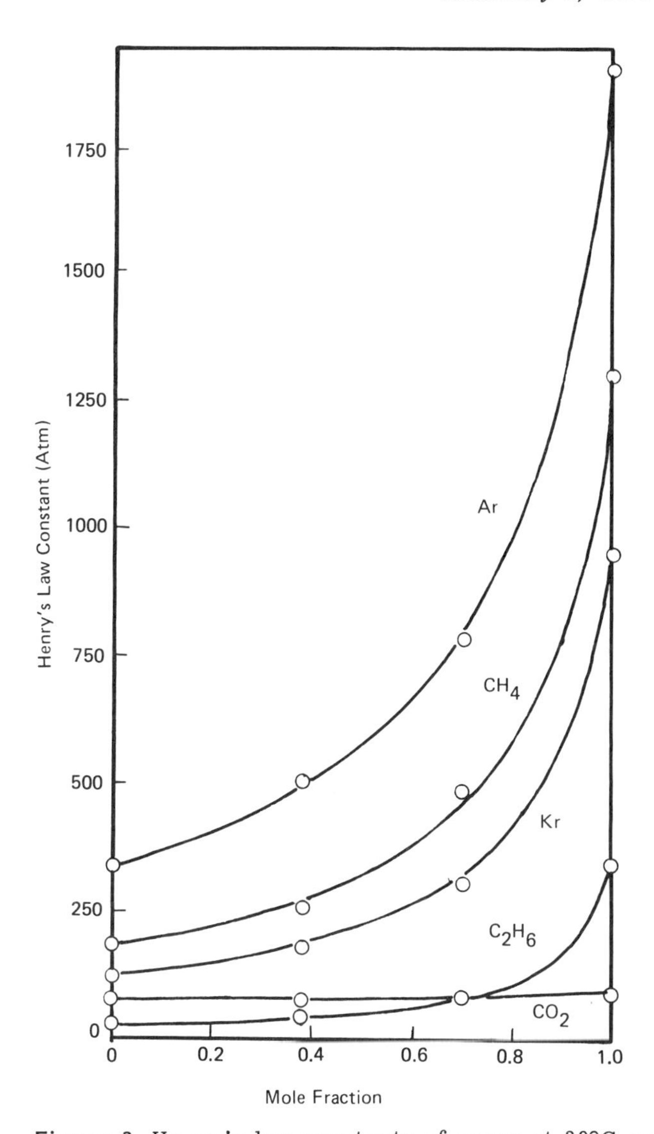

Figure 2 Henry's law constants of gases at 30°C as a function of the mole fraction of propylene carbonate in a binary solvent (n-hexadecane plus propylene carbonate).

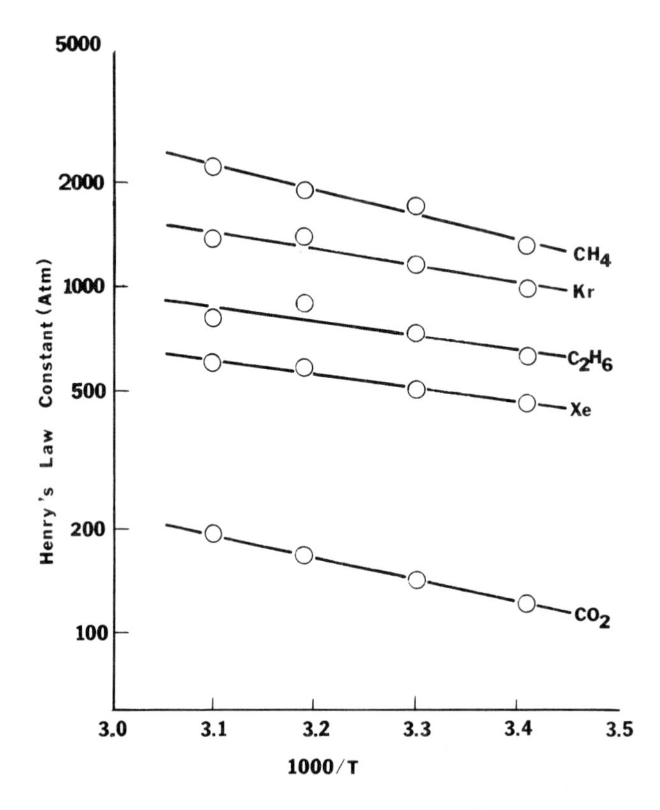

Figure 3 Temperature dependence of the Henry's law constants of several gases in a molten salt solvent.

shown in Fig. 1. These alkanes are very insoluble in the ionic solvent and should have positive heats of solution. However, the slopes of the lines for these two solutes give heats of -3.2 and -1.9 kcal/mol for methane and ethane, respectively. These abnormal values are due to the fact that the retention mechanism for these solutes is not partition, but adsorption on the surface of the molten salt. The calculated values are heats of adsorption, not heats of solution. In the case of ethane, the retention mechanism is probably a combination of partition and adsorption. Adsorption at the gas–liquid interface is a potential problem that must be recognized for very nonideal systems and for nonpolar gases in very polar solvents. Rapid equilibrium and degassing are the advantages gained by having the solvent spread as a thin film on a support; adsorption at the gas–liquid interface is a possible disadvantage.

VI. DISCUSSION

These recent examples of the direct determination of the solubility of gases in several different types of solvents illustrate the potential of this technique. The method has been shown to be as accurate as static methods for many, but not all, gas–liquid systems. The primary factor limiting the accuracy, and hence the utility, of the method is the accurate determination of the void volume or dead time of a chromatographic system.

A more reliable and accurate method for the determination of t_m is a critical need. An entirely new method, preferably a direct method, will have to be developed, because the limits have been reached for schemes based on a linearization of the $\ln(t_{Ri} - t_m)$ data. The void volume problem is not critical for most GC applications and has thus far been ignored by most chromatographers; however, this is the main limiting problem in the determination of gas solubilities.

The method is also limited for reactive gases, such as the nitrogen and sulfur oxides. The limitations can, however, usually be overcome by the use of special column materials. If the gas chemically reacts with the solvent, then no solubility method, including GLC, is useful. The requirements for very fast, accurate injection and detection systems are also a disadvantage. In the case of a mass-spectrometric detector, the disadvantage is particularly significant. The increased use and availability of quadrupole instruments as dedicated GC detectors is an indication that the universal dread of this instrumentation may be waning.

The advantages of the method are the same as the advantages of any gas-chromatographic method compared to static methods. These include speed, simplicity, and elimination of the common restrictions to 25°C and 1 atm.

ACKNOWLEDGMENTS

Acknowledgment is made to the National Science Foundation (Grant No. CHE-8207756) and to the donors of the Petroleum Research Fund, administered by the American Chemical Society, for support of this research.

REFERENCES

1. E. Wilhelm and R. Battino, Chem. Rev. *73*, 1 (1973).
2. E. Wilhelm, R. Battino, and R. J. Wilcock, Chem. Rev. *77*, 219 (1977).
3. E. R. Thomas, B. A. Newman, T. C. Long, D. A. Wood, and C. A. Ekert, J. Chem. Eng. Data *27*, 399 (1982).

4. M.J. Horvath, H. M. Sebastian, and K. C. Chao, Ind. Eng. Chem., Fundam. *20*, 394 (1981).
5. A. M. da Silva, S. J. Formosinho, and C. T. Martins, J. Chromatogr. Sci. *18*, 180 (1980).
6. E. E. Isaacs, F. D. Otto, and A. E. Mather, Can. J. Chem. Eng. *55*, 751 (1977).
7. D. Legret, D. Richon, and H. Renon, AIChE J. *27*, 203 (1981).
8. D. Legret, D. Richon, and H. Renon, J. Chem. Eng. Data *27*, 165 (1982).
9. J. W. Swinnerton, V. J. Linnenbom, and C. H. Cheek, Anal. Chem. *34*, 483 (1962).
10. S. K. Shoor, R. D. Walker, and K. E. Gubbins, J. Phys. Chem. *73*, 312 (1969).
11. K. E. Gubbins, S. N. Carden, and R. D. Walker, J. Gas Chromatogr. *3*, 98 (1965).
12. B. A. Cosgrove and J. Walkley, J. Chromatogr. *216*, 161 (1981).
13. J. C. Leroi, J. C. Masson, H. Renon, J. F. Fabries, and H. Sannier, Ind. Eng. Chem. Process Des. Dev. *16*, 139 (1977).
14. D. Richon and H. Renon, J. Chem. Eng. Data *25*, 59 (1980).
15. D. Richon, P. Antoine, and H. Renon, Ind. Eng. Chem. Process Des. Dev. *19*, 144 (1980).
16. H. P. Kollig, J. W. Falco, and F. E. Stancil, Environ. Sci. Technol. *9*, 957 (1975).
17. F. I. Stalkup and R. Kobayashi, AIChE J. *9*, 121 (1963).
18. K. T. Koonce and R. Kobayashi, J. Chem. Eng. Data *9*, 494 (1964).
19. K. T. Koonce, H. A. Deans, and R. Kobayashi, AIChE J. *11*, 259 (1965).
20. L. D. van Horn and R. Kobayashi, J. Chem. Eng. Data *12*, 294 (1967).
21. K. Asano, T. Nakahara, and R. Kobayashi, J. Chem. Eng. Data *16*, 16 (1971).
22. J. F. Parcher, J. Chromatogr. *251*, 281 (1982).
23. S. Ng, H. G. Harris, and J. M. Prausnitz, J. Chem. Eng. Data *14*, 482 (1969).
24. C. C. Chappelow and J. M. Prausnitz, AIChE J. *20*, 1097 (1974).
25. J. Y. Lenoir, P. Renault, and H. Renon, J. Chem. Eng. Data *16*, 340 (1971).
26. W. Hayduk, E. B. Walter, and P. Simpson, J. Chem. Eng. Data *17*, 59 (1972).
27. M. B. King and H. Al-Najjar, Chem. Eng. Sci. *32*, 1241 (1977).
28. O. R. Rivas and J. M. Prausnitz, Ind. Eng. Chem. Fundam. *18*, 289 (1979).

29. F. Riedo, D. Fritz, G. Tarjan, and E. sz. Kovats, J. Chromatogr. *126*, 63 (1976).
30. M. S. Wainwright and J. K. Haken, J. Chromatogr. *184*, 1 (1980).
31. R. J. Smith, J. K. Haken, and M. S. Wainwright, J. Chromatogr. *147*, 65 (1978).
32. L. S. Ettre, Chromatographia *13*, 73 (1980).
33. M. S. Wainwright, J. K. Haken, and D. Srisukh, J. Chromatogr. *179*, 160 (1979).
34. W. E. Sharples and F. Vernon, J. Chromatogr. *161*, 83 (1978).
35. J. F. Parcher and D. M. Johnson, J. Chromatogr. Sci. *18*, 267 (1980).
36. M. S. Wainwright, J. K. Haken, and D. Srisukh, J. Chromatogr. *188*, 246 (1980).
37. M. S. Wainwright and J. K. Haken, J. Chromatogr. *256*, 193 (1983).
38. J. K. Haken, M. S. Wainwright, and D. Srisukh, J. Chromatogr. *257*, 107 (1983).
39. S. Masukawa, J. I. Alyea, and R. Kobayashi, J. Gas Chromatogr. *6*, 266 (1968).
40. S. Masukawa and R. Kobayashi, J. Gas Chromatogr. *6*, 257 (1968).
41. S. Masukawa and R. Kobayashi, J. Chem. Eng. Data *13*, 197 (1968).
42. S. Masukawa and R. Kobayashi, J. Gas Chromatogr. *6*, 461 (1968).
43. A. Yudovich, R. L. Robinson, and K. C. Chao, AIChE J. *17*, 1152 (1971).
44. Y. Hori and R. Kobayashi, J. Chem. Phys. *54*, 1226 (1971).
45. T. Nakahara, P. S. Chappelear, and R. Kobayashi, Ind. Eng. Chem. Fundam. *16*, 220 (1977).
46. E. A. Mason and W. E. Rice, J. Chem. Phys. *22*, 843 (1954).
47. J. F. Parcher and M. Selim, Anal. Chem. *51*, 2154 (1979).
48. P. J. Lin and J. F. Parcher, J. Chromatogr. Sci. *20*, 33 (1982).
49. W. Jennings, *Gas Chromatography with Glass Capillary Columns*, Academic Press, New York, 1978, p. 76.
50. F. Khoury and D. B. Robinson, J. Chromatogr. Sci. *10*, 683 (1972).
51. D. C. Locke, in *Advances in Chromatography*, Vol. 14 (J. C. Giddings, E. Grushka, J. Cazes, P. R. Brown, eds.), Marcel Dekker, New York, 1976, Chap. 4.
52. R. J. Laub and R. L. Pecsok, *Physiochemical Applications of Gas Chromatography*, Wiley, New York, 1978, Chaps. 2 and 3.
53. J. R. Conder and C. L. Young, *Physicochemical Measurement by Gas Chromatography*, Wiley, New York, 1979, Chap. 2.

54. X. Guardino, J. Albaiges, G. Firpo, R. Rodrigues-Vinals, and M. Gassiot, J. Chromatogr. *118*, 13 (1976).
55. J. K. Haken, M. S. Wainwright, and R. J. Smith, J. Chromatogr. *133*, 1 (1977).
56. C. J. Mundis, L. Yarborough, and R. L. Robinson, Ind. Eng. Chem. Process Des. Dev. *16*, 254 (1977).
57. J. S. Wilkes, J. A. Levisky, R. A. Wilson, and C. L. Hussey, Inorg. Chem. *21*, 1263 (1982).

8

Multiple Detection in Gas Chromatography

Ira S. Krull and Michael Swartz / Northeastern University, Boston, Massachusetts

John N. Driscoll / HNU Systems, Inc., Newton Highlands, Massachusetts

I. INTRODUCTION AND BACKGROUND

The use of more than a single detector response in gas chromatography (GC),* commonly referred to as multiple detection GC, has recently gained widespread popularity and acceptance as a reliable and useful analytical method for improved identification of GC analtyes. In most instances, multiple detection GC provides more compound specificity, qualitative and quantitative identification, and decreased overall analysis times than has been possible with the more traditional single detector GC approaches. In the past, analyses for compounds present in complex sample matrices has involved the use of GC with a variety of selective and/or general detectors, such as flame ionization (FID), electron capture (ECD), photoionization (PID), Hall electrolytic conductivity (HECD), and flame photometric (FPD). Even today many trace analyses depend on single detection methods in GC, as opposed to either multiple detection GC or high-performance liquid chromatography (HPLC). In part, this is due to the lower detection limits possible in GC, as well as the widespread available and familiarity of the instrumentation required. However, concomitant with excellent sensitivity in GC is the problem of sample matrix effects and interferences that often lead to misidentification of a particular analyte. As the need for lower detection limits increases, more demands are placed on the GC or detector systems to resolve interferences. At the parts per billion (ppb) level and below, positive identification by single detection GC can be difficult, if not impossible, since a large number of similar analytes may emerge from the GC column with similar retention times and chromatographic properties. Even with capillary GC columns, there is still an increased need for a more accurate and reliable method of analyte identification, and in many cases this can be provided by multiple detection GC approaches. By selecting the appropriate detector combination for a particular analyte or class of analytes, many sample interferences or extraneous GC peaks can be distinguished from those of major interest. Given all of the commercially available and research-type detectors for GC today, there are a large number of combinations possible in multiple detection GC. One could use a general and selective set of detectors, in dual detection GC, in order to determine the general complexity of the entire sample makeup, as well as those analytes which only respond to that selective detector being used. Other approaches to multiple (dual) detection GC could involve (1) general–general detectors and (2) selective–selective detectors. Of course, when one considers multiple detection GC with three or more detectors in constant operation, then the number of possible detector combinations and configurations increases still fur-

*Appendix IV contains a complete listing of abbreviations and acronyms.

ther. As we discuss below, the unique combination of general–selective–selective detector combinations provides an enormous amount of qualitative and quantitative data that can lead to significantly improved analyte identification.

Economic factors and overall analysis times are also of prime importance in considering different analytical techniques for a particular application. When proper agreement is made between the detectors to be used and the particular analytical application, analysis times can often be cut significantly. For most laboratories, any savings in analysis time means an economic advantage, and indeed multiple detection GC often realizes both of these goals. Thus dual detection GC can, at times, provide just as much specific analyte identification information as alternative approaches, such as mass spectrometry (MS), for compound identification, but now at greatly reduced costs for the basic instrumentation and maintenance or operation.

We describe here the basic operating principles and instrumentation for both series and parallel multiple detection in GC, general and selective detectors that have been used in the past, data acquisition and manipulation for improved analyte identification, specific applications from the current literature, and suggestions for further developments and utilizations of multiple detection GC. However, in order to better understand that which follows, we have summarized in tabular form some of the basic principles and approaches needed in this field. Table 1 summarizes most of the widely available and applied general- and selective-type detectors in GC, with some brief indication of how these work. Table 2 summarizes the two or three basic schematic arrangements for multiple detection GC, with two or three detectors used, as well as some of the possible combinations of general and selective detectors that have been or remain to be developed for specific GC applications. The information contained in these particular tables has been derived from recent texts and literature references [1–11].

In all multiple detection GC work we are deriving additional qualitative and quantitative analyte information on the basis of the number and types of detectors being used for each specific analyte. Multiple GC chromatograms will therefore be generated, and the amount of data that needs to be accumulated, stored, recorded, manipulated, and interpreted is much greater than for single detection GC. Some systematic approach must therefore be incorporated to enable the analyst to derive as much information as possible from each individual GC injection run, and to easily and accurately utilize that data for later purposes and interpretation. This can take the form of multiple linear chromatograms, as in routine single detector GC, or it can take the form of laboratory computer data handling, storing, and final calculating within run or post-run. A number of suitable laboratory computer type data-handling stations or systems have evolved within the past 5 years that permit the GC analyst to accumulate data from mul-

Table 1 Summary of the More Widely Employed Detectors in Gas Chromatography

Detector name	Abbreviation	Operating principles	General–selective
Thermal conductivity	TCD	Differences in thermal conductance of analyte versus carrier gas	General
Flame ionization	FID	Formation of analyte ions by burning of compound in flame	General for compounds that contain carbon/hydrogen
Electron capture	ECD	Analyte capture of ions or electrons formed by radiation of carrier gas molecules	Selective for electron-capturing analytes (halogens, nitro compounds, unsaturated compounds)
Photoionization	PID	Photochemical formation of ions from analyte molecules by photoinduced ionization	Selective for those compounds with suitable ionization potentials below that of lamp in use
Alkali flame ionization	AFID	Burning of analytes in presence of inorganic salt bead to form ions then collected/counted	Selective for nitrogen- and phosphorous-containing compounds
Hall electrolytic conductivity	HECD	Thermal generation of inorganic ions from analytes, determination of conductance of solution with ions	Selective for compounds that can release organic/inorganic ions by thermal cleavage (organohalogens, N-nitroso, N-nitro, etc.)
Mass spectrometer	MS	Formation of parent and fragment ions by electron impact or chemical ionization impact	General and selective for all compounds that can be volatilized–ionized or ionized–volatilized
Thermionic ionization	TID	Formation of ions by thermal removal of electrons, collection and counting of ions	General for all compounds forming ions thermally, selective for many classes of compounds as function of thermal temperature being used

Microwave-induced plasma emission	MIP	Excitation of analyte and emission of light as function of elements	Selective for those elements capable of being excited and emitting at measurable wavelengths
Flame photometric	FPD	Flame excitation of analyte and emission of light dependent on particular elements present	Selective for sulfur- and phosphorus-containing compounds, dependent on wavelength-selective filters used
Ultraviolet	UV	Absorbance of light by specific chromophore within analyte	Selective for those analytes which can absorb UV radiation incident
Thermal energy analysis	TEA	Thermal fragmentation to release a molecule (NO, NO_2) which is then detected by chemiluminescence	Selective for N-nitroso or nitro compounds
Atomic absorption spectroscopy (flame or graphite furnace AA)	FAA GFAA	Thermal atomization, followed by absorbance of light at specific wavelength from hollow cathode light source	Selective for specific elements, especially metals
Inductively coupled plasma emission spectroscopy	ICP	Radiofrequency-induced thermal atomization, excitation, ionization, with final emission of light at wavelengths unique for each element	Selective for specific elements, especially metals, some nonmetals
Direct current plasma emission spectroscopy	DCP	DC thermal atomization, excitation, ionization, with final emission of light at wavelengths unique for each element	Selective for specific elements, especially metals, some nonmetals

Table 2 Summary of the Basic Schematic Arrangements of Detectors Possible in Multiple Detection GC

Arrangement type	Detector arrangement	Specific example
Parallel (side by side) dual or multiple	General–general	Thermal conductivity–flame ionization
	General–selective	Thermal conductivity–electron capture
	Selective–selective	Photoionization—microwave induced plasma
Series (upstream–downstream) dual	General–general	Thermal conductivity–flame ionization
	General–selective	Thermal conductivity–electron capture
	Selective–selective	Photoionization–electron capture
Series–parallel	General–general (series) General (parallel)	Thermal conductivity–flame ionization Mass spectrometry
	General–selective (series) General (parallel)	Thermal conductivity–electron capture Mass spectrometry
	General–general (series) Selective (parallel)	Thermal conductivity–flame ionization Mass spectrometry or photoionization
	General–selective (series) Selective (parallel)	Thermal conductivity–electron capture Mass spectrometry or photoionization

Series–parallel (Continued)	Selective–general (series) Selective (parallel)	Photoionization–flame ionization Electron capture
	Selective–general (series) General (parallel)	Photoionization–flame ionization Thermal conductivity or mass spectrometry
	General–selective (series) Selective (parallel)	Thermal conductivity–electron capture Mass spectrometry or photoionization
	Selective–general (series) Selective (parallel)	Photoionization–flame ionization Electron capture
	Selective–general (series) General (parallel)	Photoionization–flame ionization Thermal conductivity or mass spectrometry
	Selective–selective (series) General (parallel)	Photoionization–electron capture Flame ionization
	Selective–selective (series) Selective (parallel)	Photoionization–electron capture Mass spectrometry or flame photometric

tiple detecton GC runs and manipulate such data according to predetermined software programs for specific end purposes and identification. We propose to describe here some of the better commercial laboratory computer systems which are fully compatbile and interfaceable with multiple detection GC. Such data-handling systems go beyond the older, more traditional reporting integrator or disk integrator-type chromatographic recorders used in the mid to late 1970s. Clearly, this type of data-handling approach is highly desirable, if not absolutely essential, for optimal, efficient employment of multiple detection GC on any semi- or fully routine basis in analytical chemistry.

The final sections of this review deal with that which has already been described in the literature with regard to dual or multiple detection in GC, along with specific detector combinations and arrangements and final analyte applications. This should be a comprehensive summary and interpretation of all that has been published in the open literature up to the very end of 1983, but not necessarily all that has been reported in the patent literature to date. We apologize in advance for any inadvertent omissions of specific literature references, and urge our colleagues to forward copies of any such papers to our attention for future incorporation.

II. INSTRUMENTAL ARRANGEMENTS AND OPERATING PROCEDURES IN MULTIPLE DETECTION GAS CHROMATOGRAPHY: DATA ACQUISITION AND MANIPULATION-INTERPRETATION

A. Instrumental Arrangements and Operating Procedures in Series Multiple Detection in Gas Chromatography

As mentioned above, there are basically three generally used arrangements for multiple detection gas chromatography (Fig. 1), wherein with dual detectors these can be placed one behind the other, termed here and elsewhere a series arrangement (upstream—downstream). In this orientation, each detector sees approximately the same amount or concentration of each analyte peak as they elute from the GC column, assuming that the first (upstream) detector does not destory (nondestructive) more than about 1% of the analyte molecules by its detection/operating methods. Thus, if we were, purely for pedagogical reasons, to consider a dual series arrangement of the same two nondestructive detectors, such as two thermal conductivity ones, then the two chromatographic peaks observed should be approximately identical in peak height, shape, and total area. This assumes that there is no effective dead volume effect on the analyte peak as it elutes from the first (upstream) detector into the second (downstream) one. In actual practice, there will always be some extra dead volume effect, depending on the particular upstream detector, and this will therefore affect the final peak shape, height, and width seen on the second, downstream detector chromatogram.

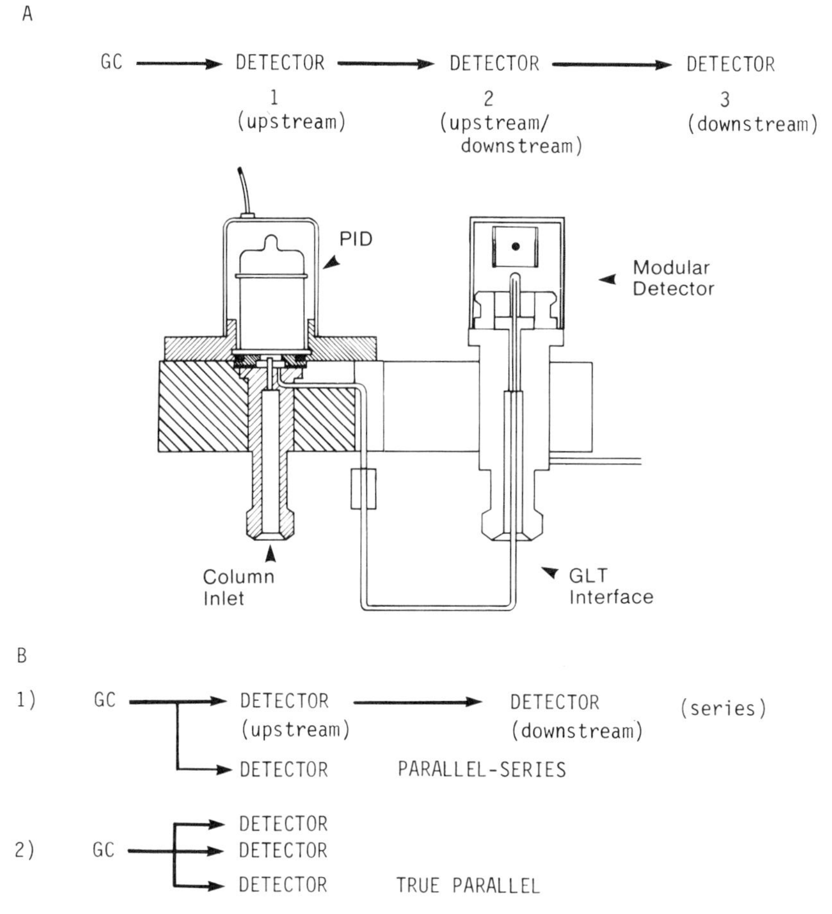

Figure 1 Schematic diagram of possible multiple detector arrangements: (A) series (upstream-downstream) multiple detection, and schematic diagram of a typical series (upstream-downstream) arrangement of two detectors for GC. PID = photoionization detection; GLT = glass lined tubing; modular detector could be any other GC detector (ECD, FID, NPD, etc.). (B) parallel adjacent (side-by-side) multiple detection GC.

There is at least one significant advantage in using the series orientation of two or more detectors, in that there is no effluent stream splitting used here, so that all of the mass of each analyte peak elutes from the separation column into the first and subsequent detectors. Thus minimum detection limits (MDLs) are not adversely affected by the stream splitting that is required in the parallel adjacent orientation of two or more detectors, as mentioned below. Because the sample peaks are not split postcolumn, all of the analyte material is allowed to enter first the upstream detector, be detected, and then pass through into the second, downstream detector. Those commerical instruments which offer multiple detection capabilities usually will employ the series detector arrangement, especially since this requires little additional hardware or splitting adapters.

A major requirement in series multiple detection GC is that the upstream detector(s) cannot be of the destructive type, that is, it must be able to detect the analytes of interest, and pass almost all of the detected materials into the downstream detector(s) unscathed. Therefore one could not satisfactorily use as the upstream detector an FID, since this would burn all of the analyte material and allow none of the original analyte to pass to the downstream detector(s). Satisfactory upstream detectors might be thermal conductivity, photoionization, ultraviolet, or any similar detector that does not destroy more than about 1% of the original analyte in the detection process. Those detectors that would not be suitable upstream would include flame photometric, microwave-induced plasma emission, Hall electrolytic conductivity, flame ionization, thermal energy analysis, or any other detector that effectively destroys all or most of the original analyte of interest.

In general, with two detectors in series, both being nondestructive types, the upstream one should have the smaller effective dead volume, so that MDLs are affected as little as possible on the downstream detector. If one is going to use one destructive- and one nondestructive-type detector, then no matter what their differences in effective dead volume, the nondestructive type must go upstream. With regard to general- or selective- type detectors, it does not really matter whether these are upstream or downstream, as long as the first (upstream) detector is nondestructive. If both are nondestructive, then it does not really matter whether the upstream is general or selective, as long as the effective dead volume of the upstream is less than that of the downstream.

Any extra connections between the two series detectors must be as minimal in effective dead volume as possible, since this would have a direct effect on the MDLs in the second, downstream detector. Such connections therefore must be gas tight, as short and tight as possible, and heated to at least the maximum temperature being used in the first, upstream detector. Any cold spot in the connecting hardware between the two detectors would lead to deposition of part or all of the analyte, which could result in no peak observed on the downstream detector or a very broadened, distorted downstream detector peak that is not due to any malfunction of this detector per se. Vast differences in peak shapes between the two detector chromatograms should and/or could immediately suggest that a cold spot or two is present between the two detectors or within the downstream, second detector in series. Another possibility would be slow adsorption–desorption of the analyte(s) in the upstream detector, perhaps due to a cold spot within that detector and not elsewhere.

The question always arises: When does one use two general detectors, two selective ones, or one selective and one general type? This obviously will depend on the nature of the analyte being studied and the reaons for using multiple detection GC. To some extent, we suggested some of the reasons already, but perhaps it would be good to expand somewhat now. Multiple detection GC does not improve detectability or minimum detection limits; therefore if this is the goal, then one should not be considering this particular GC approach. If, on the other hand, improved analyte identification is the major goal of the analysis, in lieu of perhaps the more involved mass-spectral type of analysis, then surely multiple detection GC should be strongly considered. This improved analyte determination, not detection, results from the presence or absence of the analyte peak on one or more of the multiple detector chromatograms. If one has an organohalogen compound, then it should appear on any general detector as well as on a selective detector that responds to such organics (Hall electrolytic conductivity, electron capture, microwave-induced plasma, etc.). The use of a series orientation of a general and a selective detector for such compounds should therefore lead to two chromatograms, wherein the analyte of interest appears, hopefully isolated from all other sample components, on both chromatograms. And, the relative peak heights for the suspected analyte from the sample matrix, for the absolute amount or concentration injected, must be the same, versus an internal standard, as for an authentic standard of this analyte injected under identical GC detector operating conditions. One can then do all sorts of manipulative data-handling operations on these two sets of chromatographic data, as indicated below, but the most important thing to always remember is that the suspected analyte must provide identical results chromatographically and detector-wise as the authentic standard. This means that if, on the general-type detector, there is another material coeluting with the analyte of interest, perhaps not at all resolved, then the relative peak heights for the suspected analyte in the sample must be different than the same data for the authentic compound. This, of course, assumes that the contaminant peak eluting with the compound of interest on the general detector responds differently on the selective second detector. This is not a 100% foolproof test, since there will surely be a situation where two compounds coelute and both respond to the same relative extent on the same two detectors, but this situation would be rare, at best.

Why then use a general detector at all in multiple detection GC, series or parallel; why not just use two or more selective-type detectors? At times, selective–selective approaches will indeed provide more meaningful information with regard to specific analyte identification, but this will depend on the particular analyte of interest and the detectors available within any laboratory setting. We suggest that one of the multiple detectors should be a general, sensitive type, because this will provide the analyst with a better impression of the total complexity of that sample mixture. Selective detectors only respond to certain classes or types of compounds, and thus they will not indicate all of the possible components present. At least one of the remaining detectors should be selective for the particular analyte or analyte class of interest, and, if possible, we suggest that multiple detection GC use two selective-type detectors that operate on entirely different principles. This leads us to the opinion, and it is only an opinion, that multiple detection GC should utilize three separate detectors, one general and the other two selective. The general detector then provides an overall picture of everything that is present within that sample matrix, while the other two selective detectors indicate which particular analytes or analyte classes may be present, along with vastly simplified and easy-to-read or interpret chromatograms. In almost all cases, the selective detector chromatogram is vastly simpler in the number of peaks present, in comparison with general-type detectors.

What about the choice of which general or which selective detectors to employ, assuming that one had a completely free choice? Clearly, the principle criterion must be detection limits, for that particular analyte in that sample injected. If the MDLs are not low enough for that situation, then there is just no reason to even be considering that detector. If the MDLs for all available general detectors are the same, then the next important criteria would be destructive versus nondestructive, and then the effective dead volume, if this is to be the upstream detector. If one is choosing which two selective detectors to use in the series arrangement, one of these must be nondestructive, since there will be three in series, but the other can still be destructive, if it is placed last. If both of the selective detectors to be used are destructive, then one cannot put all three in series, and one therefore has to resort to another arrangement, parallel (side by side) or a combination of series—parallel, as below. Still, assuming that we can indeed put all three detectors, one general and two selective, in a completely series arrangement, then one should try and choose the two selective ones so that they operate on entirely different principles. This would vastly increase the overall analyte specifity or selectivity for the final three-detector analysis, since the two selective detectors together become much more selective for that particular analyte when they use different operational principles. There is just less chance of a false positive arising with the two selective detectors when they have different operational parameters or approaches that when these are more similar. Thus one would not want to use both a Hall and Coulson electrolytic conductivity detector approach for a particular analyte, if one could alternatively use, for example, a Hall ECD together with an ECD or MIP selective for chlorine.

In other words, choose the two selective detectors so that there is less chance of another sample component producing the same relative peak heights with the same GC retention time as the analyte of interest. Perhaps the ultimate in dual selective detection would be a parallel combination of MS and MIP, wherein the MS would provide a complete mass spectrum for the analyte and the MIP would respond with one or two channels for specific elements of interest, as below. This particular dual detector GC approach could not be used in the series arrangement, since both the MS and MIP are destructive-type detectors and would not pass any of the original analyte from one detector (upstream) to the next (downstream). Of course, having now read this, the reader will ask what about the recent development of inductively coupled plasma (ICP)–MS, wherein specific ions from a sample mixture are formed within the ICP and then introduced into the MS. But this is really a new form of sample ionization and introduction for mass spectrometry; it is not—at least not yet—an approach to series multiple detection in GC.

In dual detection GC, which is perhaps more common than multiple (triple) detection approaches, more people appear to be using series arrangements, but this may be due to the commercially available systems. Parallel GC requires the introduction of some sort of GC effluent splitter, sometimes commercially available, which has to be calibrated daily and have its flow rates checked daily and which can become easily clogged or broken. If detection limits are not the problem, then we recommend the parallel situation, in part because it is easier to use three detectors in the parallel–series arrangement than all three in series, and partly because it is often easy to change the effluent splitter ratio to accommodate different analytes with certain multiple detection situations. Ability to vary the ratio of analytes going to each detector provides the analyst with another experimental parameter that can, at times, be used to improve analyte determinations simply by making another injection at a different split ratio, all other parameters remaining constant, as below.

B. Instrumental Arrangement and Operating Procedures in Parallel Multiple Detection Gas Chromatography

Figure 1 summarizes the two major instrumental arrangements for multiple detection GC: (1) the already described series (upstream–downstream) situation (Fig. 1A) and (2) the parallel adjacent (side by side) detector arrangement (Fig. 1B). There is a third possible situation, which is really an extension of (2) that is, with three or more detectors in the parallel orientation, at least two of these could also be in series [Fig. 1B(1)], with the remaining detector in parallel to the first two. Alternatively, in the parallel situation, all three detectors could be parallel to each other, as in Fig. 1B(2) where none of them are upstream–downstream to one another. This latter approach is termed here the true parallel situation, while that of Fig. 1A(1) is now termed series–parallel. Where the two detectors in Fig. 1A(1) are in series, the same requirements and considerations already discussed above will apply. Thus we need only to now consider the parallel arrangement, and whether this were to involve two or more detectors in true parallel is immaterial.

Whereas the series orientation required an initial consideration of which detector could precede another, this is not of importance now; that is, since all of the multiple detectors actually see part of the GC effluent stream, independent of the others on line, they can therefore be destructive or nondestructive. Similarly, dead volumes within each detector do not affect the detector peaks or performance from any of the remaining detectors, so this too is not critical for detector placement. Again, one should ideally choose one general detector and two selective ones, for the exact same reasons that were discussed above. The most important consideration here is the nature of the GC effluent split to be used and what sort of a split ratio one requires. There are basically three materials used for either fixed- or variable-ratio splitters in GC: (1) all stainless steel or a related impervious metal, (2) glass-lined stainless steel, and (3) all glass, ceramic, or fused silica. Chemically resistant plastic-coated or all-plastic tee materials are not generally recommended, especially since they drastically change dimensions and therefore split ratios as a function of operating temperature within the GC oven. In general, an all-glass or glass-lined, fixed-ratio splitter is to be recommended for any parallel detector arrangement, for a number of reasons. The glass or glass-lined tee is impervious to almost all routinely encountered analytes and carrier gases, which is not always the case with stainless steel or other metal surfaces. Glass surfaces can also be silanized before and during a large number of analyses, and this further reduces any undesired chemical interactions between the tee surfaces and the analytes or sample components. The manufacture of glass-lined or all-glass tees leads to very reproducible parts, especially for the fixed-ratio designs, and thus it is fairly easy to replace such a tee, if needed. Though most commercially available GC effleuent splitters have been designed to only handle two detectors in the parallel orientation, it is a fairly simple matter to design and manufacture a similar glass or glass-lined tee for three or more detectors, again in parallel. With each newly purchased or in-house manufactured effluent splitter, it is clearly necessary to experimentally determine the exact split ratio for any particular combination of detectors. More importantly, it is crucial to measure the actual split ratio at the start of, during, and after all multiple detection GC studies. It is never safe to assume from day to day that the fixed-ratio splitter leads to precisely the same absolute split, and this must be experimentally determined accurately and precisely at least twice each day.

We do not recommend the use of a variable-ratio splitter, unless one is going to do isothermal GC separations, in which case the split ratio should remain constant throughout. However, if one is going to do temperature or flow programming GC separations, then the variable-ratio splitter could lead to significant changes in the actual split ratio during the entire run. Only a fixed-ratio splitter can suffice, unless one is willing and able to accurately and reproducibly determine (experimentally) the absolute split ratio at each temperature or flow rate throughout the GC run with a variable-ratio design. Even with a fixed-ratio splitter in operation, it is recommended that the absolute split ratio be experimentally determined at various temperatures in the temperature program run. If flow programming is being used, with changes in the carrier gas flow rates, then again the absolute split ratio should be determined at various points into this flow programming run. It is never satisfactory to assume that a fixed-ratio splitter provides the same exact split at the end of any GC run as it did at the very start. Such ratios must be determined at least once experimentally.

What sort of split ratio should be used with two or three detectors in parallel? This has to do with the minimum detection limits and sensitivities possible for the given arrangement of detectors, as well as on the specific analyte of interest. If both detectors have about the same approximate relative response factors (RRFs) for a given analyte, then an equal split ratio can be used, whether this be 50/50 or 33/33/33. However, if the two detectors of design for that particular analyte have vastly different response factors, then clearly a very different split ratio is desired, one that would convey more of the GC effluent and analyte of interest to that detector which has the weaker (weakest) relative reponse factor. This could lead to a split ratio of 99/1 or 99.9/0.1, depending on the particular detectors in operation and the specific analyte of interest. What is needed here is to arrive at some sort of split ratio that will eventually allow both detectors to satisfactorily detect the analyte at detector amplifier attenuations as similar to each other as feasible. Although it is also possible to have approximately equal split ratios and to work at vastly different detector attenuations, we recommend the alternative approach. As long as the split ratio is constant and accurately determinable, it seems easier and less troublesome to work at similar detector attenuations and sensitivities, rather than have to artifically set detector settings vastly different from one another. However, with very different split ratios, such as 99.9/0.1, problems can arise with regard to changes in this ratio during the temperature program, and such changes in the effluent ratio might not be readily determined. Thus, if such a split ratio were to be used, it is important to verify that this ratio remains completely constant from one run to another, especially between analyses of the sample itself and of any standard solution. Internal standards can, to some extent, alleviate the possible problems here, but only if there is no change in the split ratios within given runs and therefore from run to run.

Just as in any part of the GC system, effective dead volume must be kept to a minimum, and this holds true for the fixed-ratio splitter. For commercial splitters, the dead volume has usually been minimized, but for in-house custom-made splitters, this must be considered in the construction. Extra dead volume effects postcolumn lead to loss of resolution, peak merging, peak broadening, decreased peak heights, and worse detection limits, just as in any type of chromatography. This is especially true for capillary column GC work, and special fixed-ratio splitters are already commercially available for such multiple detection GC work.

C. Data Acquisition, Manipulation, and Final Interpretation in Multiple Detection Gas Chromatography

Multiple detection GC is capable of generating a sizable amount of data with each injection, especially for complex samples in capillary column GC separations. Whereas in the past a single pen linear chart recorder, perhaps with a built-in disk integrator, might suffice for single detection GC work, this is clearly not the case with two or more detectors operated simultaneously. It is still possible to do dual detection GC with one dual pen linear strip chart recorder, and thereby obtain two separate chromatograms for manual or semiautomatic data reduction and interpretation. However, the older Hewlett-Packard single pen reporting integrators for chromatography are not really ideal for multiple detection GC data-handling requirements. There are a number of things that one would like to have done with any "ideal" data station in multiple detection GC, and thse might include the following:

1. Acquire and print complete chromatograms for each detector used and retain the ability to reprint the same or modified chromatograms according to the program requirements as set forth by the analyst/operator
2. Determine the correct baseline for the accurate and reproducible automatic determination of peak heights and peak areas
3. Accurately determine with a high degree of precision retention times, Kovats indices, and, if desired, McReynolds constants for a given analyte peak(s)
4. Record together with each chromatogram the actual operating conditions that were used to obtain that detector chromatogram
5. With an internal standard present, determine relative response factors for each analyte peak of interest, having the volume and concentration of internal standard and sample in program
6. Normalize, according to software program, response factors for each analyte peak, on a per nanogram or microgram basis, for analyte peaks and the internal standard peak
7. Calculate normalized relative response factors for each analyte peak of interest, using data first generated above

8. Determine, using an internal standard or an external standard method, absolute amounts of each analyte present, using either peak heights or peak areas, as determined above.
9. Automatically ratio normalized relative response factors, for two or more detectors, in order to obtain detector ratios for relative response factors (RRFs)
10. Identify an unknown analyte peak in a complex sample mixture on the basis of (a) relative retention time versus the internal standard versus similar analyses of standards of analyte and the internal standard injected separately; (b) normalized relative response factors for two or more detectors; and (c) ratios of normalized relative response factors, as determined above
11. Use a memory bank of relative response factors and normalized relative response factors for possible analytes of interest, and to then compare the data generated from a new sample with this memory bank, using data produced as above
12. Do blank subtractions, base-line normalizing, and peak resolving where needed, for qualitative and quantitative determinations in complex environmental type samples

Clearly, in order to understand more fully the above data-handling requirements, we should more fully explain what some of the above terms signify [12]. Any final data manipulations are fully dependent on accurately determining the split ratio throughout the temperature programmed analysis, the absolute amounts of each compound injected as authentic standards, the total volume and volume injected of the sample solution, specific recorder and detector attenuation settings, and specific chromatographic separation parameters. If, for example, we use o-nitrotoluene (o-NT) as the internal reference compound, then all other detector responses can then be referenced to o-NT as being 1.00 on each detector. The choice of an internal standard is dependent on the particular analyte(s) of interest, chromatographic conditions, and the detectors utilized. Ideally, one would like to have an internal standard that responds similarly on all or most of the detectors being used to the analyte(s) being studied, and, an internal standard that has similar chromatographic retention times and peak shapes as the analyte(s) of interest. This would then allow one to use either peak heights or peak areas for all final calculations, though we have tended to use peak heights more than areas, perhaps because manual calculations have thus far been employed. Relative response factors (RRFs) are determined directly by measuring peak heights and absolute amounts of each compound reaching that detector. The ratio of peak heights (millimeters or centimeters) divided by the amount in nanograms or micrograms reaching the detector then provides normalized response factors on a per nanogram or microgram basis (mm/ng, mm/μg, cm/ng, cm/μg, etc.). Normalized relative response factors are simply obtained by using the relative response factor for the internal standard, now o-nitrotoluene (o-NT), as 1.00, referencing all other RRFs to that value. Naturally, such calculations, manual or computer assisted, are based on detector responses measured or corrected at the same exact attenuation settings on the detector amplifier and recorder or reporting/computing integrator. Detector ratios (ECD/PID, FID/MIP, etc.) for a particular compound are calculated by first determining the RRFs and normalized RRFs for that compound on each detector, and then simply taking the arithmetic ratio of the appropriate detector RRFs. For example, if 5-nitroindan had an ECD RRF of 2.32 and a PID RRF of 1.18, then the ECD/PID ratio of RRFs would be 2.32/1.18 = 1.97. Similarly, one could calculate RRFs for two compounds on the same detector, for example, PID/PID or ECD/ECD, as well as calculating the ECD/PID ratios for two compounds and then taking the ratios of these ratios, for example, ECD/PID/ECD/PID or ECD/PID divided by ECD/PID for the two analytes of interest [12]. All of these calculations have already been described in the literature, they are trivial to perform, but they require time and effort on the part of the analyst involved. Clearly, if all of the chromatographic detector data could be automatically recorded, perhaps by a dedicated laboratory computer data station (work station) for that multiple detector GC, then separate chromatograms could be produced and stored, and all of the above and more data manipulations could be performed according to any prescribed software program for that laboratory computer. Such instrumental capabilities would mean that the analyst would then be free to perform other more useful functions, and that the data manipulations and interpretations would be done more accurately, faster, more precisely, more reproducibly, and at far less cost and manual energy input.

The next part of this chapter describes some of the more interesting and innovative laboratory work stations designed for modern chromatography data acquisition, storage, manipulation, and final printout in tabular or chromatographic fashion. It would be impossible to summarize all of the commercially available laboratory computers now on the market for chromatography, but we have tried to present a cross section of those that seem to offer exactly what multiple detection GC may require today and tomorrow.

We felt that this might be an opportune time to describe some of the commercially available laboratory computer data stations amenable to multiple detection GC requirements and interfacing. This mini-review of laboratory computers available today for this sort of application is not meant to be completely exhaustive but, rather, to give the reader more of a feel for what can now be done in GC data acquisition, handling, storing, retrieval, and eventual manipulation with possible interpretation. Much of this information comes via recent commercial literature by manufacturers of the laboratory computer hardware and software. Such materials can be obtained by the reader by writing directly to the firms, and we have provided complete names and addresses in Appendix I. Clearly, custom software could be produced in a laboratory already having the hardware computer facilities and expertise in program writing. But, rather than describe how to write one's own software, we felt it might be more useful to describe what is commercially available, and how such systems could fit in with multiple detection GC requirements.

Current data-handling systems available for use in multiple detection gas chromatography can be divided into two major groups: (1) computing or reporting integrators and (2) stand-alone computer systems or direct computer links via the appropriate hardware. Computing integrators perhaps represent the bottom end of the spectrum as far as cost, capabilities, and flexibility, while hardware and software designed and marketed specifically for use with personal computers such as the Apple and IBM PC perhaps represent the other end of this spectrum. That is, these latter devices are the state of the art in data acquisition, processing, and control. We begin by first describing some of the commercially available computing integrators, their capabilities, requirements, and applications. We then proceed to discuss some of the more sophisticated systems available today for use in multiple detection GC.

1. Computing Integrators

a. The Spectra-Physics (Appendix I) Model SP4270 Computing Integrator. The Spectra-Physics SP4270 computing integrator (Fig. 2) is an easy-to-use, low-cost unit that utilizes a modular design to permit an optional expansion to dual detector operations. It comes standard with 16K of random access memory (RAM), programmable in BASIC, and an 8-in. printer/plotter. Also available as an option, at an additional cost, is an RS232 interface module that allows the integrator to be interfaced to various hardware such as a cathode ray tube (CRT) and/or data storage devices such as disk drives. Along with dual channel capability, other obvious differences between this type of integrator and the older Hewlett-Packard types are the SP4270's programmability and the ability to reprocess the data once generated. However, inherent in the design of the SP4270 is the fact that only one of the two channels operating will be plotted as a chromatogram; the remaining channel supplies the operator with the processed data only, and like the Hewlett-Packard integrator, it cannot replot the data as chromatograms. This necessitates a duplicate injection, if the second channel detector signal or a review of either of the chromatograms is desired. The data from both channels are still in the memory and can be accessed in any way the operator sees fit (ratioing, normalization, statistical evaluation, etc.). Indeed, the entire method is accessible, and the operator/analyst need only know a little BASIC programming to tailor the method and final report to particular requirements. However, without some type of an external storage device,

Figure 2 Spectra-Physics Model SP4270 computing integrator. (Courtesy of Spectra-Physics Corporation, San Jose, Cal.)

any methods developed or data generated would be lost if the SP4270's power were interrupted in any way, owing to the volatile nature of its memory. While the SP4270 membrane-type keyboard may be conducive to the laboratory environment, it is a rather cumbersome way to implement any extensive method development programs necessary for dual detector applications, particularly if the instrument or integrator is powered down for any reason. If this is the case, a CRT and/or a storage device of some type may become a necessity for anyone interested in serious method development or programming for optimal performance of the multiple detector setup. Spectra-Physics also produces a somewhat more sophisticated computing integrator in the SP4200 (Fig. 3). It has, in addition to the already-mentioned capabilities of the SP4270, a full ASCII keyboard and an alphanumeric LED display and is also capable of dual channel operation. The price of this system, and even that of the SP4270, with the necessary options, may not seem attractive in light of the other systems yet to be discussed, unless the laboratory already possesses the Spectra-Physics main-frame Labnet system with which these types of integrators were meant to communicate. They do, however, offer an alternative to the even higher-priced personal computer (PC)-based systems, and for most laboratories, this is of prime consideration.

Figure 3 Spectra-Physics Model SP4200 computing integrator. (Courtesy of Spectra-Physics Corporation, San Jose, Cal.)

b. SICA Chromatogram Processor Model 7000A. Systems Instruments Corporation, America (SICA) (Appendix I), offers their Chromatogram Processor Model 7000A (Fig. 4), which is a two-channel reduction system that stores the chromatograms from each channel (detector) before processing for quantitative results. By storing the data points, both chromatograms can be plotted on the same printout, either end to end or side by side, for direct comparison of the two detector outputs. The SICA 7000A comes standard with 48K of read-only memory (ROM) and 64K of RAM, an $8\frac{1}{2}$-in. printer/plotter, a full membrane-type keyboard, and an eight-digit alphanumeric light emitting diode (LED) display. The 7000A has somewhat more capability than the SP4270 and is perhaps directly comparable to the SP4200, while also being somewhat rigid in its method development and data treatment. While the 7000A is not fully programmable like the Spectra-Physics integrator, the operator/analyst can interact with the 7000A in such a way as to tailor the final report to a particular format. Both chromatograms can be viewed and replotted as necessary, as would be required for base line correction, so that reinjection is not necessary. An attractive feature of the SICA 7000A is the employment of a nonvolatile-type memory so that data and/or methods that have been developed are not lost if the instrument is powered down for any

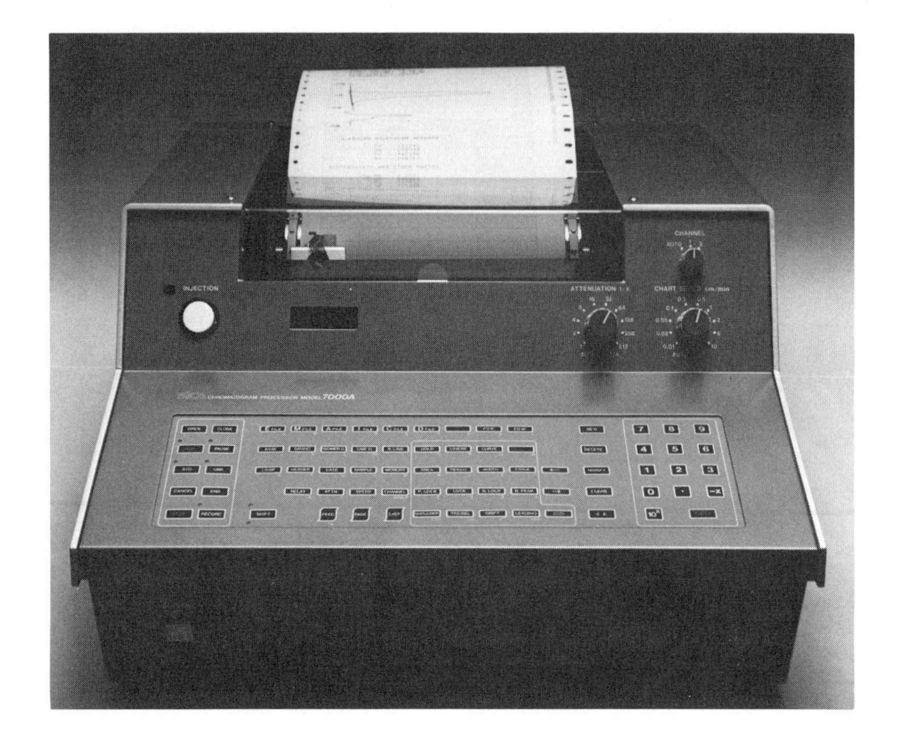

Figure 4 SICA Chromatogram Processor Model 7000A. (Courtesy of Systems Instruments Corporation, America, Dover, Mass.)

reason. Like the Spectra-Physics integrators, the 7000A also has available the option of an interfaced disk storage device, but at a much higher cost. This option is not as much of a necessity as is the case with the Spectra-Physics integrators, owing to the nonvolatile nature of the 7000A's memory. The 7000A also has the capability of communicating with a main-frame system through an optional RS232 port, and, without the disk storage system, the 7000A is priced quite competitively compared to the other computing integrators on the market. This fact justifies its consideration as a means of data handling and reduction in multiple detection GC.

2. Personal Computer Based Systems

a. The Cyborg ISAAC Chromatography Workstation. The availability of inexpensive micro- and personal computers (PCs) has recently provided a very economical means of obtaining and processing data for a mutliple detector GC system. However, it is still necessary to obtain the appropriate hardware and software to complete the interface from the analytical instrument (detector) to the computer of choice. Such an interface has been designed for the Apple Corporation (see Appendix I) Personal Computer, called the ISAAC unit (Integrated System for Automated Acquisition and Control, Cyborg Corporation; see Appendix I). The ISAAC Chromatography Workstation's (Fig. 5) unique hardware and menu-driven Chromatex software allow the operator/analyst to perform complete chromatographic data analysis in one system. Data can be acquired and observed with two real-time chromatograms on a high-resolution CRT, analyzed, stored to disk, printed with high-resolution graphics, recalculated, and then compared to previously stored data. The system's hardware and software are also designed to function as a general-purpose laboratory computer work station, allowing growth of the system with Cyborg's ISAAC expansion module. ISAAC's chromatography software Chromatext consists of interactive menu-driven displays designed to provide ease of use for data collection and establishing reliable peak processing parameters from two channels (detectors) simultaneously. Chromatext has been structured to allow the operator/analyst to create methods that are

Figure 5 ISAAC model 91A general-purpose laboratory work station with Apple computer, color monitor, and two disk drives. (Courtesy of Cyborg Corporation, Newton, Mass.)

specific for the data-handling requirements, such as those encountered in multiple detection GC. There is, therefore, no need to learn any programming for data acquisition, real-time graphics, and report generation, although the capability does exist. Indeed, the ISAAC Chromatography Workstation can also function as a general lab interface system. The optional, easy-to-use Labsoft data acquisition and control programming language permits the operator/analyst to develop specific application programs, such as detector normalization, ratioing, and qualitative identification of analytes, thereby exploiting the advantages of multiple detection GC. In addition, ISAAC's modular design permits expansions to automate a variety of laboratory instruments with specific instrument input modules that enhance ISAAC to function as a general laboratory work station. Use of the ISAAC hardware and software also does not preclude the use of the underlying PC for uses such as word processing, data base management, and a variety of other uses for which the PC was originally intended. While the cost of the PC is decreasing everyday, with the abilities becoming more and more sophisticated and flexible, a system such as the ISAAC becomes very appealing when one considers the alternative in a mainframe computer system at nearly two to three times the cost.

b. Anadata Chromcard II. An alternative to the Cyborg ISAAC system is the Chromcard II chromatography data system (Fig. 6) produced by Anadata, Inc. (see Appendix 1). The Chromcard system is also based on the popular Apple PC; it consists of two printed circuit boards that simply plug into the back of the Apple in available expansion slots. It also comes with software supplied on disks in the Apple Disk Operating System (DOS) 3.3 format. Again, all of the contributed software normally available for the Apple is still supported. Operation of the Chromcard system is very similar to the Cyborg system, in that it also uses a menu-driven format to prompt the operator/analyst through the setup, data acquisition, and report generation routines. The data and chromatograms generated are acquired via two channel operation in real time and can be observed on the CRT, stored to disk, printed, recalculated, and externally evaluated similarly to the Cyborg ISAAC system. In the Chromcard system, however, the operator must "toggle" between the two channels for CRT output so that both of the channels can be observed. The operator/analyst can switch from one channel to the other at any time, and the screen will rapidly playback the data acquired, while observing the other channel, before slipping into real-time display. Post analysis reintegration and editing are also possible; the analysis can be rerun, without a duplicate injection, from disk, using any new integration or method parameter. Like ISAAC, Chromcard allows the operator/analyst to edit the stored data graphically and manually draw in a base line if the automatic selection is inappropriate. Input to the Chromcard system is accomplished either directly from the chromatograph's recorder

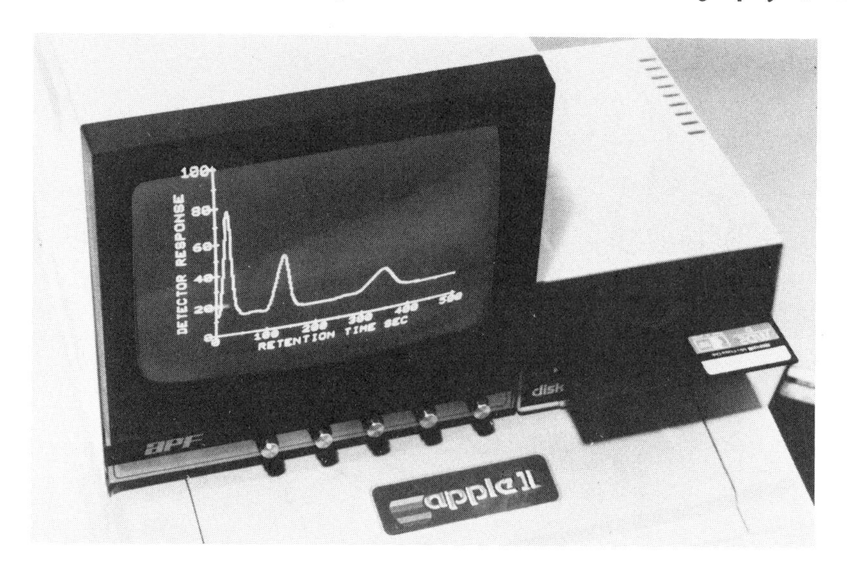

Figure 6 Anadata Chromcard II chromatography work station. (Courtesy of Anadata, Inc., of Glen Ellyn, Ill.)

output, taking advantage of the built-in scale expansion provided by the instrument attenuator, or it can operate from the 1-V integrator output, where the signal is independent of the attenuator setting. This results in valid data even if the peaks go off scale. Although it does not possess the capability of instrument control by a digital-to-analog output, Chromcard is still fully compatible with automatic operation, such as that required for autosamplers. The Chromcard system is somewhat less expensive than the Cyborg system, yet it possesses many of the same capabilities, while experiencing a high degree of commercial success in multiple detector areas.

c. *Nelson Analytical Model 3000 Multi Instrument Data System.* At the top of the line in PC-based systems is the Nelson Analytical (see Appendix I). Model 3000 Multi-Instrument Data System (Fig. 7). Based on the IBM PC, the 3000 provides some of the most advanced and sophisticated data-handling capabilities for multiple detection GC available today. Compared to the Apple, the IBM PC is a somewhat more sophisticated and higher-priced system, with the added capability of high-resolution color graphics. With the 3000, the operator/analyst can acquire data from six chromatographs with two detectors on each, simultaneously, even with each chromatograph connected to an autosampler, and run off-line programs unrelated to chromatography at the same time. During the course of an actual run, an interface stores the raw data, point by point, in buffer memory within the interface

Figure 7 Nelson Analytical Model 3000 Multi-Instrument data system. (Courtesy of Nelson Analytical, Inc., Cupertino, Cal.)

itself. At or near the end of a run, the data for all channels are transferred to the memory of the computer. That data are then reduced, reported, and stored in more permanent disk files for later recall. The 3000's chromatography software runs in the IBM PC DOS operating system, and like the previously mentioned systems, it is menu driven for tailored report and method generation. Postrun reanaysis by different parameters is, of course, also possible, in addition to various other capabilities. A replotted chromatogram can be quickly and easily obained according to new specifications, and chromatogram comparisons, overlays, ratioing, and difference calculations are all possible via the available software in accordance with the capabilities of the aforementioned systems. Up to eight chromatograms with vertical and horizontal scale expansions can be processed on one CRT screen or on continuous "pages" to permit easier viewing. While somewhat higher priced, the 3000 represents a significant opportunity for many laboratories to obtain a versatile and advanced laboratory system that is capable (if not overly so) of multiple detection data-handling requirements in GC.

III. SERIES MULTIPLE DETECTION GAS CHROMATOGRAPHY

There is a significant advantage in utilizing the series orientation in multiple detection GC in that no effluent stream splitting is necessary. Therefore all of the mass of each analyte peak elutes from the separation column into the first and subsequent detectors. Analysis of the entire sample by both detectors provides greater sensitivity than that obtained by effluent splitting, and the problem of ratio control in the splitter is eliminated. The series orientation is even more advantageous if the analyst is utilizing high-resolution capillary column GC, where effluent splitter dead volume can become significant and degrade system performance and efficiency. The series orientation has minimum requirements; as long as the upstream detector is nondestructive and preferably the detector with the lowest internal dead volume, the analyst need merely choose the proper detector combination for the application and, finally, install the detectors appropriately, as outlined earlier. What follows is a summary of what has been reported and what is currently being accomplished in the field of multiple detection GC with series orientation, with specific detector combinations and applications.

Various combinations of detectors in series orientation have been used to elicit a wide range of information from many different sample matrices. Perhaps one of the more significant combinations in recent years has been the coupling of a PID and an ELCD* to determine purgeable aromatics in the presence of chlorinated compounds in water samples [13–15]. The analysis of trace volatile organic hydrocarbons and aromatics in water is performed by the purge and trap method, according to the Environmental Protection Agency's (EPA) methods 601 and 602, respectively. Screening for both compound classes by these methods requires two concentrators and one or two GCs, or it requires a time-consuming switch from one chromatographic system to the other, which greatly increases overall analysis times. This has been overcome now by chromatographing both the halocarbons and aromatics on a single column with a PID and a Hall or Coulson electrolytic conductivity detector (PID/ELCD) in series. This combination provides a highly sensitive detector (PID), and a highly sensitive and selective detector (ELCD). As can be seen from Fig. 8, using the 10.2-eV lamp, the aromatic compounds along with some unsaturated halocarbons exhibit similar responses via the PID, while only the chlorinated compounds respond to the ELCD [14]. Table 3 summarizes the responses one might observe in somewhat more detail. Kirshen and Wood showed that the ELCD/PID response ratios can be used for qualitative identification of some additional compounds that respond to both detectors [13]. In

---

*ELCD-Electrolytic Conductivity Detector refers to a Hall or Coulson Detector

Table 3 Series PID/HECE Responses and Relative Retention Times for Organohalogens and Aromatics

Compound	Relative Retention time on column[a]			Detector response[b]	
				PID	HECD
Chloromethane	0.069	0.177	ND[f]	–	+
Bromomethane	0.086	0.294	ND	–	+
Vinyl chloride	0.100 (0.097)	0.177 (0.170)	ND	+	+
Dichlorodifluormethane	0.100	0.101	ND	–	+
Chloroethane	0.129	0.411	ND	–	+
Dichloromethane	0.199	0.497	ND	–	+
1,1-Dichloroethylene	0.360 (0.356)	0.323 (0.318)	ND	+	+
1,1-Dichloroethane	0.459	0.657	ND	–	+
trans-Dichloroethylene	0.516 (0.512)	0.448 (0.444)	ND	+	+
cis-Dichloroethylene	0.516 (0.512)	0.620 (0.621)	0.710 (0.704)	+	+
Chloroform	0.550	0.620	ND	–	+
1,2-Dichloroethane	0.591	0.889	0.433	–	+
1,1,1-Trichloroethane	0.650	0.680	ND	–	+
Carbon tetrachloride	0.670	0.525	0.426	–	+
Bromochloromethane	0.703	0.796	ND	–	+
1,2-Dichloropropane	0.760	1.00	1.00	–	+
Trichloroethylene	0.797 (0.796)	0.680 (0.678)	0.646 (0.630)	+	+

Benzene	0.813	0.870	0.630	+	–
Dibromochloromethane	0.820	0.972	ND	–	+
1,1,2-Trichloroethane	0.820	1.08	1.77	–	+
2-Bromo-1-chloropropane (ISTD)	0.860	1.12	ND	–	+
Dichloroiodomethane	0.902	1.04 (1.04)	2.43 (2.16)	+	+
Bromoform	0.912	1.12	2.48	–	+
1,1,1,2-Tetrachloroethane	0.912	1.12	1.77	–	+
Tetrachloroethylene	0.981 (0.983)	0.796 (0.796)	1.16 (1.21)	+	+
1,1,2,2-Tetrachloroethane	0.981	1.33	2.99	–	+
a,a,a-Trifluorotoluene	1.00	1.00	1.00	+	+
Toluene	1.02	1.12	1.21	+	–
Chlorobenzene	1.08 (1.08)	1.12 (1.12)	2.37 (2.41)	+	+
Ethylbenzene	1.18	1.28	1.91	+	–
Bromobenzene	1.23 (1.23)	ND	3.42 (3.50)	+	+
Isopropylbenzene	1.31	ND	2.37	+	–
m-Xylene	1.41	1.36	2.23	+	–
Styrene	1.41	ND	2.86	+	–
o-Xylene	1.49	1.41	2.37	+	–
p-Xylene	1.49	1.36	2.07	+	–
o-Chlorotoluene	1.60 (1.58)	ND	3.23 (3.30)	+	+

Table 3 (continued)

Compound	Relative retention time on column[a]			Detector response[b]	
	A[c]	B[d]	C[e]	PID	HECD
p-Propylbenzene	1.58	ND	2.78	+	–
p-Chlorotoluene	1.71 (1.72)	ND	3.23 (3.30)	+	+
m-Dichlorobenzene	1.71 (1.72)	ND	3.88 (3.98)	+	+
o-Dichlorobenzene	1.80	ND	5.28 (5.43)	+	+
p-Dichlorobenzene	1.83	ND	3.42 (3.50)	+	+

[a]Relative retention times are relative to the internal standard, a,a,a-trifluorotoluene, using the appropriate detector. When two numbers are given, the first number represents the relative retention time for the HECD.
[b]Compound causes a response (+) or not (–) on the indicated detector.
[c]Primary analytical column: 1.8 m by 2 mm inner diameter glass packed with 1% SP1000 on 60/80 Carbopack B, held at 60°C for 10 min during trap desorption, then temperature programmed at 7°C/min for 10 min, then 12°C/min to a final temperature of 200°C.
[d]Halocarbon confirmatory column: 1.8 m by 2 mm inner diameter glass packed with n-octane on Porasil C, held at 50°C for 4 min during trap desorption, then temperature programmed at 4°C/min to a final temperature of 140°C.
[e]Aromatics confirmatory column: 1.8 m by 2 mm inner diameter glass packed with 5% SP1200/5% Bentone 34 on 100/120 Supelcoport, held at 60°C for 4 min, then programmed at 3°C/min to 110°C.
[f]ND, not determined.
Source: Ref. 15.

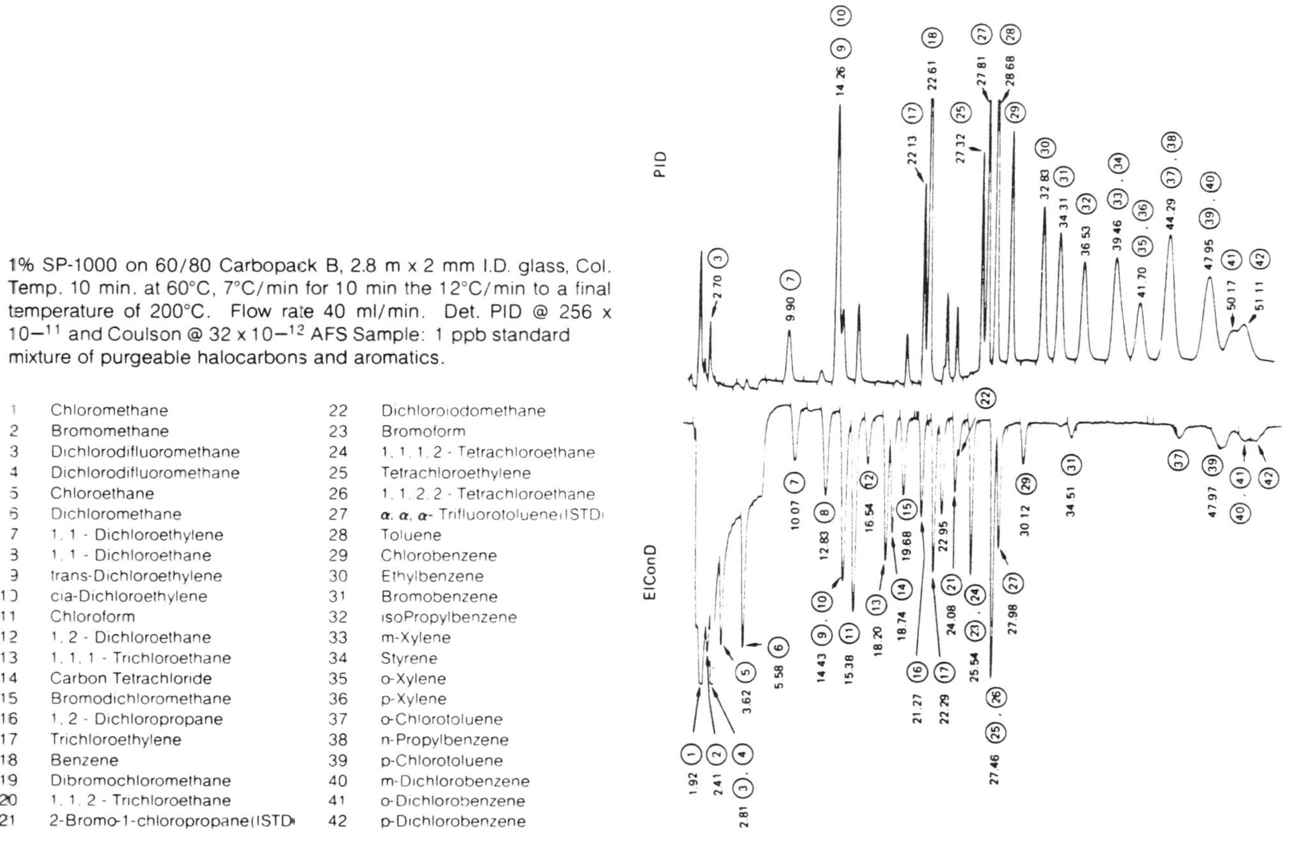

Figure 8 Purgeable chlorinated and aromatic compounds analyzed by series GC PID/ELCD (Courtesy D. Coulson & HNU Systems Inc., Newton, Massachusetts.)

addition, they undertook a very informative study to determine the contribution of the series orientation of these detectors to peak broadening, with overall results summarized in Table 4. This was determined by measuring the ELCD peak widths at half-height when using the ELCD only versus using the PID and ELCD in series. As can be seen, peak broadening is exhibited by all of the hydrocarbons, but not to any substantial degree. The broadening is attributed to dead volume and adsorption effects in the PID and transfer conduits.

Coulson [14] has applied the PID ELCD in series for analysis of chlorinated aromatics and chloroalkenes in water (Fig. 9). This permits analysis by EPS methods 601 and 602 via a single injection. Sulfur gases at low ppm levels have been analyzed using a PID (11.7 eV)/Far UV combination. The higher energy 11.7 eV lamp is necessary in the PID for detection of COS. The far UV absorbance detector which has a universal response will also detect both SO_2 and O_2 which have ionization potentials greater than 12 eV. The PID with a 9.5 eV lamp provides considerable selectivity as seen by the selective detection of 10 ppm of DDT and DDE in the presence of 1000 ppm of toxaphene (Fig. 10). This analysis requires GC-MS analysis for the detection of DDT or DDE when Toxaphene is present. The advantage here is that the complex mass spectrometer is not required and lower levels of these pesticides can be determined in this matrix.

Another advantageous combination of detectors in series is the coupling of PID and FID to provide data that can then be used to classify hydrocarbons according to structural groups. Driscoll et al. [16] exploited this particular combination to aid in the detection and identification of aromatic and aliphatic hydrocarbons in complex mix-

Table 4 Peak Broadening with PID/HECD in Series

	Peak width at half height, HECD		
Compound	HECD	PID/HECD	Percentage difference
$CFCl_3$	9.95	10.8	8.5
1,2-DCE=	11.4	11.2	1.8
1,2-DCE	10.15	10.2	0.5
1,1,1-TCE	9.6	10.4	8.3
$CHBrCl_2$	12.2	12.3	0.8
t-1,3-DCP=	11.45	11.55	0.9
$CHBr_3$	14.05	15.55	10.6

Source: Ref. 13.

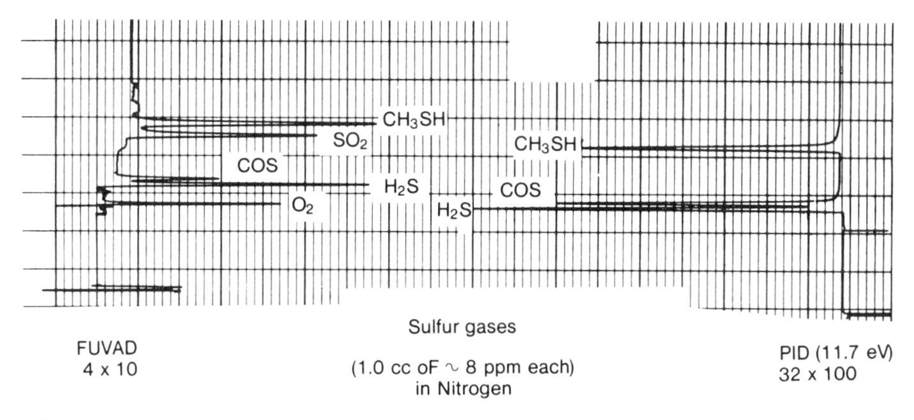

Figure 9 Analysis of sulfur gases by GC PID (11.7)/Far UV absorbance (Courtesy HNU Systems Inc., Newton, Massachusetts.)

tures. By using response ratios in the form of relative molar response per mole of carbon (RMR), it was possible to distinguish between these two groups of compounds very readily. The FID was used to measure the relative levels of a particular hydrocarbon, regardless of the degree of unsaturation; a 10-fold difference was observed between alkenes and aromatics. Those compounds with a PID/FID ratio, normalized to Benzene = 35, of about 10 to 20, could be identified as alkenes, those with a ratio of 25 to 50 as aromatics, and those with ratios less than 8 as alkanes. Initially, simple hydrocarbon mixtures were evaluated to demonstrate the viability of the overall technique, and later work extended the technique to high molecular weight heteroatom molecules, such as chlorinated and sulfur based pesticides. Table 5 shows the PID/FID response ratios used for the identification, along with the compound type. A practical application of the approach was demonstrated with the successful analysis of aromatics in a light hydrocarbon feedstock used in a synthetic natural gas plant. Fig. 11 is a chromatogram comparing the magnitude of dual detector responses for this sample, and Table 6 gives the peak identity and the PID/FID response ratios for the hydrocarbon peaks. It is also noted that these types of detectors (PID/FID) are independent of the design or manufacturer of the particular unit used, since different designs were tested with very similar overall results.

Nutmagul et al. [17] extended applications of the PID/FID combination by utilizing a somewhat modified PID adapted for capillary work to analyze atmospheric samples. PID/FID ratios normalized to toluene were used to distinguish between saturated and unsaturated hydrocarbons as outlined before by Driscoll et al. [16]. Figure 12 illustrates the chromatograms obtained from a 500 ml air sample evaluated by capil-

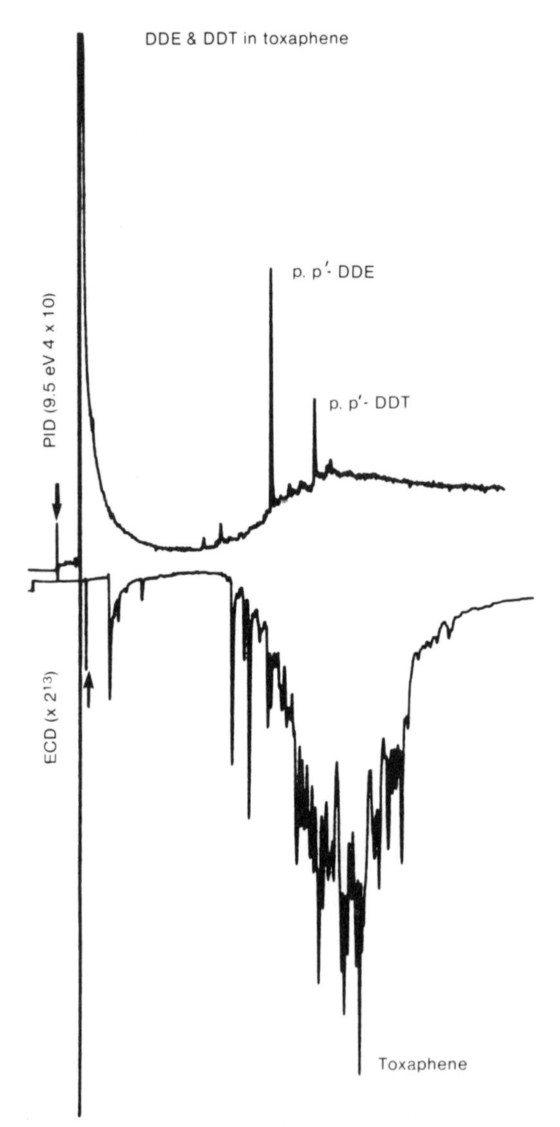

ANALYTICAL CONDITIONS

Sample: 10 ppm p, p'-DDE, 10 ppm p, p'-DDT and 1000 ppm Toxaphene
Column: 20 M OV-101 Fused Silica Capillary
Column Oven Temp.: 150°C to 250°C at 8°C/min.
0 min. initial hold and 16 min. final hold
Split Ratio: 60/1

Figure 10 Analysis of DDT, DDE in the presence of Toxaphene by series GC-PID/ELCD (Courtesy HNU Systems Inc., Newton, Massachusetts.)

Table 5 Normalized Relative Response Ratios for Simple Hydrocarbons and Pesticides

Compound	Type of structure	Normalized PID/FID ratio*
Hexane	Alkane	1
Cyclohexane	Alkane	9
Octane	Alkane	4
cis-2-Octene	Alkene	22
Benzene	Aromatic	35
Toluene	Aromatic	41
p,p'-DDT	Aromatic	70
o,p'-DDD	Aromatic	70
o,p'-DDT	Aromatic	68
p,p'-DDE	Aromatic	60
Heptachlor epoxide	Alkene	23
Aldrin	Alkene	23
α-BHC	Alkane	4
β-BHC	Alkane	4
γ-BHC	Alkane	5
Chlorobenside	Aromatic	50
Endosulfan I	Alkane	4
Endosulfan II	Alkane	8
Tetrasul	Aromatic	59

*These PID/FID ratios have been normalized to a PID/FID ratio of 35 for benzene and have been recalculated for this paper.
Source: Ref. 16.

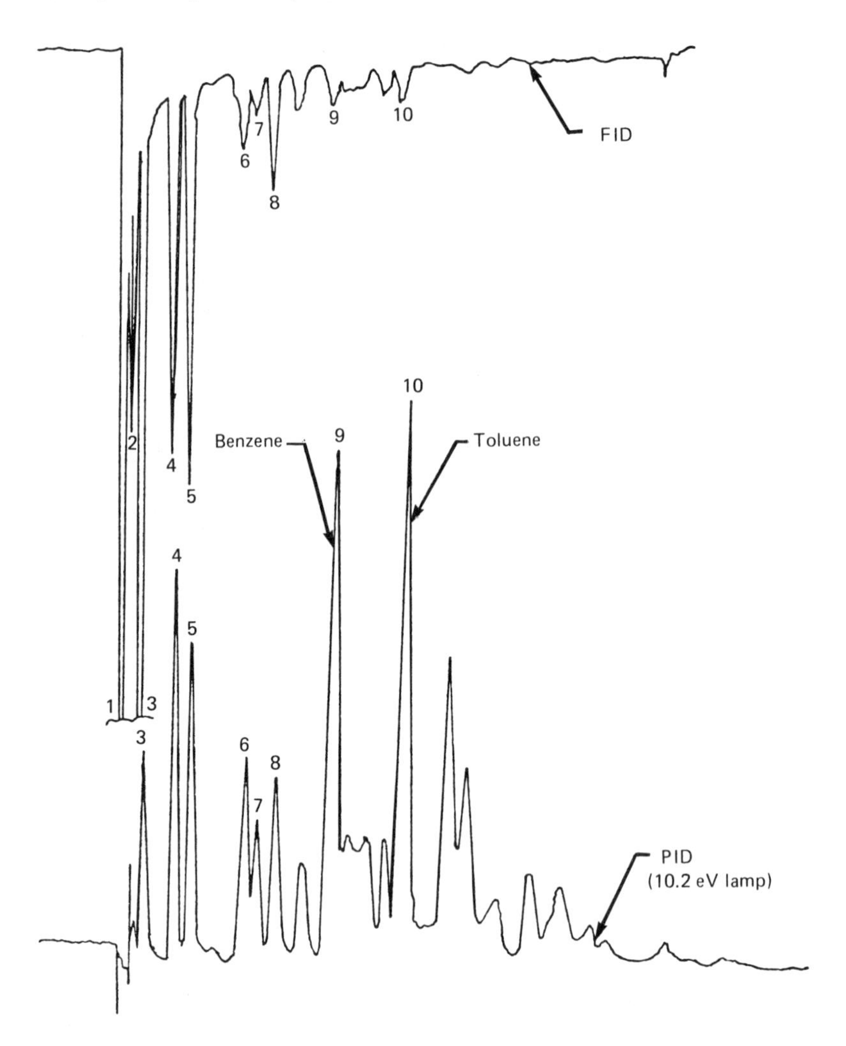

Figure 11 Comparison of the PID and FID for a light hydrocarbon synthetic natural gas feedstock (conditions: 3 ft X 1/8-in. inner diameter Porapak Q column, temperature programmed from 0 to 190°C at 8°C/min; N_2 flow rate 36 ml/min, 0.5 ml of liquid petroleum gas injected with high-pressure valve). (From Ref. 16.)

Table 6 Light Hydrocarbon Natural Gas Feedstock Analysis

Peak number	Compound	Normalized PID/FID ratio*	Retention time (min)
4	Isopentane	2	1.33
5	n-Pentane	1.5	1.68
	Neohexane	6	2.17
6	2-Methylpentane	3	2.28
7	3-Methylpentane	5	3.06
8	n-Hexane	3	3.38
	Methylcyclopentane + 2,2'-dimethylpentane	3	3.86
9	Benzene	35	4.56
	Unknown	28	4.81
	Unknown	10	5.58
10	Toluene	24	5.92
	Unknown	23	7.24
	Unknown	16	7.75

*Normalized to PID/FID ratios = 35 for benzene and recalculated for this paper.
Source: Ref. 16.

lary dual detector GC. Table 7 lists the identified peaks and compounds of Fig. 12, along with the PID/FID ratios used for the identification. Utilizing the data of Table 7 and Fig. 12, it was possible to further characterize the real air sample constituents by calculating the hydrocarbon distribution. Using the data in Table 6 a PID/FID ratio of 35 for normalization, Nutmagul's value for toluene is 23 compared to a value of 24 in Table 6.

Kapila and Vogt [18] also used this particular combination to again monitor trace amounts of aromatic hydrocarbons by analyzing an extract of a composite tar sample obtained from a pilot coal gasification plant. With PID/FID in series and the above response ratio methods, eluted peaks from coal gasification samples were classified as either aromatic or aliphatic hydrocarbons. Although no specific data were provided, Fig. 13 displays results obtained from such a sample. In addition,

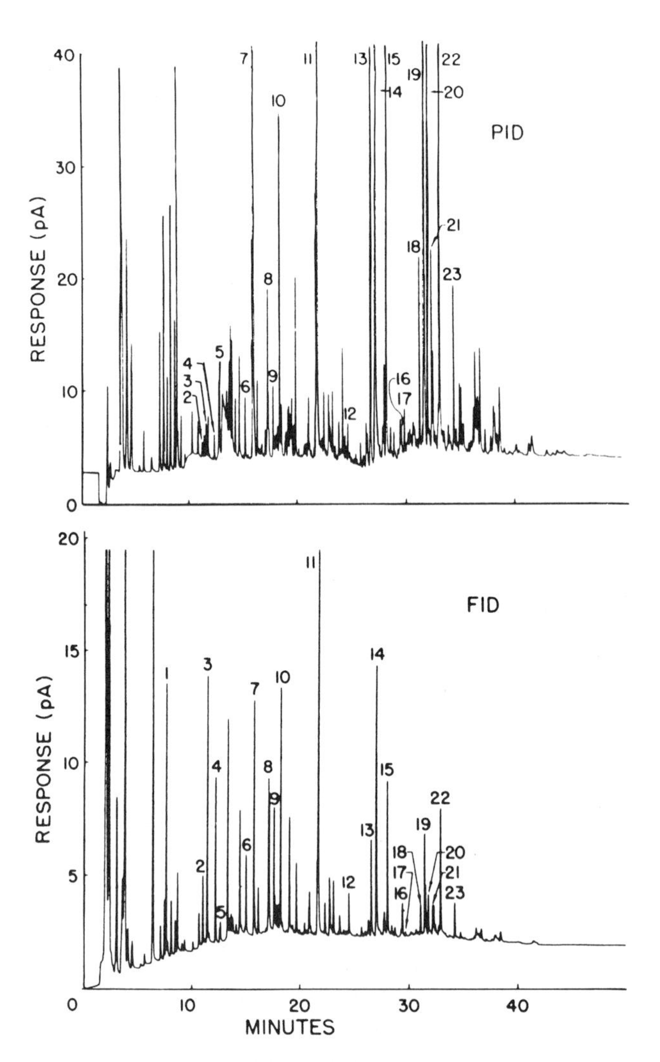

Figure 12 PID (top) and FID (bottom) chromatograms of a real air sample collected at 1700 LT in a small parking lot on April 2, 1983. The sample aliquot size was 500 ml of air and the UV lamp intensity for the PID was 70% full scale. (From Ref. 17. Copyright 1983 American Chemical Society.)

Table 7 Hydrocarbon Compounds Determined in Real Air Sample, Including the Observed PID/FID Ratios Normalized to Toluene

Peak number	Compound	PID/FID normalized to toluene (X100)
1	n-Butane	—
2	2,3-Dimethylbutane	4
3	2-Methylpentane	3
4	3-Methylpentane	3
5	1-Hexene	75
6	2,4-Dimethylpentane	15
7	Benzene	129
8	2,3-Dimethylpentane	40
9	3-Methylhexane	12
10	2,2,4-Trimethylpentane	28
11	Toluene	100
12	n-Octane	22
13	Ethylbenzene	84
14	p- and m-Xylene	116
15	o-Xylene	82
16	n-Nonane	25
17	Isopropylbenzene	134
18	n-Propylbenzene	103
19	p-Ethyltoluene	76
20	1,3,5-Trimethylbenzene	158
21	o-Ethyltoluene	105
22	1,2,4-Trimethylbenzene	123
23	1,2,3-Trimethylbenzene	95

Source: Ref. 17.

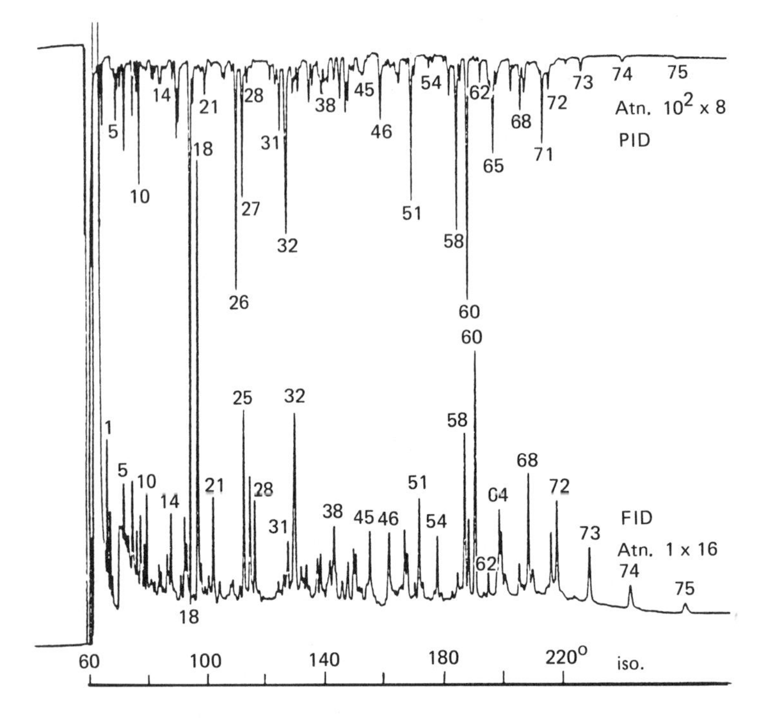

Figure 13 Chromatograms of a neutral fraction of a tar sample from a coal gasification pilot plant. (From Ref. 16.)

efficiency of the series combination of PID/FID was evaluated in terms of theoretical plates and is presented in Fig. 14. A less than 6% loss in efficiency was realized in terms of the height equivalent to a theoretical plate measured as detector response to hexadecane via the series approach.

Still another application of the series approach utilizing the PID coupled with another detector was that of Townes and Driscoll [19]. A series orientation of PID/NPD was used for the detection and identification of amines, with two different lamp energies employed to obtain the PID data. It was shown that with the PID (10.2 eV)/NPD, the PID provided a general response and the NPD a selective response for nitrogen-containing compounds, while the PID (3.8 eV), NPD, the response ratios were indicative of the type of amine present (primary, secondary, or tertiary). A demonstration of the selectivity of the NPD over the more universal response on the PID (10.2 eV) is illustrated in Fig. 15, where hydrocarbon peaks two and three are not detected on the NPD. It should be noted in Fig. 15 that as the lamp energy was reduced to 8.3 eV, the PID response become more selec-

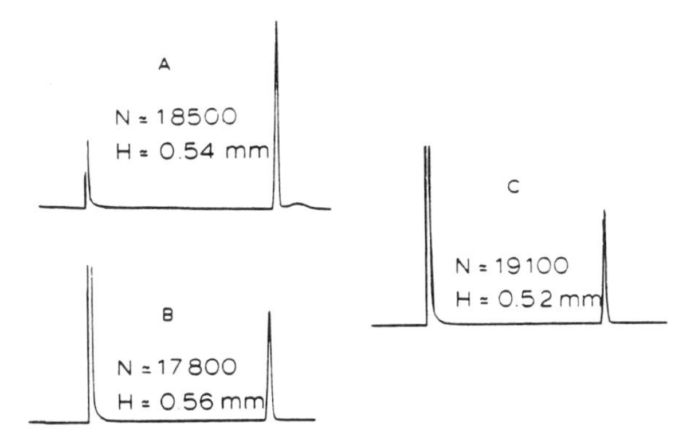

Figure 14 Efficiency of the chromatographic system expressed in terms of theoretical plates: (A) photoionization detection, (B) flame ionization detection in series with PID, and (C) flame ionization detection direct from the column. (From Ref. 18.)

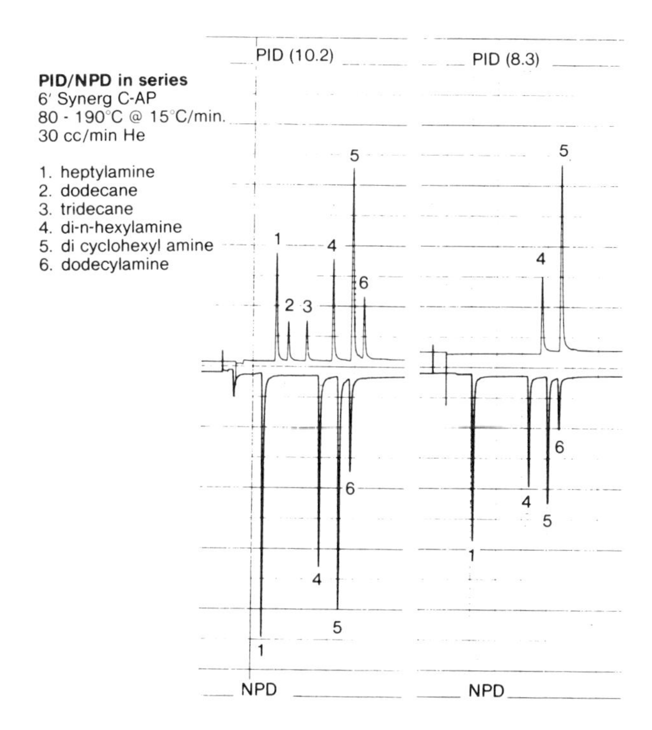

Figure 15 PID/NPD chromatograms for hydrocarbons and amines. (From Ref. 19.)

tive. Hence the PID (8.3 eV) becomes a more selective detector than the NPD. Primary amines, on the basis of ionization potentials, do not respond and are therefore not detected on the PID (8.3 eV), while the secondary and tertiary amines are detected, as further illustrated in Fig. 16. Therefore, using the corresponding PID/NPD response ratios at various PID lamp energies, nitrogen content and amine classification can be obtained, as in Table 8. It should also be noted, however, that the 8.3 eV lamp has a lower intensity than the 10.2 eV lamp, and therefore, the final detection limits are somewhat lower. There exists, then, a trade-off (10.2 versus 8.3 eV) between the increased sensitivity, or the ability to differentiate amine classes, that must be considered.

The final PID series application that will be presented here is that by Conron et al. [20] and Jaramillo et al. [21], consisting of the detection and identification of halogenated hydrocarbons by PID/ECD in series. This particular application is perhaps unique when compared to other detector combinations, in that it is the only combination so far mentioned in the literature that offers a dual detector series approach that is totally nondestructive. Although the PID response is essentially independent of the degree of chlorination, as shown by

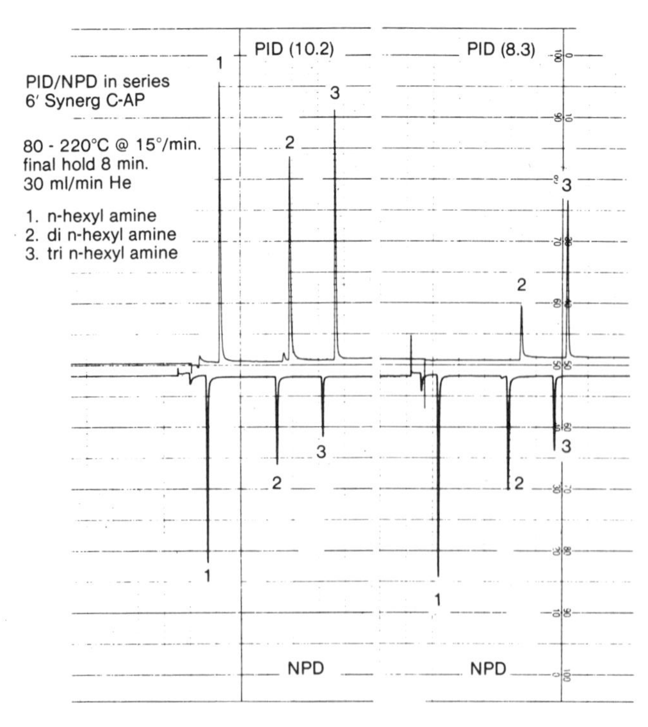

Figure 16 PID/NPD chromatograms for primary, secondary, and tertiary amines. (From Ref. 19.)

Table 8 PID/NPD Detector Response Ratios for Amines and Detector Combinations

	Ratios	
	PID (10.2)/NPD	PID (8.3)/NPD
n-Hexylamine	1.6	—[a]
Dodecylamine	2.6	—[a]
Di-n-hexylamine	2.3	0.01
Dicyclohexylamine	3.2	0.03
Tri-n-hexylamine	4.0	0.04
Benzylamine	3.4	—[a]
Dibenzylamine	6.4	0.004
Tribenzylamine	9.8	0.03

[a]No response for PID (8.3).

Detector Combinations

PID (10.2)/NPD	Semiqualitative	All peaks/those containing N
	Structural identification	PID/NPD ratios are indicative of N content: <1; more than 1N; e.g., di- or triamine $>1, <2.5$; primary aliphatic amine >3; primary aromatic amine
PID (8.3)/NPD		Confirming amine classification 1°—no PID response but positive NPD response 2° or 3°—positive PID and NPD response (see text) with appropriate ratios
Optional PID/FID	Additional confirmation of amine skeletal structure	
		Aliphatic 15–20 Cycloaliphatic 25–30 Aromatic 40–50

Source: Ref. 19.

Langhorst [22], it is more sensitive to these types of compounds than the FID, making it more attractive for this type of application. The ECD response can vary by four orders of magnitude as the degree of chlorination is changed, and when used in series with the PID additional qualitative information can be obtained, as indicated in Table 9. For aromatic compounds with less than three chlorine atoms, the PID will be more sensitive, but with four or more chlorine atoms, the ECD will be more sensitive. PID/ECD ratios can then provide a unique means for sample identification of compounds such as PCBs, DDE, DDT, and toxaphene. As before, additional lamp energies were also employed to further enhance the specificity and selectivity of the method.

Another series application applied to the analysis of halogenated compounds was the arrangement of ECD/FID reported by Sodergren [23]. The detection system was used to study the degradation and fate of persistent pollutants in aquatic ecosystems. Usually these pollutants are closely associated with lipids, and therefore it is an advantage to be able to study the occurrence and the amount of both lipids and, for example, organochlorine residues. The particular detector combination was chosen on the basis of the advantage of simultaneously detecting compounds with different properties in the same sample. Figure 17 shows the response of the detectors to just such a sample. Organochlorine residues and methyl esters of fatty acids were simultaneously detected. The ECD was not affected by the fatty acids, and the FID showed no response for the organochlorine compounds at the concentrations used here.

Perhaps the most novel approach in terms of series orientation detector design has been that developed by Poy [24]. In this application, a new line of detector heads with the ionization cells coupled in series was developed to, in effect, accomplish ECD/FID detection within a single detector termed an EC + FID. Figure 18 shows the basic elements and functions of this particular detector. The EC + FID head is installed on the base of a conventional ionization detector without mod-

Table 9 Comparison of PID (10.2 eV)/ECD Response Ratios for Chlorobenzene

Number of chlorine atoms	1	2	3	4	5	6
PID (10.2 eV)[a]/ECD ratios	b	4.0	1.1	0.16	0.04	0.03

[a]PID has approximately the same response (within 20%) for all chlorobenzenes.
[b]Very little response on the ECD
Source: Ref. 20.

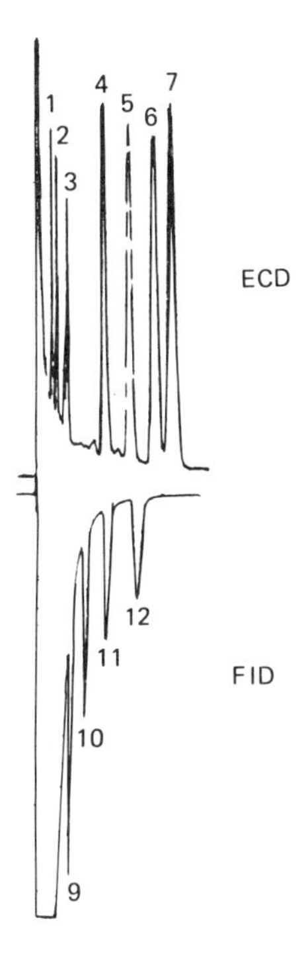

Figure 17 Simultaneous ECD/FID responses to a mixture of organochlorine pesticides and methyl esters of fatty acids (peaks: 1, lindan; 2, BHC; 3, aldrin; 4, p,p'-DDE; 5, dieldrin; 6, p,p'-DDD; 7, p,p'-DDT; 8, lauric acid; 9, myristic acid; 10, palmitic acid; 11, stearic acid; and 12, oleic acid). (From Ref. 23.)

ification, and is heated and controlled electronically by conventional means. The overall performance of the series-coupled system was shown to be comparable with that of single detection. The main thrust of this article was the evaluation of the new detector design; applications were used for the evaluation similar to those outlined earlier for the same dual detector combination via two separate series detectors [23].

Herres et al. [50] in 1983 reported an interesting interfacing of GC with a series arrangement of Fourier transform infrared spectroscopy (FTIR) and FID for the analysis of certain volatile organics in fruits. Gas-chromatographic separations were performed using a fused wall-coated, open tubular capillary column with temperature programming. Following the GC separation step, the column effluent was transferred at elevated temperatures to the light pipe of the GC–

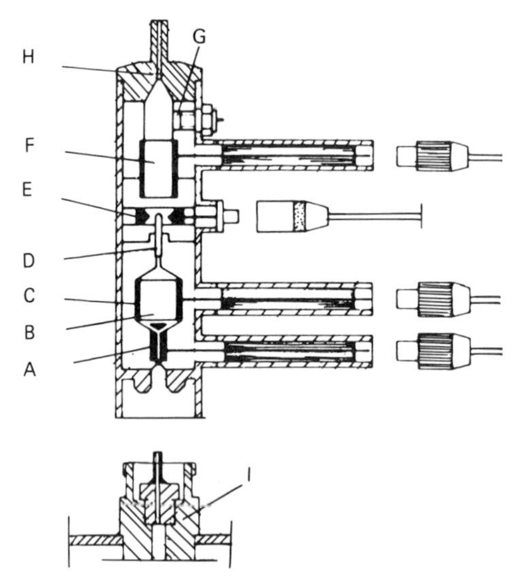

Figure 18 Schematic diagram of the DANI EC + FID model 68/51 dual detector: (A) collection electrode, (B) ECD section, (C) cathode, (D) burner of FID section, (E) polarization electrode, (F) collector, (G) flame igniter, and (H) chimney and combustion inlet. (From Ref. 24.)

infrared (IR) interface, part of the overall FTIR spectrometer (Bruker IRFS 85). The total effluent was then passed back into the GC oven to the FID system for generation of the second series chromatogram. A number of series IR/FID chromatograms are provided in this article, illustrating the utility of a selective IR detector followed by a totally general FID unit. In addition, this particular series dual detector approach had the ability to provide individual vapor phase IR spectra from a typical GC–IR run for various individual peaks.

IV. PARALLEL MULTIPLE DETECTION IN GAS CHROMATOGRAPHY

There have been relatively few prior extensive reviews of this aspect of multiple detection in GC, but certainly that of Pigliucci et al. should be mentioned [25]. These authors have discussed a variety of selective detectors that could be used, especially the nitrogen–phosphorus detector (NPD), flame photometric detector (FPD), and electron capture detector (ECD). A single NPD could be used in either a series or parallel orientation to monitor simultaneously for both sulfur and phosphorus emission from each analyte, thus providing two separate chromatograms from a single injection with a single detector, using two separate light pipes. This review article also summarized various configurations of multiple columns and/or multiple detectors that could be used with commercial gas chromatographs. With a particular Perkin-Elmer model GC system and their detectors, a wide number of possible combinations of detectors could be installed and used simultaneously. With two ECDs, one FPD with two light pipes, and either an FID or NPD in place, five different detector channels could be monitored simultaneously with each injection per analyte. We have intentionally underplayed the possible use of multiple columns with multiple detectors in GC, but certainly this is always possible. Thus, with a single injection, two separate columns of differing resolving capabilities and multiple detectors in parallel at the end of each column, an even larger number of chromatograms and detector printouts could be obtained per injection of sample. It is also possible, given the appropriate GC instrumentation, to have two separate columns hooked up to two separate injection ports, with various detectors at the end of each column, and with simultaneous injections onto each column, additional chromatographic and detector information could be readily obtained. Surely the easier approach is to use a single injection, with a single column or dual columns in place, injection splitting at known ratios to each column, all followed by multiple detectors at the end of each column [25]. Various applications of these approaches have been presented and discussed in this earlier review, as in the multiple detection analysis of oil samples, pesticide mixtures, polychlorinated biphenyls, and others. Typical chromatograms for multiple detection are presented, but relatively little is indicated in the way of data handling, detector response ratios (quantitative), data manipulation, and ancillary techniques suggested above.

Another earlier review is that of Grob [26], who emphasized the glass capillary column in GC, with several pages devoted to multiple detection methods and applications. Of course, in capillary GC with multiple detection, effluent splitters, tees, and connecting lines can significantly affect total system dead volume, chromatographic efficiencies and resolutions, and even more attention must be devoted to the hardware used here. As Grob pointed out, "valuable preliminary information from a first separation can be obtained when the column effluent is split onto detectors with markedly differing specificity." Mentioned here are several earlier references to the use of efficient capillary effluent splitters, such as the methods of Etzweiler and Neuner-Jehle [27,28] and applied by Bertsch and co-workers [29]. Detectors initially engineered for packed column GC applications provide excessive dead volume and band broadening when used with capillary columns, and thus many workers and companies have redesigned and engineered detectors specifically for capillary GC interfacing. Another approach to multiple detection in GC is discussed in this review by Grob, namely, that which uses multicolumn systems for two-dimensional separations. Thus effluents from the first column can be split to two separate subsequent capillary columns, each of these being followed by single or multiple detection. At the same time, it has been possible to take the effluent from the first GC column, split this with one fraction going to another different second column, with the remainder directed to a general or selective detector or series of detectors. This particular multiple detection—multiple column GC arrangement is depicted schematically in Fig. 19 [26]. A number of specific applications of multiple column/detection capillary GC have been provided by Grob, such as that for diesel oil and cigarette smoke. Though this particular review is almost 10 years old, it is still an excellent introduction to capillary GC, column switching, multiple column analysis, and, of course, multiple detection approaches.

Figure 19 indicates the ability to monitor chromatographic profiles of injected samples in between a first and second separation step, using either a general (FID) or selective-type detector after the first split (parallel). When all of the eluent from the first column is entirely transferred to the second column, with a different packing material or capillary column in place, the next detection step (general or selective) then provides another chromatographic profile for the same sample initially injected into the first column in series. Thus, with a single injection, two different GC profiles or chromatograms are produced, with either general or selective detection possible after each separation, or multiple detection (series or parallel) after each column.

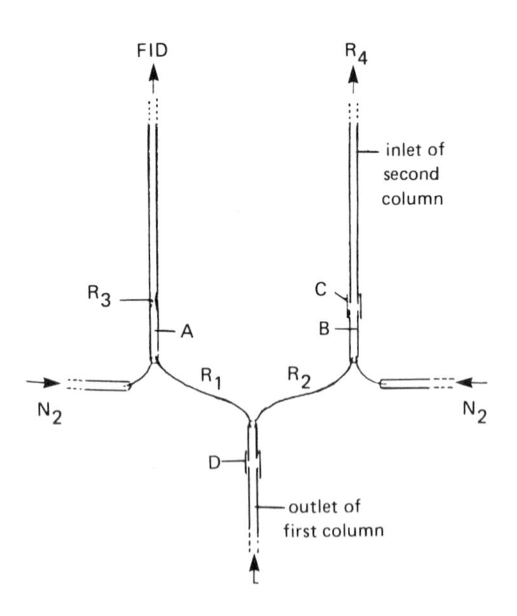

Figure 19 Splitting device for controlled transfer of column effluents to FID and to second column, respectively. (From Ref. 26.)

Just as detectors can be placed either in series or parallel after the GC column, so too can columns be placed in series, as in Fig. 19, or in parallel, as discussed above. Such arrangements therefore provide the analyst with the ability to obtain a large amount of chromatographic and detector information with a bare minimum number of injections in a minimum amount of time. One should always keep in mind the opportunity of using multiple detection GC with various combinations of columns, and there is no limitation in these approaches to using just a single or even two columns, in series or parallel, other than those inherent limitations built into the instrumentation available to the analyst.

In a more recent review of multidetection GC systems with capillary column separations, Gagliardi et al. [30] have described the Carlo Erba high-resolution GC instrumentation and technology. Figure 1 illustrates in schematic fashion the particular detector arrangement utilized in this GC instrument, wherein the capillary column is followed by a gas makeup tee (glass-lined tubing) and then a fixed-ratio splitter leading to three detectors. The first parallel set of detectors are the FID and ECD, with appropriate gas makeup lines indicated, and another selective detector has been placed in series behind the ECD, this being a flame photometric detector (FPD). Additional hydrogen makeup gas is supplied in between the ECD and FPD, again using glass-lined tubing in this connector tee. This is termed a parallel–series arrangement by these authors, as discussed further above. The fixed-ratio splitter used between the end of the capillary column and the first two detectors was a glass-lined stainless steel splitter, commercially available from a number of suppliers (Carlo Erba, Varian, Scientific Glass Engineering, J & W Scientific, Precision Sampling).* The glass-lined outlet splitter was deactivated with a 0.1% solution of Carbowax 20M in methylene chloride, and then heated at 250°C in a stream of inert gas [30]. Other approaches have been described in the literature for deactivating glass surfaces used as splitters or connections in GC, such as by the use of silylation reagents [31,32]. In one particular approach recommended for deactivating glass columns or glass-lined metal tubing, a 20% dimethyldichlorosilane–80% toluene solution is left in contact with the glass surface for a set period of time, after the surface has been washed and treated with various inorganic and organic solvents or solutions. Other recommended recipes (see Appendix II) have been described in the literature for deactivating glass surfaces, and this is very important to keep in mind in all multiple detection GC interfacing and eventual applications. A large number of applications of this triple detection system are described in this particular review, such as the analysis of

*Appendix III gives a list of these commercial suppliers and their addresses.

beer samples, the head-space vapor of wine samples, different water samples, and river or waste water. Figure 20 is a triple detection analysis of a beer sample analyzed with a glass capillary column followed by ECD, FPD, and FID. Different degrees of selectivity are evident on the ECD and FPD, while the FID provides a much more general, total picture of the constituents present in this sample.

Bashall [33] of Carlo Erba wrote another minireview in 1984 also dealing, to some extent, with multiple detection GC. This particular application utilized ECD and FID in series, for a series of homologous hydrocarbons and chlorinated pesticides and environmental pollutants. Actually, this particular work should have been discussed previously, but since it referred to the same GC instrumentation as in the previous reference, we have included its discussion here.

An excellent discussion of platinum—iridium (Pt—Ir) effluent splitters for multidetector GC and pneumatic solute switching applications has been provided by Anderson and Bettsch [34]. Practical details are provided to produce low dead volume Pt—Ir to glass connections and various components for effluent splitting. It is shown that the column effluents from a glass capillary column can be split in any desired ratio and maintained constant, regardless of column flow rate. Band broadening in the splitter remains small, even for very low flow rates of carrier gas. In general, variable-ratio splitters do not always provide for constant split ratios as a function of oven temperature during a temperature programmed run. Thus variable-ratio splitters are probably better used with isothermal separations, wherein it is demonstrated that the split ratio remains constant throughout the analysis. Then another ratio could be used with the same or different isothermal temperature conditions, if this were desired. However, if one desires to use temperature programming approaches, then it is far wiser and more reproducible to use a fixed-ratio splitter, wherein it is proven that the ratio is indeed constant at all parts of the programmed run. This must be demonstrated at least once each day with that particular splitter in place, which is readily done just by measuring gas flows at the outlet of each parallel detector at the beginning, middle, and end of the temperature program run. Although rare, even fixed-ratio splitters could change their physical characteristics as a function of temperature, thereby altering the final split ratio. Therefore it is safer to demonstrate the overall constancy of the split ratio throughout the temperature program run, rather than to simply assume it is constant because of the fixed nature of the splitter. If the fixed-ratio splitter is of a design that is symmetrical to both parallel detectors, then in principle there should be no change in the split ratio as a function of varying the oven temperature. The demonstration of this for that particular splitter and program run removes any possible doubt.

There are several possible materials that have been and could be used in splitters for multiple detection, fixed or variable ratios, and

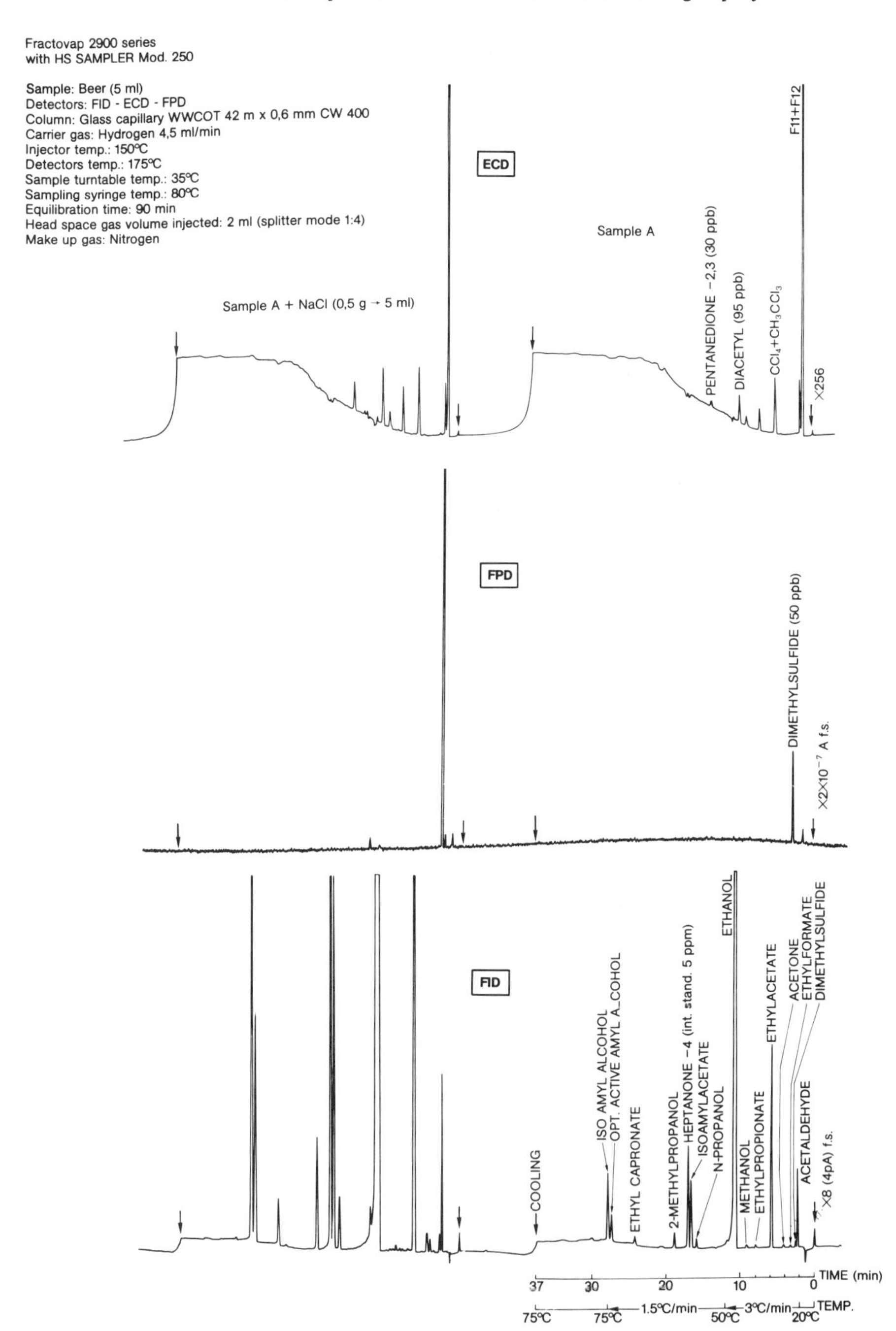

Figure 20 Multiple detection capillary GC analysis of a beer sample using FPD, ECD, and FID. (From Ref. 30.)

these include (1) glass-lined stainless steel, (2) all-glass tubing, (3) all-metal tubing (stainless steel or nickel), (4) Pt–Ir, and (5) quartz tubing. In general, all-glass or quartz tubing offers advantages over all metal tubing, especially with regard to inertness, ease of deactivation and cleaning, reshaping, and repairs. Glass-lined stainless steel tubing provides rigidity, safety in operations, and protection of the glass lining against breakage, together with inertness and the ability to be manipulated and bent to various shapes. Most commercially available splitters today are either all glass or glass-lined stainless steel, and most analysts interested in multiple detection GC seem to be moving in one or both of these directions for splitters and connectors. Some attempts have been made to use polymeric materials or Teflon-lined metal tubing, but these do not appear to be as inert or stable as the glass linings.

In view of the fact that several reviews of this field have already appeared, at least through 1975, most of the specific references to parallel multiple detection that follow date from 1976 to the present time. We have tried to be comprehensive in the coverage of the literature, but if we have inadvertently omitted any specific references from 1976, we apologize to the authors responsible and the current readers. Our own publications are included, but only as these have intentionally used parallel multiple detection for specific applications or demonstration purposes.

Though most parallel approaches have used simple two-way splitting for dual detection results, a number of references have dealt with three- and four-way splitting arrangements for truly multiple detection results. One of thse, by Hrivnac et al. [35], has used capillary GC columns with a four-way splitter illustrated in Fig. 21. This used

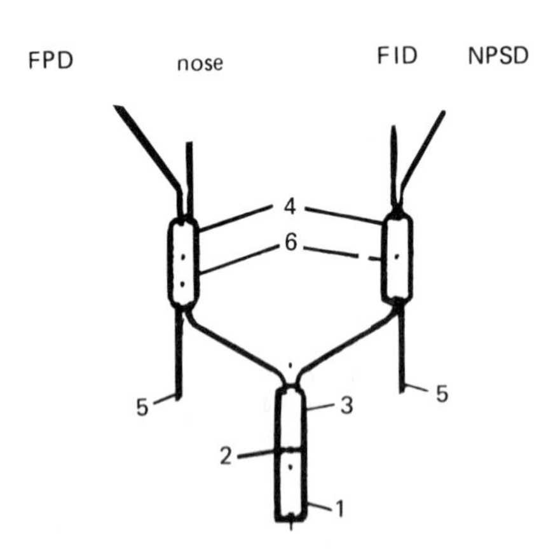

Figure 21 Schematic view of a four-way splitting arrangement: (1) column outlet, (2) PTFE shrink tubing, (3) initial splitting, (4) final splitting, and (5) purge gas inlet. (From Ref. 35.)

an initial two-way splitter followed by two additional two-way splitters which then led the effluent to three separate detectors and a fourth stream to vent for possible organoleptic determinations on part of the original effluent. The three conventional detectors included an FID serving as the monitoring device, a nitrogen–phosphorus detector (NPD), and finally a sulfur-selective flame photometric detector (FPD). Thus, in a typical GC analysis, simultaneous detection of sulfur and nitrogen compounds was possible with the selective detectors, together with a general chromatogram for all volatile materials present in the same sample. If desired, additional organoleptic determinations could be made for individual eluting analytes, avoiding exposure to any possibly toxic materials. Applications of this particular multiple detection approach were desired for meat flavor additives [35]. A detailed description and construction procedures for this particular four-way splitter are provided in this article. One particular novel method was the use of heat-shrinkable tubing to seal the openings of the splitter, an approach that is simpler, if not as rugged, as an all-glass or glass-lined metal tubing approach.

An excellent article appeared in 1977 by Bachmann and colleagues [36] describing the simultaneous connection of an FID and an ECD to a capillary column. The connection of the two detectors was by use of an all-glass splitter, fixed ratio, with a separate support gas stream introduced to the ECD just prior to the splitter. These workers also make use of response ratios for the two detectors employed as an aid in identifying individual analytes. We have already discussed this approach to determining relative response factors, normalized relative response factors, and finally dual detector response ratios, as above. Similar approaches are described in this particular reference for a series of simple chlorinated hydrocarbons using isothermal GC separations on a capillary OV-101 column at 70°C.

Bertsch et al. [37] have described a heart cutting technique in capillary GC applied to sulfur compounds in cigarette smoke, with final parallel dual detection by FID and FPD (sulfur mode). The sulfur-specific FPD was used to detect sulfur-containing compounds in cigarette smoke, while the FID was used solely as a more general, nonselective detector. As these authors have also noted, "the combination of a specific and a universal detector was chosen to facilitate component selection, but other detector combinations are equally applicable." The splitting tees used here were all metal, stainless steel, fixed ratios, and silanized beforehand using standard procedures. In general, we have not found silanization to be as efficient or successful an approach with metal-lined surfaces as with glass-lined ones. This may be why more and more workers have tended to invoke the glass-lined parts today as opposed to metal ones. Dual detector chromatograms are also provided here, now taking certain portions of the sample mixture, heart cutting, to the parallel detectors for evaluation.

Bjorseth and Eklund [38] have described the use of parallel ECD/FID with capillary columns for the analysis and identification of polynuclear aromatic hydrocarbons. The ECD/FID ratios were determined for 46 separate PNA compounds, and such ratios varied from 0.02 to 117, with relative standard deviations better than ±20% for 10 replicate analyses. Such results could therefore provide additional evidence in identifying PNAs in environmental samples. The ECD/FID ratio combined with retention time represents a way of assigning chromatographic peaks with a relatively high degree of certainty. However, certain precautions must be used in applying dual detector response ratios, and it is mandatory to use the same exact chromatographic conditions when analyzing mixtures of standards and real-world samples. The ECD/FID ratio also depends on the split ratio and long-term variations in the detector performance. Impurities in the sample may also contribute to the ECD response of a particular PNA compound. Thus correct ECD/PID ratios may be used to support the identification of a particular peak, but a higher value might indicate the presence of an impurity in the peak or an entire misidentification. This article also provides typical dual detector chromatograms for urban air particulates from Oslo, Norway. All of these capillary GC separations involved temperature programming, in view of the complex mixture of analytes present in such real-world samples.

Lopez-Avila [39] has described a dual detector capillary GC method for the trace analysis of sludge extracts, including the use of FID, ECD, FPD, and a thermionic nitrogen–phosphorus detector (TSD). Of interest in this particular article is a detailed discussion of the various merits and disadvantages of T-versus Y-shaped effluent splitters in capillary GC. The author came to the conclusion that Y splitters have been found to be superior over the more conventional T splitters for a number of valid reasons. Only dual detector responses were obtained here, these being FID/ECD, FID/TSD, FID/FPD, and ECD/TSD. Split ratios were about 1:1 or 1:6, these ratios being determined experimentally by measuring the flow rates of the combined carrier gas and makeup gas at the very exits of the detector towers. This is indeed the only truly valid and virtually infallible method of determining split ratios, with all restrictions and detectors in place at operating temperatures and flow rates. The splitting ratio of the effluent splitter was found to vary between 0.98 and 1.06 when measured at room temperature. In comparing the FID with more selective detector chromatograms, the cleanliness of the selective traces, together with base line resolved peaks in these same chromatograms, demonstrates that extensive sample cleanup is not always necessary when selective detectors are employed. The correct use of detector selectivity can be used to simplify the analysis and provide information that a particular peak contains sulfur, nitrogen, halogen, and so on. In many cases, this type of information could be missed if the samples were only analyzed by GC/MS, which in the total ion mode becomes a totally general-type detector. This particular study also used different splitting ratios for the same analyses. This showed that the splitter does not have any significant effects on peak shape and peak ratio linearity.

McCarthy et al. [40] described the application of dual parallel detection in capillary GC for the analysis of various petroleum samples (oil). Simultaneous detection was achieved using an FID and the Hall electrolytic conductivity detector (HECD) operated in either the sulfur- or nitrogen-specific mode. The HECD could be made element specific for halogen-, sulfur-, or nitrogen-containing compounds; however, no use was made here of dual detector response ratios as a method of analyte identification or confirmation, as in many related studies. Figure 22 illustrates the capillary effluent splitter used in this work, with provision for the addition of a makeup gas or diverting of the solvent front. Typical chromatograms obtained from each detector are provided for various petroleum residue samples. The results suggest that element-selective detection in combination with general detection by an FID can provide, at times, information that nicely complements and assists analyte identification provided by traditional GC/MS methods. As these workers conclude, the use of multiple detection in GC offers some significant advantages, especially for complex samples, and it should be more widely used by others interested in complex analyte identifications and applications.

In 1981 Becher [40] described an analogous application of capillary GC with parallel dual detection using FID and NPD. The specific interest was in the trace analysis of amines in workplace atmospheres using nitrogen-selective detection via NPD. The capillary column effluent was split by means of a Pt–Ir tubing, according to the work of Etzweiler and Neuner-Jehle [27,28] discussed above. The FID/NPD splitting ratio was held constant at 1.05, measured at room temperature, but apparently not during the actual temperature programmed analysis. Actually, as we have mentioned several times already, though fixed-ratio splitters should be constant at any temperature, unless this is actually demonstrated experimentally, there is no way of knowing this for certain. Hence the particular NPD/FID ratios described in this work may be questionable, although still usable in comparing standards and samples analyzed under the same conditions with the same splitter in place. These workers concluded that simultaneous flame ionization–nitrogen-selective detection offers a significant advantage in the determination of aromatic amines in complex environmental-type samples. Also, with the FID, a large solvent peak can cause interference with rapidly eluting analytes, but with the NPD, the solvent peak is negligible. It is also true that aromatic amines respond better on the NPD than the FID, thus providing both improved analyte identification and sensitivity.

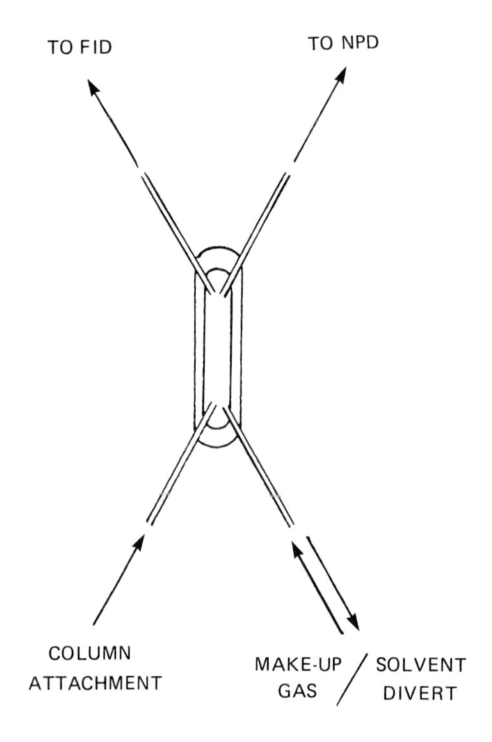

Figure 22 Schematic diagram of a fixed-ratio splitter used for dual detection capillary GC. (From Ref. 40.)

A recent article in *American Laboratory* by Parliment [42] has discussed the use of capillary column GC with three parallel detectors using a unique fixed-ratio (1:1:1) three-way splitter assembly [43]. The approach was applied for the analysis of nitrogen- and sulfur-containing compounds present as volatile aromas in foods or fragrances. The three detectors employed were an FID, an NPD selective for nitrogen, and an FPD selective for sulfur. The actual splitter assembly was constructed from a stainless steel cross to which four 0.5-mm inner diameter glass-lined stainless steel arms were silver soldered and bent. Three parallel arms were designed to fit directly into the heated detector zone of the gas chromatograph and to extend to the respective flame tips of each detector. The capillary column then fitted into the fourth arm of the assembly, and 30 ml/min of helium makeup gas was added through a tee [43]. This is one of the few articles that describes a determination of the efficiencies of the overall GC system with and without the splitter in place, using a particular test compound (tridecane). The incorporation of the effluent splitter caused a reduction in the number N of effective theoretical plates of 2475 to 2150 per meter. Less than 13% loss of efficiency was therefore observed with this particular splitter assembly. The linearity of each detector was then determined over a wide range of concentrations or masses injected for a test compound (trimethylthiazole). Finally, the entire GC multiple detection system was applied to a large number of samples, standard mixtures of flavor chemicals, jasmine essence, horseradish oil, and others. Figure 23 illustrates typical capillary chromatograms for each of the three detectors employed, wherein the FID shows the most complex chromatogram for all substances present, and the other two selective detectors respond to either nitrogen- or sulfur-containing materials. This particular work was a very nice illustration of the capabilities of multiple detection capillary GC, although improved analyte identification would have been possible by the use of relative response ratios for two or even three detectors. No relative response ratios were apparently determined or reported in this particular study.

In 1982 Ramdahl et al. [44] published an article dealing with the use of dual detection capillary GC in the analysis for nitrated polycyclic aromatic hydrocarbons (PAHs). These workers employed simultaneous FID and NPD selective for nitrogren as the detection methods, with a commercial Pt–Ir fixed-ratio splitter maintained at a split ratio of 1.13 (FID/NPD), measured at room temperature. Again, as above,

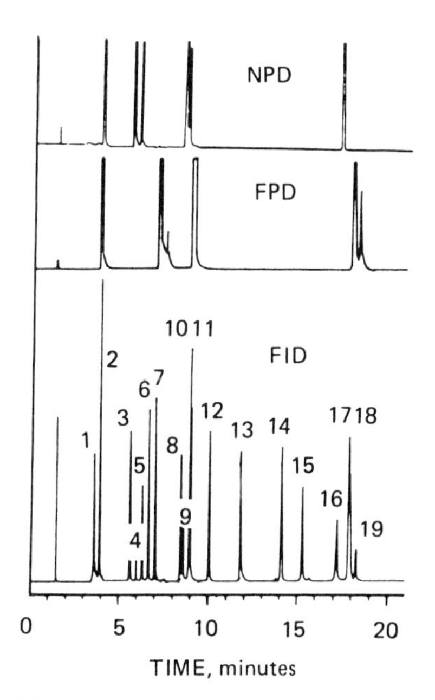

Figure 23 Capillary GC chromatograms of flavor chemical mixture. (From Ref. 42.)

there is the question of not having demonstrated the actual split ratio in effect during the GC analyses, here using temperature programming for all runs. Tables are presented for the determined FID, NPD, and NPD/FID response factors and ratios for various nitropyrenes, relative to an internal standard such as benzo(e)pyrene or 1-nitropyrene. A number of capillary chromatograms are also presented, both single and dual detection, for standard mixtures of nitropyrenes. real tar samples, carbon black extracts, and so on. In general, these authors concluded that "the use of both general and nitrogen selective detectors simplifies complex chromatograms and facilitates identification of trace amounts of nitro-PAHs in environmental samples."

Krull et al. [45–47] have described the use of parallel multiple detection in packed column GC for the analysis of various boron-containing compounds, here making use of a general FID and a very element-specific microwave-induced plasma emission detector (MIP). The column effluent splitter was of a fixed-ratio design, prepared in-house, glass-lined stainless steel, which split the effluent about 50:50 between the FID and MIP detectors. Following an open-split Henneberg-type interface between the GC and the MIP, the emission beam from this detector was then split 50:50 with an in-line beam splitter. This permitted the simultaneous monitoring of both carbon and boron emission wavelength lines of design and permitted three separate chromatograms from each sample injection. For those compounds which contained both carbon and boron, elemental response ratios could then be obtained, knowing the elemental formula for standard compounds. This elemental ratio information for standards could then be used to confirm retention time identification of a particular analyte present in a complex biological sample matrix. Compounds such as steroidal carboranes and catechol boronate esters were studied with this GC FID–MIP approach, which also allowed for the quantitative analysis of such analytes as well as qualitative identification. A large number of papers have discussed GC MIP, often with dual wavelength or multiple wavelength monitoring, but we have not felt that these were really multiple detector/detection GC references, and therefore these have not been included in this particular review [48].

In 1982 Cox and Earp [49] published an excellent article demonstrating the use of capillary GC with parallel simultaneous PID and FID for trace level organics in ambient air samples. Retention time data and normalized relative response ratios are presented for 143 organics, such as alkenes, aromatics, aldehydes, ketones, organochlorines, and organosulfur compounds. Splitter ratios (FID/PID) of 1:1 or 3:1 were used for most of this work, but there is no discussion of how such splitter ratios were actually measured or when. The fixed-ratio splitter used here was of a commercial design, using glass-lined stainless steel tubing silver soldered into a low dead volume union. Two lengths of narrow inner diameter, fused silica tubing were cemented into a two-hole vespel ferrule which butted against one end of the glass-lined tubing and butted against the two-hole ferrule. The ratio of flow to each detector was then controlled by adjusting the length of splitter tubing. The basis of the splitter function is the pressure drop across each of the splitter tubes. It is important that the splitter tubing be of a smaller inner diameter than that of the GC column, but not that much smaller as to cause significant back pressure and decreased carrier gas flow rate. Figure 24 illustrates a typical set of PID and FID chromatograms for a 38-component mixture of hydrocarbons, using an initial cryogenic sample trapping just prior to the GC PID/FID run [49]. On the basis of the relative response ratio data obtained here, the authors concluded that it is possible to estimate chemical classes for sample components. However, whenever possible, PID/FID data should be used in conjunction with GC retention time and related data. The ability to screen for volatile compounds within the classes studied makes this entire approach useful for screening complex samples, such as air above hazardous waste disposal sites.

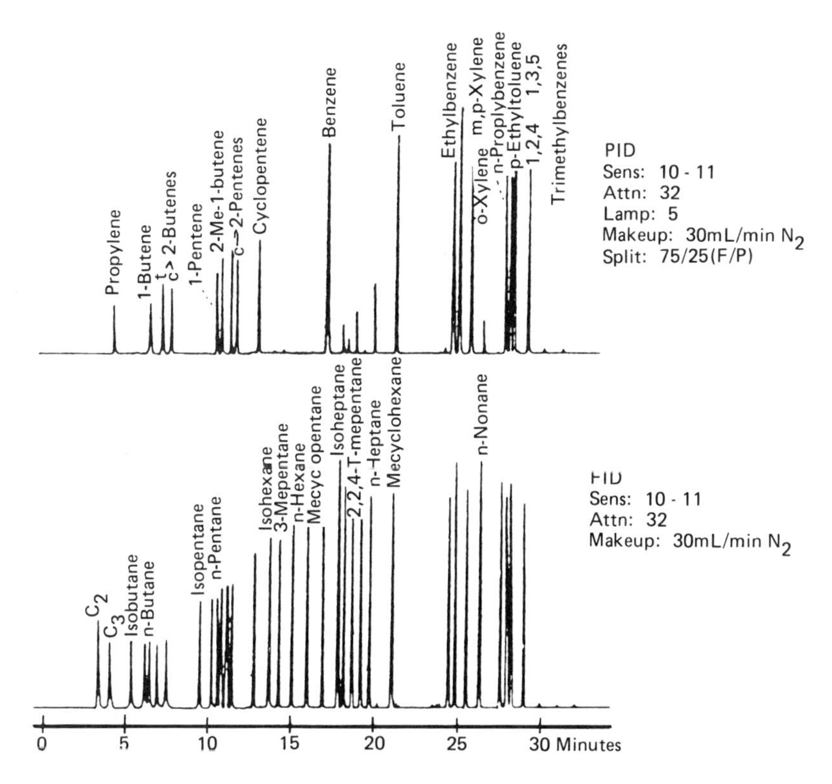

Figure 24 Analysis of a 38-component mixture of hydrocarbons by the cyrogenic GC PID/FID method. (From Ref. 49.)

In 1983 Krull and co-workers [12] described the use of dual detection GC with packed glass columns for the trace analysis and characterization of a large number of organic nitro compounds, including explosives, drugs, industrial chemicals, propellants, and environmental pollutants (nitro-PAHs and PAHs); GC retention times together with relative response factors and ratios of ECD/PID response factors were reported. Detection limits for ECD and PID, relative response factors, and ratios of RRFs were derived form results for mixtures of organic nitro compounds separated by packed column GC with temperature programming. The fixed-ratio splitter was constructed in house from commercially available parts, including a glass-lined tee and glass-lined stainless steel capillary tubing. The fixed-ratio splitter was monitored before, during, and after various days' analyses to ensure that the actual eluent split to each detector was reproducible and well defined. Temperature programming did not change the eluent splitting ratio using this particular type of fixed-ratio GC splitter, but this is not necessarily true for all possible designs of fixed-ratio splitters in multiple detection GC. A large number of dual detector GC chromatograms were presented here for various mixtures of organic standards, and RRFs and ECD/PID ratios for aliphatic nitro compounds, aromatic nitro compounds, nitro-PAHs, and PAHs were determined in this study. Using ECD/PID ratios it was indeed possible to differentiate between the PAHs and their nitro analogs, since such ratios differed by several orders of magnitude in all cases (see Table 10).

Finally, in 1983 Sandra et al. [51] described the construction of a new polyimide effluent splitter for parallel multiple detection in capillary GC applications. A number of detector arrangements were used here, such as ECD/FID, NPD/FID/sniffing, ECD/NPD/FID, and MS/FID/ECD/NPD/sniffing, all using this type of three-way effluent splitter, along with provisions for a makeup gas line. Multiple detector chromatograms are also presented, especially for trimethylsilyl derivatives of steroids and biological applications involving such analytes. This type of splitter was fixed ratio, 1:1 or 1:1:1, and could be used for a large number of detector arrangements and applications. The choice of general and selective detectors used here was excellent, in that both element- and mass-selective ones provided excellent analyte identification possibilities. However, the article was somewhat brief in describing how effluent split ratios were determined and how reproducible these ratios were within any day or from day to day, and the overall accuracy, precision, and reproducibility of this particular effluent splitter were never really described or discussed in depth. Though a polyimide effluent splitter may have certain advantages over glass-lined metal or all-glass ones, this was a difficult conclusion to arrive at from just this one article. Perhaps more work needs to be done in this area in the coming years.

Table 10 Relative Response Factors and ECD/PID Ratios for PAHs and Their Nitro-PAH Analogs[a]

Compound	ECD	PID	ECD/PID[b]
o-Nitrotoluene	1.00	1.00	1.00
Indan	7.18×10^{-5}	0.54	1.34×10^{-4}
5-Nitroindan	2.32	1.18	1.97
Naphthalene	8.01×10^{-6}	0.286	2.80×10^{-5}
2-Nitronaphthalene	4.73	1.25	3.78
Fluorene	—[c]	0.762	—[c]
2-Nitrofluorene	3.48	0.75	4.64
Anthracene	6.73×10^{-3}	0.520	8.83×10^{-3}
9-Nitroanthracene	2.50	1.38	1.81
Pyrene	1.53×10^{-2}	0.388	3.94×10^{-2}
3-Nitropyrene	2.18	0.370	5.89

[a]GC condition as indicated in Ref. 12.
[b]All calculations were made using peak heights and not peak areas. ECD and PID responses were first normalized to that of o-nitroluene as 1.00 (cm/ng) knowing amounts injected, detector attenuations, and peak heights obtained; analyzed as mixtures of PAHs or nitro-PAHs with o-nitroluene present.
[c]Not possible to obtain ECD response for fluorene at microgram levels or above.
Source: Ref. 12.

V. THE FUTURE OF MULTIPLE DETECTION GAS CHROMATOGRAPHY

Despite the fairly widespread commercial availability of both general and selective detectors, together with a reasonable number of commercial fixed- and variable-ratio splitters for packed column and capillary GC work, the percentage of articles discussing and employing multiple detection is probably less than 5% of the total number of GC papers published today. There are some obvious reasons for this, such as the fact that most people do not see the need to use multiple detection, and they may not want to bother installing splitters in their GC units. However, we suspect that as more and more commercial GC units become available with multiple splitters already installed, more and more analysts will come to utilize these approaches. Thus we would predict that the number of articles using such approaches will increase somewhat in the next few years, but it may never reach 25 to 50% of all GC articles. Future areas for development will utilize different combinations of general—selective or selective—selective detectors, since such arrangements seem to provide more meaningful and reliable analyte criteria than other combinations. More and more such detectors will become commercially available for capillary interfacing, and improved fixed- and variable-ratio splitters will be introduced by various GC manufacturers or supply houses dealing in chromatography. More data are needed with regard to relative response factors versus internal standards, and especially ratios of RRFs for various detector combinations and new classes of analyte compounds. Such data will eventually be compiled, as are retention time data, Kovats indices, McReynolds constants, and related chromatographic information of value.

As more and more analysts recognize how little additional effort is required to utilize multiple detection GC, and how very much more useful and reliable data can be generated with no additional sample injections, clearly more and more applications will begin to appear using these approaches. There is no limit to which detectors can be used, or which class of analytes would prove suitable, or which real-world samples could be better analyzed. The use of element-selective detectors, such as the MIP, inductively coupled plasma (ICP), direct current plasma (DCP), atomic absorption (FAA or GFAA), flame photometric, and nitrogen—phosphorus, will become more and more widely used in multiple detection GC, especially for inorganic, organometallic, and heteroatom organic compounds. Analysts should and will rely more and more on such specific detectors, in combination with each other or general detectors, for improved analyte identification in complex sample matrices, biological, industrial, and environmental in origin.

VI. SUMMARY AND CONCLUSIONS

Although multiple detection in GC has always been available, it is only within the past decade or so that it has become really popularized and more widely employed on a semiroutine basis. It is capable of providing unique information related to individual, resolved analytes in complex samples with no additional analytical time, effort, or money beyond the initial instrumental investments. There are no additional analyses required beyond the one that would be needed for any GC single detector run/analysis. Though some sensitivity may be lost in both the series and parallel modes of operation, unless one is working at trace or ultratrace levels, such losses of detection limits are usually not very serious or significant. In any event, it is always necessary to minimize dead volume in any GC connection, splitters, detector fittings, injection ports, and so on, and judging from what has been reported for some of the fixed- or variable-ratio parallel splitters already studied, losses of column efficiencies have not been serious. In the series mode, this is even less of a problem, as long as one places the nondestructive minimum dead volume detector first in line. By all means, loss of column efficiencies should be determined, and further minimized if this shows a problem present. We have also discussed the possible configurations of various parallel splitters for multiple detection, the nature of the materials already described and available, commercial suppliers of such splitters (see Appendix II), and suitable physical dimensions. Ways to measure splitter ratios are also important, for both fixed- and variable-ratio splitters, and this must be done by anyone using such approaches. It is not feasible or valid to assume that the split represented by the manufacturer will hold at all temperatures of any given GC run. This must be determined at the various temperatures to be actually used for that particular analysis; otherwise there is no experimental evidence to support detector splitting ratios. A fixed- or variable-ratio splitter measured at room temperature may or could show a ratio that is very different from what will be in effect at higher operating temperatures. Reproducibility of the split ratio is immensely important in using relative response ratios and detector ratios obtained for standards versus actual samples. Though an internal standard may, to some extent, get around a lack of reproducibility in the split ratio, it is still much wiser to demonstrate split ratio numbers and reproducibility. Data obtained without an internal standard and without monitoring split ratios are questionable, at best.

There is a tremendous amount of creativity yet possible in utilizing various combinations of general and selective detectors in GC, and the use of multple element determinations with a single detector (MIP, ICP, FPD, etc.) should become more and more valuable in confirming a compound's identity, especially when it is present in a very complex sample matrix with many GC peaks, and/or when a GC/MS instrument is not available or unsuitable. Despite what many analysts wish to believe, GC/MS approaches do not solve all problems in analytical chemistry, and there is still a great deal that multiple detection GC has to offer both the organic and inorganic analyst. It is our hope that as more and more chemists realize the opportunities inherent in these approaches, more and more wise use of such methods will be reported and applied in the coming years.

ACKNOWLEDGMENTS

We appreciate the opportunity to review the field in multiple detection gas chromatography, and the original invitation provided by Jack Cazes, editor of *Advances in Chromatography*. Our own involvement in this field evolved, in part, as a result of collaborations between Northeastern University and HNU Systems, Inc. The work described involving GC ECD/PID of organic nitro compounds and explosives would not have been possible without the direct financial support and instrumentation donations provided by HNU Systems, Inc., to Northeastern University, as part of formal research and development contracts between these two organizations. Some of the work performed at Northeastern University formed the M.S. theses of Randy Hilliard and Michael Swartz, graduate students in the Departments of Criminal Justice and Chemistry at Northeastern University.

This is contribution number 191 from The Barnett Institute at Northeastern University.

Appendix I. Commercial Suppliers of Chromatography Data Stations and/or Laboratory Computer Systems for Analytical Chemistry

Spectra-Physics Corporation Autolab Division	3333 N. First Street, San Jose, California 95134
Systems Instruments Corporation America (SICA)	106 Centre Street, Dover, Massachusetts 02030
Apple Computer Corporation	20525 Mariani Avenue, Cupertino, California 95014
Cyborg Corporation	55 Chapel Street, Newton, Massachusetts 02158
Anadata, Inc.	516 North Main Street, Glen Ellyn, Illinois 60137
Nelson Analytical, Inc.	10061 Bubb Road, Cupertino, California 95014
International Business Machines Corporation (IBM)	1000 N.W. 51st Street, Boca Raton, Florida 33432

Appendix II. Method of Deactivation of Glass Columns or Glass-Lined Metal Tubing for Multiple Detection Gas Chromatography*

Using a faucet aspirator pump and rubber tubing, pull air through the glass objects to be treated. Then individually aspirate the following liquids through the column, taking the column or glass-lined tubing out of the liquid to empty it before continuing (all solvents used must be of high analytical purity):

1. Methylene chloride
2. Methanol
3. Acetone
4. Water
5. 50% HCL—50% water left in column, squeeze rubber tubing to stop suction, leave in contact with glass surface 5 min
6. Water
7. Acetone
8. Methanol
9. Methylene chloride
10. 20% dimethyldichlorosilane—80% toluene left for 5 min in contact with glass surface, as above
11. Methanol
12. Methylene chloride
13. Methanol
14. Methylene chloride

Then dry glass surfaces by aspirating air for 2 to 3 min and purging with nitrogen for several minutes to completely dry. The column or glass-lined or all-glass tubing is then ready to use and should be placed within the GC instrument as soon as feasible and immediately used after further temperature conditioning, as desired.

*This particular method of deactivation was provided by HNU Systems, Inc., 160 Charlemont Street, Newton Highlands, Massachusetts 02161, and has been specifically recommended by them for treating their glass columns for Permabond or Synerg C packing materials, prior to actual packing.

Appendix III. Commercial Suppliers of Gas Chromatography Effluent Splitters

Firm	Mailing address
Varian Associates, Inc. Varian Instrument Group	611 Hansen Way, Palo Alto, California 94303
J&W Scientific, Inc.	3871 Security Park Drive, Rancho Cordova, California 95670
Scientific Glass Engineering, Inc.	2007 Kramer Lane, Suite 100, Austin, Texas 78758
Precision Sampling Corporation	P.O. Box 15119 or Box 15886, Baton Rouge, Louisiana 70895
The Anspec Company, Inc.	122 Enterprise Drive, Ann Arbor, Michigan 48107
Supelco, Inc.	Supelco Park, Bellefonte, Pennsylvania 16823-0048
Analabs, Unit of Foxboro Analytical	80 Republic Drive North Haven, Connecticut 06473
Alltech Associates, Inc.	2051 Waukegan Road, Deerfield, Illinois 60015
Applied Science, Inc.	2051 Waukegan Road Deerfield, Illinois 60015
Carlo Erba Strumentazione Sp.A.	P.O. Box 10364, Milan I-20100, Italy

Appendix IV. Abbreviations and Acronyms

Abbreviation	Full definition
AA	Atomic absorption spectroscopy
AFID	Alkali flame ionization detector
CRT	Cathode ray tube (computer monitor)
COS	Carbonyl sulfide
DCE	Dichloroethane
DCE=	Dichloroethylene
DDE	2,2-bis-(p-chlorophenyl)-1,1-dichloroethylene
DDT	1,1,1-trichloro-2,2-bis[p-chlorophenyl]ethane
DCP	Direct current plasma emission detector
DOS	Disk operating system
ECD	Electron capture detector
ELCD	Electrolytic conductivity detector
EPA	Environmental Protection Agency
FAA	Flame atomic absorption spectroscopy
FID	Flame ionization detector
FPD	Flame photometric detector
FTIR	Fourier transform infrared spectroscopy
FUVAD	Far ultraviolet absorbance detection
GC	Gas chromatography
GFAA	Graphite furnace atomic absorption spectroscopy
GLT	Glass lined tubing
HECD	Hall electrolytic conductivity detector
HPLC	High-performance liquid chromatography
ICP	Inductively coupled plasma emission detector
IR	Infrared spectroscopic detection
ISAAC	Integrated Systems for Automated Acquisition and Control
ISTD	Internal standard
LED	Light emitting diode

Appendix IV. (continued)

Abbreviation	Full definition
MDL	Minimum detection limit
MIP	Microwave induced plasma emission detector
MS	Mass spectrometric detector
NPD	Nitrogen-phosphorus detector
o-NT	o-Nitrotoluene
PAH	Polycyclic aromatic hydrocarbon
PC	Personal computer
PCBs	Polychlorinated biphenyls
PID	Photoionization detector
PNA	Polynuclear aromatics
RAM	Random access memory
RMR	Relative molar response
ROM	Read-only memory
RRF	Relative response factor
TCD	Thermal conductivity detector
TEA	Thermal energy analysis
TCE	Trichloroethane
TCE=	Trichloeothylene
TID	Thermionic ionization detector
TSD	Thermionic nitrogen-phosphorus detector
UV	Ultraviolet

REFERENCES

1. J. A. Perry, *Introduction to Analytical Gas Chromatography. History, Principles, and Practice*, Marcel Dekker, New York, 1981.
2. R. L. Pecsok, L. D. Shields, T. Cairns, and I. G. McWilliam, *Modern Methods of Chemical Analysis*, 2nd ed., Wiley, New York, 1976, Chap. 7.
3. H. H. Willard, L. L. Merritt, Jr., J. A. Dean, and F. A. Settle, Jr., *Instrumental Methods of Analysis*, 6th ed., D. van Nostrand, New York, 1981, Chap. 16.

4. H. F. Walton and J. Reyes, *Modern Chemical Analysis and Instrumentation*, Marcel Dekker, New York, 1973, Chap. 12.
5. H. Hachenberg, *Industrial Gas Chromatographic Trace Analysis*, Heyden, London, 1973, Chap. 1.3.
6. R. R. Freeman, *High Resolution Gas Chromatography*, 2nd ed., Hewlett-Packard, Avondale, Penn., 1981.
7. I. S. Krull, in *Sample Preparation Technology Manual* (G. L. Hawk and J. N. Little, eds.), Zymark Corporation, Hopkinton, Mass., 1982, p. 13.
8. S. M. McCown and C. M. Earnest, *Am. Lab. 10*, 33 (1978).
9. W. A. Aue, *J. Chromatogr. Sci.* 13, 329 (1975).
10. L. S. Ettre, J. Chromatogr. Sci. *16*, 396 (1978).
11. D. J. David, *Gas Chromatographic Detectors*, Wiley, New York, 1974.
12. I. S. Krull, M. Swartz, R. Hilliard, K. -H. Xie, and J. N. Driscoll, J. Chromatogr. *260*, 347 (1983).
13. N. A. Kirschen and R. P. Wood, *Varian Application Note*, Varian Instruments Group, Palo Alto, Cal., 1981.
14. D. R. Coulson, Stanford Research Institute, Newton Highland, Mass., (Priv. Comm. to J. Driscoll, 1982).
15. B. Kingsley, C. Gin, D. Coulson, and R. Thomas, in *Water Chlorination, Environmental Impact and Health Effects*, Vol. 4 (R. L. Jolley, ed.), Ann Arbor Science, Ann Arbor, Mich., 1981, Chap. 41.
16. J. N. Driscoll, J. Ford, L. F. Jaramillo, and E. T. Grubber, J. Chromatogr. *158*, 171 (1978).
17. W. Nutmagul, D. R. Cronn, and H. H. Hill, Jr., Anal. Chem. *55*, 2160 (1983).
18. S. Kapila and C. R. Vogt, J. HRC & CC *4*, 233 (1981).
19. B. D. Townes and J. N. Driscoll, Am. Lab. *14*, 56 (1982).
20. D. W. Conron, B. D. Townes, and J. N. Driscoll, paper presented at the 1982 Pittsburgh Conference on Analytical Chemistry and Applied Spectroscopy, Atlantic City, N.J., March 1982.
21. L. F. Jaramillo, J. N. Driscoll, and D. W. Conron, paper presented at the 1981 Pittsburgh Conference on Analytical Chemistry and Applied Spectroscopy, Atlantic City, N.J., March 1981.
22. M. Langhorst, J. Chromatogr. Sci. *19*, 98 (1981).
23. A. Sodergren, J. Chromatogr. *160*, 271 (1978).
24. F. Poy, J. HRC & CC *2*, 243 (1979).
25. R. Pigliucci, W. Averill, J. E. Purcell, and L. S. Ettre, Chromatographia *8*, 165 (1975).
26. K. Grob, Chromatographia *8*, 423 (1975).
27. F. Etzweiler and N. Neuner-Jehle, Chromatographia *6*, 503 (1973).

28. N. Neuner-Jehle, F. Etzweiler, and G. Zarske, Chromatographia *6*, 211 (1973).
29. W. Bertsch, F. Shumbo, R. C. Chang, and A. Zlatkis, Chromatographia *7*, 128 (1974).
30. P. Gagliardi, G. R. Verga, and F. Munari, Am. Lab. *13*, 82 (1981).
31. *Pierce Handbook and General Catalog*, 1983–84, Pierce Chemical Co., Box 117, Rockford, Ill. 61105.
32. *Instructions for Use of Permabound Supports*, HNU Systems, Inc., 160 Charlemont St., Newton Highlands, Mass. 02161.
33. A. Bashall, Am. Lab. *16*, 98 (1984).
34. E. L. Anderson and W. Bertsch, Jr., J. HRC & CC *1*, 13 (1978).
35. M. Hrivnac, W. Friechknect, and L. Cechova, Anal. Chem. *48*, 937 (1976).
36. K. Bachmann, W. Emig, J. Rudolph, and D. Tsotsos, Chromatographia *10*, 684 (1977).
37. W. Bertsch, F. Heu, and A. Zlatkis, Anal. Chem. *48*, 928 (1978).
38. A. Bjorseth and G. Eklund, J. HRC & CC *2*, 22 (1979).
39. V. Lopez-Avila, J. HRC & CC *3*, 545 (1980).
40. L. V. McCarthy, E. B. Overton, M. A. Maberry, S. A. Antoine, and J. L. Laseter, J. HRC & CC *4*, 164 (1981).
41. G. Becher, J. Chromatogr. *211*, 103 (1981).
42. T. H. Parliment, Am. Lab. *14*, 35 (1982).
43. T. H. Parliment and M. D. Spencer, J. Chromatogr. Sci. *19*, 435 (1981).
44. T. Ramdahl, K. Kveseth, and G. Becher, J. HRC & CC *5*, 19 (1982).
45. I. S. Krull, S. W. Jordan, S. Kahl, and S. B. Smith, Jr., J. Chromatogr. Sci. *20*, 489 (1982).
46. S. W. Jordan, B. L. Karger, I. S, Krull, and S. B. Smith, Jr., Chem. Biomed. Environ. Instrum. *12*, 263 (1982–83).
47. S. W. Jordan, I. S. Krull, and S. B. Smith, Jr., Anal. Lett. *15*, 1131 (1982).
48. I. S. Krull and S. Jordan, Am. Lab. *12*, 21 (1980).
49. R. D. Cox and R. F. Earp, Anal. Chem. *54*, 2265 (1982).
50. W. Herres, H. Idstein, and P. Schreier, J. HRC & CC *6*, 590 (1983).
51. P. Sandra, M. Verzele, and E. Vanluchene, J. HRC & CC *6*, 504 (1983).

Author Index

Numbers in parens are reference numbers and indicate that an author's work is referred to although the name may not be cited in text. Italic numbers give the page on which the complete reference is listed.

Subject Index